FLUID MECHANICS

*Also from Macmillan*

Engineering Thermodynamics
Theory, worked examples and problems
G. Boxer

Introduction to Engineering Fluid Mechanics,
Second Edition
J. Fox

Principles of Engineering Thermodynamics
E. M. Goodger

# FLUID MECHANICS

## Theory, worked examples and problems

H. C. Lowe

*Principal Lecturer,*
*Department of Mechanical and Civil Engineering,*
*North Staffordshire Polytechnic*

M

*First published 1978 by*
THE MACMILLAN PRESS LTD
*London and Basingstoke*
*Associated companies in Delhi Dublin*
*Hong Kong Johannesburg Lagos Melbourne*
*New York Singapore and Tokyo*

*Printed in Hong Kong*

British Library Cataloguing in Publication Data

Lowe, H C
Fluid Mechanics.
1. Fluid mechanics
I. Title
532 QC145.2

ISBN 0-333-24023-5

# CONTENTS

# PREFACE

The purpose of this book is to present the fundamental laws of fluid mechanics and illustrate them by application to a wide range of worked examples and problems.

The level is suitable for first and some second year fluid mechanics work on degree and diploma courses in mechanical and civil engineering. The book should also be useful to students preparing for CEI examinations.

The presentation is based on a logical teaching order in preference to grouping the examples under convenient topic headings. A feature of the book is the rigorous treatment of fundamentals in the introductory theory to each chapter. This is necessary because of the wide range of principles involved in fluid mechanics and my belief that the assumptions, on which the principles are based, should be examined. It is also important to consider the range of validity of derived relationships before they are applied to the worked examples.

The actual derivation of the relationships is not considered appropriate in this book and the practice of deriving basic theory in the examples is generally avoided, although there are occasions when it is desirable to link a particular aspect of derived theory with numerical work. It is assumed that the reader will cover the derivations in the lectures of an engineering course, or refer to the books listed in the bibliography.

I attach great importance to the link between certain topics covered in both fluid mechanics and a parallel course in thermodynamics. For this reason, thermodynamic concepts are summarised in Chapter 1. The principle of conservation of energy is treated in Chapter 4 and the important relationship between the Bernoulli equation and the steady-flow energy equation is discussed in detail.

In some chapters, the general three-dimensional flow case is treated, even though the subsequent discussion and examples are restricted to one or two-dimensional flow. This treatment is included for the benefit of students proceeding to an advanced study of fluid mechanics.

Vector methods are used in certain topics, particularly those in the chapter on the dynamics of fluids. Vectors are now taught at an early level and a modern book on fluid mechanics should include vector treatment where appropriate. However, the reader who is unfamiliar with vectors, or the student who prefers to avoid them, should have little difficulty in following the theory and worked examples. Scalar equations are always given, in addition to vector equations, and the majority of the examples are solved using scalars.

It is customary in books to denote vector quantities by boldface type. This is impossible in the present work and vector quantities are denoted by an underscore. For example, F denotes a scalar force and $\underline{F}$ denotes a vector force.

The notation adopted for velocity is particularly important because it is necessary to distinguish between resultant vector velocities in two and three-dimensional flow, and mean velocities used in one-dimensional flow. The symbol used for velocity, throughout this book, is u. The resultant vector velocity at a point is $\underline{u}$, with u, v and w as velocity components in the x, y and z-directions, respectively. The mean velocity in duct or channel flow (V/A) is denoted by $\bar{u}$. Other velocities are defined by an appropriate subscript e.g. $u_r$ and $u_\theta$ are radial and tangential velocity components, respectively.

The symbols used in this book conform wherever possible to the recommendations of the British Standards Institution and the Royal Society. All problems are set in SI units and the general method of numerical solution involves the substitution in physical equations of base SI units (i.e. kg, m, N, etc.) and multiples of ten. This method obviates the need for any special techniques such as the Stroud convention.

My thanks are due to the Director of the North Staffordshire Polytechnic and the Council of Engineering Institutions for permission to include questions from their examination papers. I accept sole responsibility for the solutions to the questions.

I wish to express my gratitude to Mrs Kay White for her expert typing of the manuscript and her patient interpretation of my rough notes.

Finally, I wish to thank my wife and children for their encouragement and help during the writing of the manuscript.

Stafford
May 1978

H. C. Lowe

# SYMBOLS AND UNITS

| *Symbol* | *Quantity* | *Preferred Units* |
|---|---|---|
| A | area | $m^2$ |
| $\underline{A}$ | area (vector) | $m^2$ |
| $\underline{a}$ | linear acceleration | $m/s^2$ |
| a | radius of circular cylinder | m |
| a | sonic velocity | m/s |
| B | width of liquid surface | m |
| b | breadth of weir, spillway, etc. | m |
| C | a constant | |
| C | Chézy coefficient | |
| $C_c$ | contraction coefficient | |
| $C_D$ | drag coefficient | |
| $C_d$ | discharge coefficient | |
| $C_f$ | mean skin coefficient | |
| $C_L$ | lift coefficient | |
| c | velocity (celerity) of wave propagation | m/s |
| $c_f$ | local skin friction coefficient | |
| $c_p$ | specific heat at constant pressure | kJ/kg K |
| $c_v$ | specific heat at constant volume | kJ/kg K |
| D | diameter | m |
| $D_e$ | equivalent diameter (4A/P) | m |
| d | differential operator | |
| E | internal energy | kJ |
| $E_k$ | kinetic energy | kJ |
| $E_p$ | potential energy | kJ |
| e | specific internal energy | kJ/kg |
| F | force | N |
| $\underline{F}$ | force (vector) | N |
| $F_D$ | drag force | N |
| $F_L$ | lift force | N |
| Fr | Froude number $u/\sqrt{(gL)}$ | |
| f | friction factor | |
| f | frequency of vortex shedding | $s^{-1}$ |
| g | acceleration of free fall | $m/s^2$ |
| H | enthalpy | kJ |
| $H_o$ | Bernoulli head $(p/\rho g + \overline{u}^2/2g + z)$ | m |
| $H_s$ | specific head $(\overline{u}^2/2g + y)$ | m |

| | | |
|---|---|---|
| $h$ | specific enthalpy | kJ/kg |
| $h$ | depth below free surface of liquid | m |
| $K$ | a constant | |
| $K_s$ | isentropic bulk modulus | Pa ≡ (N/$m^2$) |
| $K_T$ | isothermal bulk modulus | Pa |
| $k$ | surface roughness height | m |
| $L$ | length | m |
| $\ell$ | length | m |
| ln | natural logarithm (i.e. to base e) | |
| log | logarithm to base 10 | |
| $M$ | total mass | kg |
| $M$ | Mach number $u/a$ | |
| $m$ | mass | kg |
| $\dot{m}$ | mass flow rate | kg/s |
| $m$ | strength of source or sink ($Q/2\pi$) | $m^2$/s |
| $N$ | rotational speed | rev/min |
| $n$ | rotational speed | rad/s |
| $n$ | frequency of vortex shedding | $s^{-1}$ |
| $n$ | Manning roughness coefficient | |
| $n$ | polytropic index | |
| $P$ | wetted perimeter | m |
| $P$ | power | W |
| $p$ | pressure | Pa ≡ (N/$m^2$) |
| $p_a$ | atmospheric pressure | Pa |
| $p^*$ | piezometric pressure ($p + \rho gz$) | Pa |
| $Q$ | flow per unit depth (source or sink) | $m^2$/s |
| $Q$ | quantity of heat | kJ |
| $\dot{Q}$ | heat flow rate | W |
| $q$ | heat per mass | J/kg |
| $R$ | radius | m |
| $R$ | characteristic gas constant | kJ/kg K |
| Re | Reynolds number $\rho uL/\mu$ | |
| $R_h$ | hydraulic radius ($A/P$) | m |
| $r$ | coordinate | m |
| $r$ | radius | m |
| $\underline{r}$ | radius (vector) | m |
| $S$ | entropy | kJ/K |
| $S_o$ | slope of channel bed | |
| $S_w$ | slope of liquid surface | |
| $S$ | slope of energy line | |
| $S$ | control surface | |
| $s$ | specific entropy | kJ/kg K |
| $T$ | thermodynamic temperature | K |
| $T$ | torque | N m |
| $T$ | thrust | N |
| $t$ | time | s |
| $U$ | free stream velocity | m/s |
| $\underline{U}$ | body velocity (vector) | m/s |
| $\underline{u}$ | fluid velocity (vector) | m/s |
| $\bar{u}$ | mean fluid velocity | m/s |
| $u$ | a fluid velocity | m/s |
| $u$ | fluid velocity component (x-direction) | m/s |

| | | |
|---|---|---|
| $u_a$ | axial component of velocity | m/s |
| $u_R$ | relative velocity | m/s |
| $u_r$ | radial component of velocity | m/s |
| $u_S$ | fluid velocity relative to control surface | m/s |
| $u_\theta$ | tangential component of velocity | m/s |
| $u_V$ | fluid velocity relative to control volume | m/s |
| $u_w$ | whirl component of velocity | m/s |
| V | a volume | $m^3$ |
| V | control volume | |
| $\dot{V}$ | volumetric flow rate | $m^3/s$ |
| v | fluid velocity component (y-direction) | m/s |
| $\upsilon$ | specific volume | $m^3/kg$ |
| W | work | J |
| We | Weber number $u\sqrt{(\rho L/\gamma)}$ | |
| $W_x$ | shaft or electrical work | J |
| w | work per mass | J/kg |
| w | fluid velocity component (z-direction) | m/s |
| x | coordinate | m |
| y | depth of liquid (channel flow) | m |
| y | coordinate | m |
| z | height above arbitrary datum | m |
| z | coordinate | m |

*Greek symbols*

| | | |
|---|---|---|
| $\alpha$ (alpha) | angle | |
| $\alpha$ | kinetic energy correction factor | |
| $\beta$ (beta) | isobaric expansivity (thermal coefficient of cubical expansion) | $K^{-1}$ |
| $\beta$ | momentum correction factor | |
| $\Gamma$ (gamma) | circulation | $m^2/s$ |
| $\gamma$ | ratio of specific heats or bulk moduli ($c_p/c_v = K_S/K_T$) | |
| $\gamma$ | surface tension | N/m |
| $\Delta$ (delta) | difference in value | |
| $\delta$ | a very small increase of | |
| $\delta$ | thickness of boundary layer | m |
| $\delta^*$ | displacement thickness of boundary layer | m |
| $\varepsilon$ (epsilon) | kinematic eddy viscosity | $m^2/s$ |
| $\zeta$ (zeta) | vorticity | $s^{-1}$ |
| $\eta$ (eta) | efficiency | |
| $\eta_H$ | hydraulic efficiency | |
| $\eta_o$ | overall efficiency | |
| $\theta$ (theta) | an angle | |
| $\theta$ | temperature on arbitrary scale | °C |
| $\lambda$ (lambda) | temperature lapse rate (-dT/dz) | K/m |
| $\lambda$ | friction factor = 4f | |

| | | |
|---|---|---|
| $\mu$ (mu) | dynamic viscosity | Pa s $\equiv$ (N s/$m^2$) |
| $\nu$ (nu) | kinematic viscosity ($\mu/\rho$) | $m^2$/s |
| $\Pi$ (pi) | dimensionless parameter | |
| $\pi$ | 3.14159 ... | |
| $\rho$ (rho) | mass density | kg/$m^3$ |
| $\Sigma$ (sigma) | summation | |
| $\sigma$ | normal stress | Pa |
| $\tau$ (tau) | shear stress | Pa |
| $\tau_o$ | shear stress at boundary | Pa |
| $\phi$ (phi) | a function of | |
| $\phi$ | velocity potential function | $m^2$/s |
| $\psi$ (psi) | stream function | $m^2$/s |
| $\omega$ (omega) | angular velocity | rad/s |

*Suffixes*

| | |
|---|---|
| c | critical |
| d | delivery |
| R | relative to |
| s | suction |
| S | relative to control surface |
| V | relative to control volume |
| x, y, z | component of vector quantity in x, y, z directions |
| o | stagnation conditions |
| 1, 2 | at inlet or outlet of control volume or machine rotor |
| $\infty$ | at a large distance upstream from body |

# 1 FUNDAMENTAL CONCEPTS

Fluid mechanics is the science which deals with the application of the fundamental principles of general mechanics to fluids at rest and in motion.

## 1.1 CHARACTERISTICS OF FLUIDS

Substances exist as solids or fluids. A *solid* can withstand both normal and tangential stresses without continuous motion taking place. The characteristic of a *fluid* is that it is continuously and permanently deformed by a shear stress no matter how small the stress may be. Thus a shear stress cannot exist in a static fluid.

### 1.1.1 The Continuum

In fluid mechanics we are concerned with macroscopic behaviour and average values of properties. We therefore suppose that a fluid is a *continuum* or a continuous distribution of matter with no voids. This supposition is valid in most engineering situations where the molecular mean free path is small compared with the smallest significant length in the problem being studied. It would be invalid in situations such as high speed, high altitude flight or vacuum work. In such cases kinetic theory must be applied.

### 1.1.2 Density and Specific Volume

The *density* $\rho$ of a fluid is the mass per unit volume under continuum conditions.

The *specific volume* $\upsilon$ is the volume per unit mass ($\upsilon = 1/\rho$).

## 1.2 DIMENSIONS AND UNITS

Properties and forms of energy in transition can be described in terms of fundamental measurable quantities called *dimensions*. For example, *length* and *time* are mutually independent primary dimensions denoted by symbols L and T, respectively. These two dimensions together with a third selected from *mass* (M) or *force* (F) are sufficient to describe all quantities of interest in fluid mechanics.

*Units* are numerical standards selected to give an indication of the magnitude of a quantity described qualitatively by dimensions. Systems of units are chosen quite arbitrarily. A given quantity will always have the same dimensions but the numerical value of the quantity in terms of units will differ according to the system of units used.

### 1.2.1 Dimensions

For convenience, a dimensional equation will be designated by use of the symbol ($\equiv$). For example, distance $\equiv$ L simply means that distance has the dimension length. Similarly

$$\text{Area} \equiv L^2, \ \text{Volume} \equiv L^3, \ \text{Velocity} \equiv LT^{-1}, \ \text{Density} \equiv ML^{-3}$$

The relationship between force and mass is given by Newton's second law which states that force, F, is proportional to the product of mass, m, and acceleration, a. Therefore $F \propto ma$. We may now apply the principle of dimensional homogeneity which states that the dimensions of every term in an equation must be the same.

$$F \equiv M \times LT^{-2} \equiv MLT^{-2}$$

With this relationship any quantity having dimensions F, L and T can be expressed in terms of dimensions M, L and T. For example

$$\text{Work} \equiv FL \equiv ML^2T^{-2}$$

### 1.2.2 The International System of Units (SI)

In SI units the magnitudes of seven physical quantities have been arbitrarily selected and declared to have unit value. The seven base units are listed below.

| *Quantity* | *Name of unit* | *Unit symbol* |
|---|---|---|
| mass | kilogram | kg |
| length | metre | m |
| time | second | s |
| thermodynamic temperature | kelvin | K |
| electric current | ampere | A |
| luminous intensity | candela | cd |
| amount of substance | mole | mol |

The units of all other quantities encountered in fluid mechanics are derived from the base units.

| *Physical quantity* | *Name of SI unit* | *Symbol for SI unit* | *Definition of SI unit* | *Equivalent form of SI unit* |
|---|---|---|---|---|
| energy | joule | J | $m^2\ kg\ s^{-2}$ | N m |
| force | newton | N | $m\ kg\ s^{-2}$ | $J\ m^{-1}$ |
| pressure | pascal | Pa | $m^{-1}\ kg\ s^{-2}$ | $N\ m^{-2}$, $J\ m^{-3}$ |
| power | watt | W | $m^2\ kg\ s^{-3}$ | $J\ s^{-1}$ |
| frequency | hertz | Hz | $s^{-1}$ | |

Some named non-SI units which are decimal multiples of SI units, and are still in use, are given below.

| *Physical quantity* | *Name of unit* | *Symbol for unit* | *Definition of unit* |
|---|---|---|---|
| volume | litre | $\ell$ | $10^{-3}m^3 = dm^3$ |
| mass | tonne | t | $10^3 kg = Mg$ |
| Celsius temperature | degree Celsius | °C | K |
| force | dyne | dyn | $10^{-5}$N |
| pressure | bar | bar | $10^5$Pa |
| energy | erg | erg | $10^{-7}$J |
| kinematic viscosity | stokes | St | $10^{-4}m^2\ s^{-1}$ |
| dynamic viscosity | poise | P | $10^{-1}$Pa s |

## 1.3 FORCE

In the section on dimensions it was shown that force and mass are related by Newton's second law of motion. Force is defined by

$$F = ma \qquad (1.1)$$

The unit of force is derived using this equation. The *newton* is that force which gives to a mass of one kilogram an acceleration of one metre per second per second.

### 1.3.1 Body Forces

It is necessary at this stage to distinguish between a *body* force, which normally acts at the centre of mass of a solid or fluid body, and a *surface* force which can act on any element of area on the surface or within the body. Examples of body forces are weight, centrifugal and buoyancy forces. Examples of surface forces are pressure force and viscous force. Any problem involving static or dynamic equilibrium is solved by a consideration of body and surface forces.

### 1.3.2 Weight

*Weight* is a body force exerted by a gravitational field and it must be expressed in force units i.e. newtons. The local acceleration due to gravity, g, is a variable quantity depending on location (latitude and altitude). At a given locality the weight, $F_W$, is found using Newton's law (equation 1.1).

$$F_W = mg$$

### 1.3.3 Surface Tension

At a liquid-gas or liquid-liquid interface, the binding force between molecules results in a force which is tangential to the surface and normal to an imaginary line drawn on the surface. If the imaginary line has a length $\delta L$, which is small enough to make the line straight compared with the surface curvature, and the line force is

$\delta F_s$, the surface tension $\gamma$ is defined

$$\gamma = \frac{\delta F_s}{\delta L} \qquad (1.2)$$

## 1.4 SURFACE STRESSES

Consider a small element of area drawn in a substance (figure 1.1). A force of magnitude $\delta \underline{F}$ is exerted across the area due to molecular effects and this force may be resolved into two components: a normal force $\delta F_N$ and a tangential or shear force $\delta F_T$.

*Stress* is defined as the limit of $\delta F/\delta A$ as the area A approaches $\delta A'$, the smallest area for which the substance can be considered a continuum.

Normal stress is defined

$$\sigma = \lim_{\delta A \to \delta A'} \left(\frac{\delta F_N}{\delta A}\right) \qquad (1.3)$$

Shear stress is defined

$$\tau = \lim_{\delta A \to \delta A'} \left(\frac{\delta F_T}{\delta A}\right) \qquad (1.4)$$

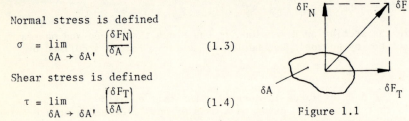

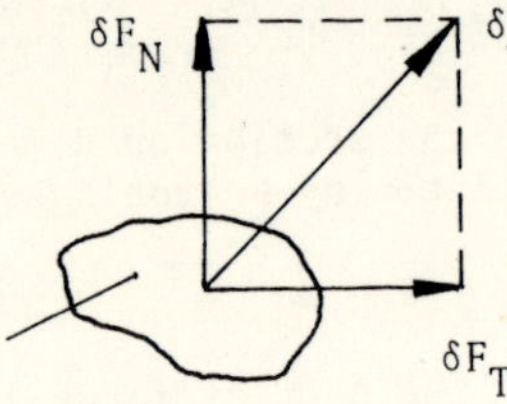

Figure 1.1

These stresses may exist in a fluid or at a solid-fluid interface.

### 1.4.1 Pressure

The normal stress in a fluid is called *pressure* p. Pascal's law states that under continuum conditions the pressure at a point in a fluid is the same in all directions. Pressure is constant at all points in a horizontal line in a stationary continuous fluid.

### 1.4 Absolute and Gauge Pressure

The *absolute* pressure of a fluid is that measured relative to a perfect vacuum. It is usually measured by the height of a vertical column of mercury supported by the pressure. If *atmospheric* pressure is measured by this method then the device is called a barometer. The *gauge* pressure of a fluid is that measured relative to atmospheric pressure. Note that a vacuum gauge reads a negative gauge pressure.

$$p_{abs} = p_{gauge} + p_{atm} \qquad (1.5)$$

### 1.4.3 Shear Stress and Viscosity

A shear force tends to cause one layer of fluid to move relative to another. In a real fluid this relative motion is resisted and the resistance is attributed to the viscosity of the fluid.

*Viscosity* is defined as the property which causes the variation in velocity between adjacent layers or laminae moving parallel to one another, by setting up shear or viscous stresses between the layers. *Dynamic viscosity* $\mu$ is defined

$$\mu = \frac{\text{shear stress}}{\text{rate of shearing strain}} = \frac{\tau}{du/dy} \tag{1.6a}$$

and $$\tau = \mu\frac{du}{dy} \tag{1.6b}$$

Equation 1.6b is known as *Newton's law of viscosity*.

It is important to note that equations 1.6 apply only to parallel (laminar) flow and to a point in the flow. A fluid which obeys equations 1.6, and for which the viscosity is independent of the rate of shear, is known as a *Newtonian fluid*. Substances which satisfy the definition of a fluid, but not equations 1.6, are known as *non-Newtonian fluids*. Note that fluids will flow with a very small shear stress but solids and plastics require a large shear stress.

For normal pressure variations, viscosity is independent of pressure for both liquids and gases. For liquids, viscosity decreases with increase in temperature. For gases, the viscosity increases with temperature.

In many fluid flow problems the ratio $\mu/\rho$ is involved. *Kinematic viscosity* $\nu$ is defined

$$\nu = \frac{\mu}{\rho} \tag{1.7}$$

1.4.4 Viscous Stresses at a Solid Boundary

Viscous stresses cannot exist in an *ideal* or frictionless fluid and if relative motion occurs between an ideal fluid and a solid surface a slip condition will exist (figure 1.2a).

If relative motion occurs between a *real* fluid and a solid boundary, a velocity profile is set up next to the surface, and viscous stresses occur (figure 1.2b). This is because it is impossible for a fluid to slip over the surface (except under very low pressure conditions) due to irregularities which are large compared with the size of fluid molecules. Hence, some fluid is trapped and some held by adsorption. If the main body of fluid were to slip over the trapped fluid, then an abrupt change in velocity would occur and du/dy and $\tau$ would be infinite. Clearly, this is impossible and the fluid velocity must change continuously, over a finite thickness of fluid known as the *boundary layer*.

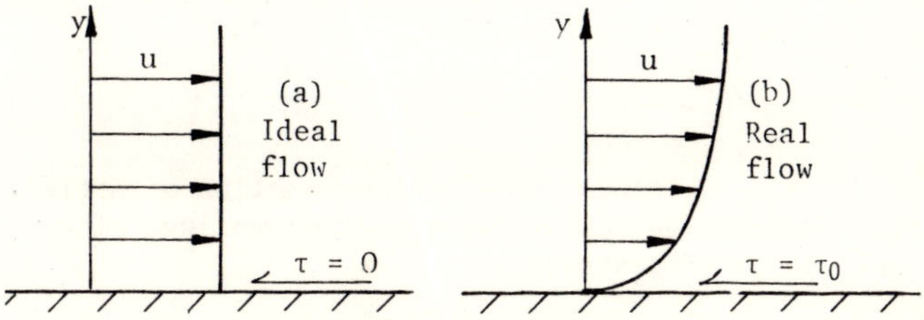


Figure 1.2

## 1.5 LAMINAR AND TURBULENT FLOW

Flow can be either *laminar* or *turbulent*. Reynolds (1883) carried out experiments in which a filament of dye was introduced into water flowing through a glass tube (figure 1.3)

(a) In laminar flow, the fluid particles move along parallel paths and there is no transverse velocity component. An injected filament of dye retains its form without diffusion.

(b) In turbulent flow, fluid particles do not remain in layers, but move in an irregular manner. The disorderly motion of the particles is superimposed on the main flow causing a continuous mixing of the fluid. An injected filament of dye rapidly diffuses throughout the flow.

Newton's law of viscosity cannot be applied if the flow is turbulent because the effective viscosity of the fluid is increased by momentum transfer due to the random motion of fluid particles (aggregates of molecules).

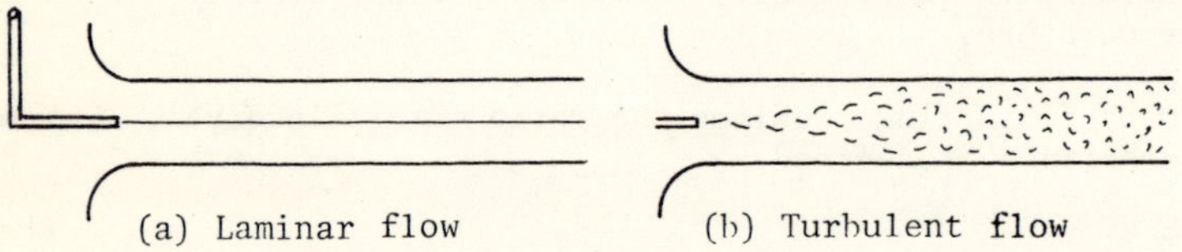


Figure 1.3

### 1.5.1 Laminar Flow with Linear Velocity Distribution

A velocity gradient will exist in the fluid between a stationary and a moving surface (figure 1.4).

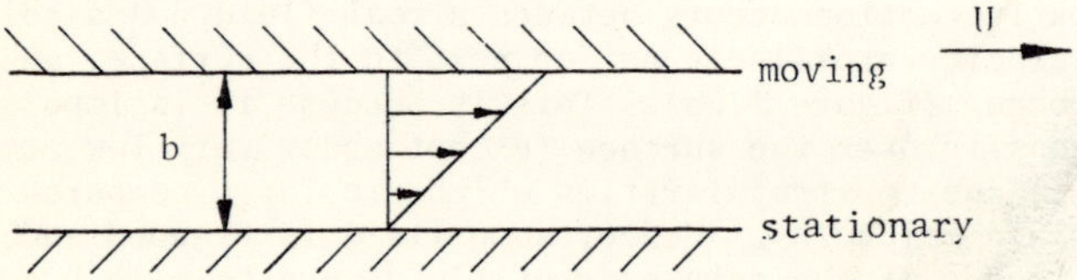


Figure 1.4

If the distance between the surfaces is small the flow may be assumed laminar and the velocity distribution linear. At any point within the fluid the shear or viscous stress is given by

$$\tau = \mu\frac{du}{dy} = \mu\frac{U}{b} \tag{1.8}$$

The moving surface produces fluid flow in the direction of motion. This type of flow is sometimes called simple *Couette* flow. The flow rate per unit depth of plate is

$$\dot{V} = bU/2 \tag{1.9}$$

## 1.6 THERMODYNAMIC CONCEPTS

The study of fluid mechanics requires a basic knowledge of thermodynamics. In this section some important concepts will be summarised.

### 1.6.1 The Thermodynamic System

A *system* is a fixed collection of matter enclosed within a specified *boundary* (real or imaginary) which separates the system from the *surroundings* (figure 1.5a). The shape and position of a system may change with respect to time.

### 1.6.2 The Control Volume

A *control volume* is a fixed region in space through which matter flows. The boundary or envelope enclosing the control volume is called the *control surface* (figure 1.5b). The shape and position of a control volume are fixed relative to the observer.

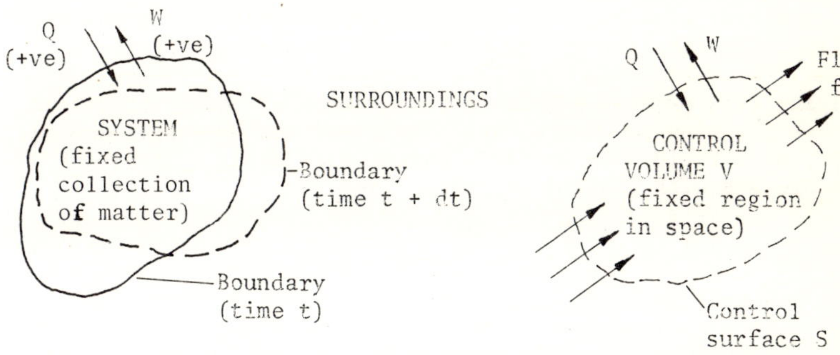


(a) Figure 1.5 (b)

### 1.6.3 Work, Heat and Temperature

*Work*, W, is defined as any action whose sole effect external to a system could be reduced to the rise of a weight. This definition covers the cases encountered in mechanics (i.e. product of force and distance moved) and also cases such as a rotating shaft or electric-current carrying conductors crossing a system boundary. Work is not a fluid property but a form of energy transfer.

*Displacement work* occurs when fluid pressure is exerted on a moving piston. For frictionless, fully resisted conditions the work done between end states 1 and 2 is given by

$$ {}_1w_2 = \int_1^2 p \, d\upsilon \qquad (1.10)$$

*Flow work* occurs when fluid enters or leaves a control volume. If the pressure at any entry or exit point is p, and the corresponding specific volume is $\upsilon$, the specific flow work, $w_F$, required to cause flow is given by

$$ w_F = p\upsilon \qquad (1.11)$$

The *temperature* of a substance is usually related to the level of molecular activity with higher temperatures corresponding to higher levels of molecular activity. The definition of temperature is abstract and it is sufficient to say here that in practice, empirical temperature, $\theta$, is measured using an arbitrary Celsius scale, but in calculations the thermodynamic temperature, T, measured in kelvin units must be used.

For practical purposes

$$T = \theta + 273.15 \qquad (1.12)$$

*Heat*, Q, is energy transferred across the boundary of a system by virtue of a temperature difference. Heat, like work, is not a fluid property but a transient quantity which can only be identified as it crosses the boundary. A process in which there is no heat transfer ($Q = 0$) is called an *adiabatic process*.

1.6.4 Conservation of Energy

The principle of conservation of energy states that in the absence of nuclear reactions energy can neither be created nor destroyed.

Applying this principle to mechanical forms of energy it can be shown that the *kinetic energy*, $E_k = \frac{1}{2}mu^2$, and the *potential energy*, $E_p = mgz$, of a rigid body or fluid particle of mass, m, are each interchangeable with mechanical work. Therefore

$$(E_{k_2} - E_{k_1}) + (E_{p_2} - E_{p_1}) = W \qquad (1.13)$$

If we now exclude kinetic, potential and other external forms of energy, we may argue that when work, $W_a$, is done on a system under adiabatic conditions, it will increase the *internal energy*, E, of the system. Thus, the change in internal energy is defined by the relationship, $E_2 - E_1 = W_a$ without having to identify the nature of internal energy. However, for fluid systems, it is possible to use molecular theories of matter to show the relationship of internal energy to molecular motion and intermolecular forces.

At this stage it is useful to introduce *enthalpy*, H, which is a composite thermodynamic property with energy units. By definition

$$H = E + pV \qquad (1.14a)$$

and $$h = e + pv \qquad (1.14b)$$

1.6.5 The First Law of Thermodynamics

If a system is subjected to heat transfer, Q, and work transfer, W, we may state

$$Q - W = E_2 - E_1 \qquad (1.15a)$$

and $$dq = de + dw \qquad (1.15b)$$

This statement is the *first law of thermodynamics*. If external effects are taken into account we have

$$Q - W = (E_{k_2} + E_{p_2} + E_2) - (E_{k_1} + E_{p_1} + E_1) \qquad (1.15c)$$

In the above equations heat transfer to a system and work transfer from a system are both positive.

### 1.6.6 The Second Law of Thermodynamics

The first law is simply an energy relationship and gives no indication of the preferred direction of energy transfer or transformation in real processes. There are several statements of the *second law of thermodynamics* but in this text we are mainly concerned with the concept of *reversibility* and the conceptual property *entropy*. A *reversible* process is one which can be stopped at any point and its path reversed so that the system and surroundings return through their original states. Real processes are irreversible due to friction, unresisted expansion or heat transfer across a finite temperature difference. In fluid flow, we are mainly concerned with irreversibility due to fluid friction (viscosity) which causes a degradation of mechanical energy to thermal energy. Therefore, frictionless (ideal) flow is described as reversible. The property *entropy* can be defined in terms of the probability of finding a system in a particular state of order or disorder. In real (irreversible) processes there is an increase in disorder and the entropy change of the system plus surroundings is always positive.

If $Q_R$ denotes the heat received or rejected reversibly, the change in *entropy*, S, for any reversible process between end states 1 and 2 is given by

$$S_2 - S_1 = \int_1^2 \frac{dQ_R}{T} \qquad (1.16a)$$

and $$ds = \frac{dq_R}{T} \qquad (1.16b)$$

From the definition of entropy (equation 1.16b) and the first law (equation 1.15b), applied to a reversible process ($dw = p\ d\upsilon$), we obtain

$$T\ ds = de + p\ d\upsilon = dh - \upsilon\ dp \qquad (1.17)$$

Equations 1.17 are applicable to both reversible (frictionless) and irreversible (real) processes.

In fluid mechanics we are particularly interested in adiabatic processes. A *reversible adiabatic* process is also an *isentropic* (s = const) process but the converse is not necessarily true.

## 1.7 PROPERTIES OF FLUIDS

Substances can exist in three different phases: *solid*, *liquid*, and *gaseous*. The term fluid describes substances in either the liquid or gaseous phases. A substance in the gaseous phase may be described as either a *vapour* or a *gas* according to its temperature and pressure.

*Evaporation*, or boiling, is a bulk liquid-vapour phase change which occurs when corresponding *saturation temperature* and *saturation pressure* values are reached; either by increasing the temperature or decreasing the pressure. This phase change is responsible for cavitation and vapour lock in liquid flow.

The term evaporation is often used to describe a surface effect whereby molecules at the liquid surface gain sufficient energy to escape. In a vessel containing liquid (e.g. a bottle of liquid petroleum gas), the rate at which the molecules escape is balanced by the rate of return. The space above the liquid then becomes saturated with vapour and the partial pressure exerted is known as the *vapour pressure*. If there are no other gases present in the space the vapour pressure is the total pressure and the space is called a *torricellian vacuum*. Vapour pressure is equal to the saturation pressure at the given temperature. It is independent of the volume of the space.

Effects similar to boiling occur if a liquid contains dissolved gases. As the pressure reduces the gases are liberated in the form of bubbles. This is sometimes called gas cavitation.

### 1.7.1 Specific Heat and Compressibility

It is now necessary to define some derivative properties.

Specific heat at constant volume $c_V$

$$c_V = \left(\frac{\partial e}{\partial T}\right)_V \simeq \frac{de}{dT} \tag{1.18}$$

Specific heat at constant pressure $c_p$

$$c_p = \left(\frac{\partial h}{\partial T}\right)_p \simeq \frac{dh}{dT} \tag{1.19}$$

Ratio of specific heats $\gamma$

$$\gamma = \frac{c_p}{c_V} \qquad \text{(constant for a perfect gas)} \tag{1.20}$$

Isobaric expansivity (coefficient of cubical expansion) $\beta$

$$\beta = \frac{1}{\upsilon}\left(\frac{\partial \upsilon}{\partial T}\right)_p = -\frac{1}{\rho}\left(\frac{\partial \rho}{\partial T}\right)_p \tag{1.21}$$

Isothermal bulk modulus of elasticity $K_T$

$$K_T = -\upsilon\left(\frac{\partial p}{\partial \upsilon}\right)_T = \rho\left(\frac{\partial p}{\partial \rho}\right)_T \tag{1.22}$$

Isentropic bulk modulus of elasticity $K_S$

$$K_s = -\upsilon\left(\frac{\partial p}{\partial \upsilon}\right)_s = \rho\left(\frac{\partial p}{\partial \rho}\right)_s \tag{1.23}$$

Ratio of bulk moduli $\gamma$

$$\gamma = \frac{K_s}{K_T} = \frac{c_p}{c_v} \tag{1.24}$$

All matter is compressible to some extent. Hence, the bulk moduli values are important for compressibility effects in both liquids and gases. For liquids $\gamma \simeq 1$. Except in high accuracy work, it is not customary to distinguish between either the bulk moduli or the specific heat values of a liquid.

## 1.8 EQUATION OF STATE

An equation which defines the state of a substance in terms of properties is called an *equation of state*. For liquids and vapours, the equation of state can only be expressed in terms of tabular or graphical data. For gases, a mathematical equation may be used. The *equation of state for a perfect gas* is

$$pV = mRT \tag{1.25a}$$

or $$p\upsilon = RT \tag{1.25b}$$

or $$p = \rho RT \tag{1.25c}$$

where R is the characteristic gas constant (287 J/kg K for air).

### 1.8.1 Gas Laws

Equation 1.25c may be applied to any process between end states 1 and 2 to give the combination law.

$$\frac{p_1}{\rho_1 T_1} = \frac{p_2}{\rho_2 T_2} = \text{constant} \tag{1.26}$$

This law may be simplified for *isothermal* (T = const) or *isobaric* (p = const) processes.

For an isentropic process

$$\frac{p_1}{\rho_1^{\gamma}} = \frac{p_2}{\rho_2^{\gamma}} = \text{constant} \tag{1.27}$$

Equations 1.26 and 1.27 may be combined for an isentropic process to give

$$\frac{p_1}{p_2} = \left(\frac{\rho_1}{\rho_2}\right)^{\gamma} = \left(\frac{T_1}{T_2}\right)^{\gamma/(\gamma-1)} \tag{1.28}$$

## 1.9 VELOCITY OF PROPAGATION OF AN INFINITESIMAL PRESSURE DISTURBANCE

A small pressure disturbance in a compressible fluid is propagated as a wave of increased (or decreased) density and pressure. The wave travels at a finite velocity or *celerity*, a, given by

$$a = \sqrt{(dp/d\rho)} \tag{1.29}$$

The velocity a is frequently called the *sonic* or *acoustic* velocity because sound travels with this velocity. The propagation occurs under isentropic conditions.

For a gas

$$a = \sqrt{(\gamma RT)} \tag{1.30}$$

For a liquid

$$a = \sqrt{(K_s/\rho)} \tag{1.31}$$

The value of the propagation velocity varies from $a \simeq 340$ m/s for atmospheric air to $a \simeq 1400$ m/s for water. For a truly incompressible fluid, $a = \infty$.

### 1.9.1 Mach Number

During fluid flow, property changes (e.g. density) are influenced by the bulk velocity but this velocity in itself is insufficient to determine the nature of flow. It is the ratio of bulk velocity, u, to propagation velocity, a, which is important. The ratio, u/a, is called the *Mach number* and is denoted by M.

For a gas

$$M = \frac{u}{\sqrt{(\gamma RT)}} \tag{1.32}$$

The flow may be described as *subsonic* ($M < 1$), *sonic* ($M = 1$), or *supersonic* ($M > 1$).

### 1.9.2 Incompressible and Compressible Flow

It is necessary at this stage to distinguish between *incompressible* and *compressible* flow. All fluids are compressible but the term incompressible flow is used to describe flow which is essentially constant density. Thus, liquid flow is always described as incompressible (except when considering 'water hammer' effects) and gas flow is treated as incompressible for $M \lessapprox 0.2$.

*Example 1.1*

Determine the dimensions of (a) pressure, p, and (b) dynamic viscosity, μ. Suggest appropriate units.

(a) By definition

$$p = \frac{\delta F_N}{\delta A} \equiv \frac{F}{L^2}$$

From section 1.2.1

$$F \equiv MLT^{-2}$$

In mass dimensions

$$p \equiv \frac{MLT^{-2}}{L^2} \equiv ML^{-1}T^{-2}$$

The SI unit for pressure is the *pascal* (symbol Pa). Pressure may be expressed in terms of Pa or N $m^{-2}$.

(b) By definition

$$\mu = \frac{\tau}{du/dy} \equiv \frac{FL^{-2}}{LT^{-1}/L} \equiv FTL^{-2}$$

In mass dimensions

$$\mu \equiv (MLT^{-2})TL^{-2} \equiv ML^{-1}T^{-1}$$

Dynamic viscosity may be expressed in force or mass dimensions. In an absolute system, $\mu$ will have the same numerical value expressed in $FTL^{-2}$ or $ML^{-1}T^{-1}$ dimensions.

The SI unit for dynamic viscosity is the *pascal second* (symbol Pa s). Alternative SI units are N s $m^{-2}$ and kg $m^{-1}$ $s^{-1}$. The cgs unit for dynamic viscosity is the *poise* (symbol P). Alternative cgs units are dyn s $cm^{-2}$ and g $cm^{-1}$ $s^{-1}$.

*Example 1.2*

Define unity brackets and by their use determine the conversion factors for dynamic viscosity in SI units to cgs and FPS units.

Unity brackets are formed from the known numerical relationships between quantities which have the same dimensions. In the SI system all unity brackets involve powers of ten. The most useful unity brackets for conversion between SI and FPS units are given below.

| *Length* | *Mass* | *Force* | *Energy* |
|---|---|---|---|
| $\left[\frac{\text{ft}}{0.3048\text{m}}\right]$ | $\left[\frac{\text{lb}}{0.4536\text{kg}}\right]$ | $\left[\frac{\text{lbf}}{4.448\text{N}}\right]$ | $\left[\frac{\text{Btu}}{1.055\text{kJ}}\right]$ |

Unity brackets can be used as multipliers to rationalise the units in a numerical equation.

Multiply dynamic viscosity in SI units by unity brackets which convert to cgs units or FPS units.

$$\frac{\text{kg}}{\text{m s}} = \frac{\text{kg}}{\text{m s}}\frac{[10^3\ \text{g}]}{[\ \text{kg}\ ]}\frac{[\ \text{m}\ ]}{[10^2\ \text{cm}]} = 10\ \frac{\text{g}}{\text{cm s}}$$

and $$\frac{\text{kg}}{\text{m s}} = \frac{\text{kg}}{\text{m s}}\frac{[\ \text{lb}\ ]}{[0.4536\ \text{kg}]}\frac{[0.3048\ \text{m}]}{[\ \text{ft}\ ]} = 0.672\ \frac{\text{lb}}{\text{ft s}}$$

Note that dynamic viscosity in lbf s/ft$^2$ and lb/ft s units will not have the same numerical value because lbf and lb are not related by an absolute system of units.

*Example 1.3*

Determine the height of the liquid column that forms in a 1 mm diameter vertical tube due to capillary action when the tube is dipped into water. Assume $\gamma$ = 74 mN/m and an angle of contact of 5°.

When a liquid-gas interface meets a solid wall, as in a tube, the meniscus becomes curved and is either convex upwards (mercury) or concave upwards (water). Surface tension exerts a force along the axis of the tube equal in magnitude to the product of the surface tension and the length of the line along which it acts. This action causes the capillary rise or fall of liquid in a tube (figure 1.6).

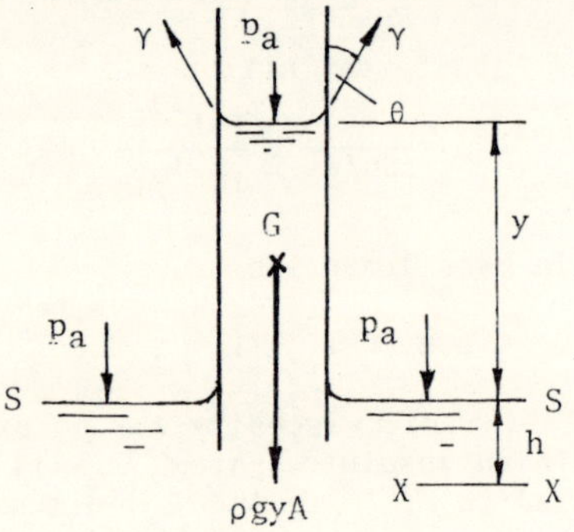


Figure 1.6

Pressure is maintained constant along any horizontal line in a stationary continuous fluid i.e. atmospheric pressure $p_a$ along line SS (or pressure $p_a + \rho g h$ along line XX). Therefore the column of water is supported by a surface tension force.

Surface tension force = Gravity body force on column

$$2\pi R\gamma \cos\theta = \rho g y \pi R^2$$

$$y = \frac{2\gamma\cos\theta}{\rho g R} = \frac{2 \times 74 \times 10^{-3} \times \cos 5}{10^3 \times 9.81 \times 0.5 \times 10^{-3}} = 30.1\ \text{mm}$$

*Example 1.4*

A force F of magnitude 500 N is applied to the small piston in an hydraulic jack. If the area, A, of the small piston is $10^3$ mm$^2$ and the area, $A_2$, of the large piston is $10^4$ mm$^2$, determine the magnitude of the mass lifted and the work done by the large piston when the small piston moves 100 mm. Assume that the pistons are at the same nominal level throughout their travel.

The concept of displacement work is used in the hydraulic jack whereby an incompressible fluid (oil or water) is displaced by a piston acting in a small cylinder and the displaced fluid acts against a load-carrying piston in a large cylinder (figure 1.7). It is assumed that velocities are low and there is no energy dissipated in heat.

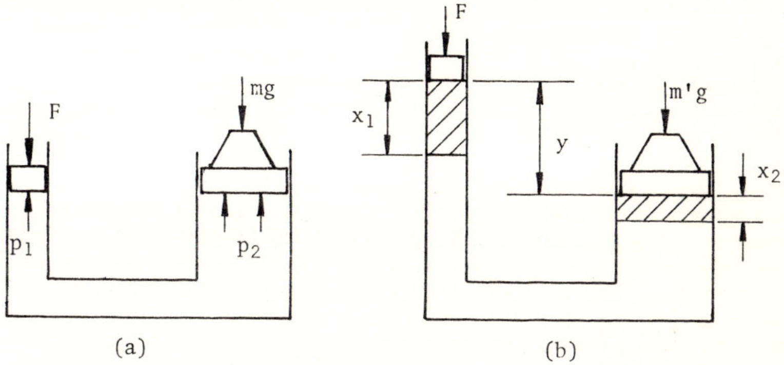


Figure 1.7

Mean pressure on face of small piston

$$p_1 = \frac{F}{A_1} = \frac{500}{10^3 \times 10^{-6}} = 500 \text{ kPa} \qquad \text{(i)}$$

Displacement work done by small piston

$$W_1 = p_1A_1x_1 = Fx_1 = 500 \times 0.1 = 50 \text{ J}$$

Mean pressure on face of large piston

$$p_2 = \frac{mg}{A_2} \qquad \text{(ii)}$$

For pistons at the same level*, $p_1 = p_2$, and equations (i) and (ii) give

$$mg = \frac{A_2F}{A_1} \qquad \text{(iii)}$$

For $A_2 > A_1$ a small force, F, can raise a large load, mg.

$$m = \frac{A_2F}{A_1g} = \frac{10^4 \times 10^{-6} \times 500}{10^3 \times 10^{-6} \times 9.81} = 510 \text{ kg}$$

For an incompressible fluid

$$A_1x_1 = A_2x_2$$

$$x_2 = \frac{A_1x_1}{A_2} = \frac{10^3 \times 10^{-6} \times 0.1}{10^4 \times 10^{-6}} = 0.01 \text{ m}$$

Displacement work done on large piston

$$W_2 = p_2A_2x_2 = 500 \times 10^3 \times 10^4 \times 10^{-6} \times 0.01 = 50 \text{ J}$$

Thus the work done on the large piston is equal to that done by the small piston if the pistons are nominally level. To be strictly accurate, the change in potential energy of the displaced volume of fluid should be taken into account since this affects both the work done on the large piston and the mass lifted. It should also be noted that, in practice, $p_2$ differs from $p_1$ because of the difference in elevation, y, of the piston faces.

*Example 1.5*

The space between two flat and parallel walls 21 mm apart is filled with water of dynamic viscosity 1.12 mN s/m$^2$. A flat plate 200 mm square and 1 mm thick is pulled through the space in such a manner that one surface remains parallel at 5 mm distance from the wall. Assuming that the velocity profiles between plate and walls are linear, determine the force and power required to maintain a plate velocity of 125 mm/s. Neglect the resistance to motion on the edges of the plate.

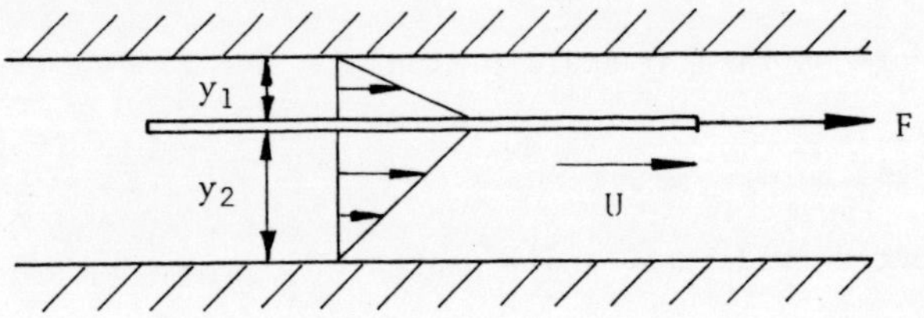


Figure 1.8

For laminar flow the shear stress at any point in the fluid is given by equation 1.6b.

$$\tau = \mu\,(du/dy)$$

Assuming Couette flow the velocity profile in the fluid is linear and the shear force on the upper surface of the plate is

$$F_1 = \tau_{o1} A = \mu(U/y_2)\ A$$

Similarly the shear force on the bottom surface is

$$F_2 = \tau_{o2} A = \mu(U/y_2)\ A$$

Total force on plate

$$F = F_1 + F_2 = \mu \frac{U}{y_1} A + \mu \frac{U}{y_2} A = \mu UA\left[\frac{1}{y_1} + \frac{1}{y_2}\right]$$

$$F = 1.12 \times 10^{-3} \times 0.125 \times 0.2 \times 0.2\ \frac{0.005 + 0.015}{0.005 \times 0.015} = 0.0015\ \text{N}$$

Power to maintain motion

$$P = FU = 0.0015 \times 0.125 = 0.000187\ \text{W}$$

This is also the rate of energy dissipation.

*Example 1.6*

A rotating cylinder viscometer is shown in figure 1.9. The space between the two cylinders is filled with the liquid under investigation and the viscosity is determined by rotating the outer cylinder at angular speed $\omega$ and measuring the torque T required to hold the inner cylinder stationary. Assuming that the clearances b and c are very small compared with the cylinder radii, derive an expression giving the dynamic viscosity in terms of the other variables.

It is assumed that the flow streamlines are circular and the clearances b and c small enough for simple Couette flow to occur. Interference effects at the junction of the base and the cylindrical surfaces are neglected. There are two distinct flow regimes in the viscometer and they must be treated separately.

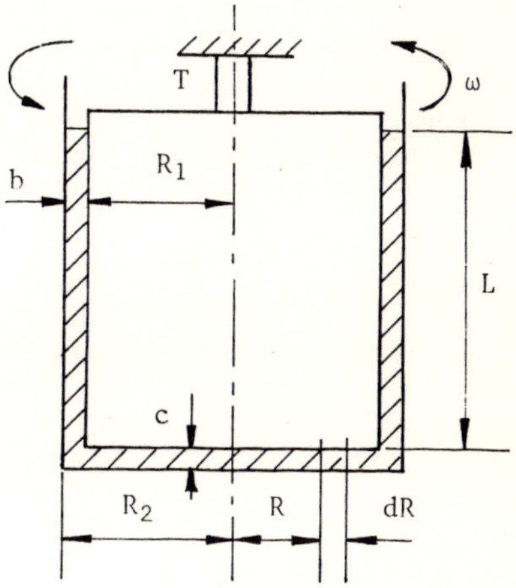

Figure 1.9

The first flow regime occurs in the annular gap between the rotating outer cylinder and the stationary inner cylinder. The tangential force acting on the surface of the inner cylinder is given by

$$F_t = \tau_o A = \tau_o 2\pi R_1 L$$

Substituting for $\tau_o$ from equation 1.8

$$F_t = \frac{\mu U}{b} 2\pi R_1 L = \frac{\mu\omega R_2 \times 2\pi R_1 L}{b}$$

The torque associated with the tangential force is

$$T_t = F_t R_1 = \frac{\mu\omega R_2 \times 2\pi R_1{}^2 L}{b} \qquad \text{(i)}$$

The second flow regime occurs in the gap between the ends of the inner and outer cylinders. The tangential surface velocity now varies and, at any radius R, it is given by $U = \omega r$. Assuming zero pressure gradient in the direction of flow between concentric streamlines distance dR apart, we may apply the simple Couette flow equation to find the tangential force acting on an elemental ring of radius R and thickness dR on the bottom of the inner cylinder. The torque associated with the tangential force is

$$dT_r = \tau_o \, dA \, R = \tau_o \, 2\pi R^2 \, dR = \frac{\mu\omega R}{c} \, 2\pi R^2 \, dR$$

The total resisting torque on the end of the inner cylinder is

$$T_r = \frac{\mu\omega 2\pi}{c} \int_0^{R_1} R^3 \, dR = \frac{\mu\omega\pi R_1^4}{2c} \qquad \text{(ii)}$$

Total resisting torque on inner cylinder

$$T = T_t + T_r = \frac{\mu\omega R_2 \times 2\pi R_1^2 L}{b} + \frac{\mu\omega\pi R_1^4}{2c}$$

therefore

$$\mu = \frac{2bcT}{\pi R_1^2 \omega (4R_2 Lc + R_1^2 b)}$$

*Example 1.7*

A Fortin barometer is contaminated by water on the mercury meniscus. If the height of the mercury column is 735 mm when the atmospheric temperature is 20 °C determine the true barometric pressure. What percentage error is involved if the torricellian vacuum is assumed to be a true vacuum? Neglect surface tension effects.

The toricellian vacuum is not a true vacuum. The space above the water meniscus will be saturated with water vapour and a vapour pressure $p_s$ will be exerted on the meniscus. The vapour pressure is equal to the saturation pressure at the temperature of the water. From tables (e.g. reference 6), at $\theta = 20$ °C, saturation pressure $p_s = 0.02337$ bar. The partial pressure of the mercury vapour is negligible.

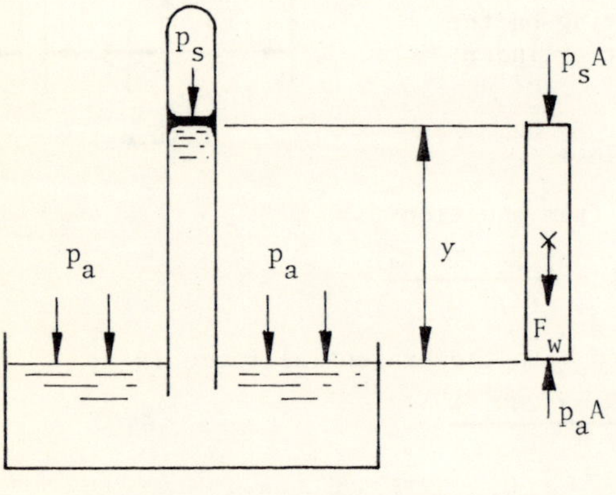


Figure 2.10
Fortin barometer

The pressure is constant and equal to atmospheric pressure $p_a$ along a horizontal line coincidental with the free surface.

Equating forces acting on the cylindrical column of water of cross-sectional area A

$$p_a A = p_s A + \rho g A y$$

$$p_a = p_s + \rho g y = 0.02337 \times 10^5 + 13.6 \times 10^3 \times 9.81 \times 0.735$$

$$= 1.004 \text{ bar}$$

If the torricellian vacuum is assumed to be a true vacuum there will be no vapour pressure exerted on the water meniscus.

$$p_a{}^1 = \rho g y = 13.6 \times 10^3 \times 9.81 \times 0.735 = 0.9806 \text{ bar}$$

$$\text{Error} = \frac{1.004 - 0.9806}{1.004} \; 100 = 2.32\%$$

(Note that this example can also be solved using the equations of fluid statics in chapter 2.)

*Example 1.8*

The density of water at atmospheric conditions of 15 °C and 1 bar may be assumed to be 1000 kg/m$^3$. Determine the change in density if the pressure is increased to 100 bar and the temperature to 100 °C. Assume $\beta = 2 \times 10^{-4}$ K$^{-1}$ and $K_T = 2 \times 10^4$ bar.

Assume an equation of state of the form

$$\rho = \phi(T, p)$$

From a mathematical theorem in calculus

$$d\rho = (\partial\rho/\partial T)_p \; dT + (\partial\rho/\partial p)_T \; dp$$

Substituting for isobaric expansivity $\beta$ and isothermal bulk modulus $K_T$

$$d\rho = -\rho\beta \; dT + (\rho/K_T) \; dp$$

Integrating between initial (suffix o) and final conditions

$$\int_{\rho_o}^{\rho} d\rho = -\int_{T_o}^{T} \rho\beta \; dT + \int_{p_o}^{p} (\rho/K_T) \; dp$$

The first integral is at constant p and the second at constant T. For liquids the density changes are small therefore we can assume $\rho = \rho_o$ in the integrations and also assume constant or mean $\beta$ and $K_T$ values over the temperature and pressure ranges involved. Therefore

$$\rho - \rho_o = -\rho_o\beta \; (T - T_o) + (\rho_o/K_T) \; (p - p_o)$$

$$= -10^3 \times 2 \times 10^{-4} \; (100 - 15) + \frac{10^3}{2 \times 10^4} \; (100 - 1)$$

$$= -12 \text{ kg/m}^3$$

Note that in this problem the liquid is simultaneously heated and compressed and the decrease in density due to temperature increase is larger than the increase in density due to pressure increase.

*Example 1.9*

Show that for adiabatic bulk compression of a liquid the contribution of temperature change to the density change is negligible. Comment on the applicability of this result to bulk compression of gases.

From example 1.8 the change in density due to bulk compression is given by

$$d\rho = -\rho\beta \, dT + (\rho/K_T) \, dp \qquad \text{(i)}$$

Thus the net density change is the sum of the contributions from pressure change and temperature change. Adiabatic bulk compression produces increases in both pressure and temperature and the relative importance of the density change due to temperature to the total density change is given by

$$r = -\frac{\rho\beta \, dT}{d\rho} \qquad \text{(ii)}$$

The first law of thermodynamics (equation 1.15b) gives

$$dq = de + p \, dv = de - (p/\rho^2) \, d\rho$$

For adiabatic bulk compression $dq = 0$. Also $c_v = (de/dT)$ from equation 1.18. Therefore

$$0 = c_v \, dT - (p/\rho^2) \, d\rho$$

$$d\rho = \frac{\rho^2 c_v \, dT}{p} \qquad \text{(iii)}$$

From equations (ii) and (iii)

$$r = -\frac{\beta p}{\rho c_v} \qquad \text{(iv)}$$

Consider the special case of water at atmospheric conditions for which $\beta = 2 \times 10^{-4}$ K$^{-1}$, $c_v = 4.186$ kJ/kg K, $\rho = 10^3$ kg/m$^3$ and $p \simeq$ 100 kPa. Numerically

$$r = \frac{2 \times 10^{-4} \times 100 \times 10^3}{4.186 \times 10^3 \times 10^3} \simeq 5 \times 10^{-6}$$

Thus the contribution of temperature change to density change is negligible for adiabatic bulk compression of a liquid and equation (i) reduces to

$$d\rho = (\rho/K_T) \, dp \qquad \text{(v)}$$

It may be deduced that the temperature rise is negligible for adiabatic compression of a liquid.

Consider now the special case of air at atmospheric conditions for which $\beta \simeq (1/300)$ K$^{-1}$, $c_v \simeq 717$ kg K, $\rho \simeq 1.2$ kg/m$^3$ and $p \simeq 100$ kPa. Numerically

$$r = \frac{100 \times 10^3}{300 \times 1.2 \times 717} \simeq 0.4$$

This result indicates that for a gas it is completely unrealistic to neglect the temperature rise which occurs during adiabatic compression.

*Example 1.10*

Determine the velocity of sound (i) in air at a temperature of 20 °C and (ii) in water of bulk modulus 2.14 $GN/m^2$. Hence determine the bulk velocity for which each flow may be assumed incompressible. For air $R = 287$ J/kg K, $\gamma = 1.4$.

The velocity of propagation of an infinitesimal pressure disturbance in an infinite fluid (velocity of sound) is given by equation 1.29.

$$a = \sqrt{(dp/d\rho)} \qquad \text{(i)}$$

The disturbance is propagated under isentropic conditions.

(i) For gases the law for an isentropic process (equation 1.27) gives

$$p/\rho^{\gamma} = \text{constant}$$

Taking logs and differentiating

$$dp/p = \gamma(d\rho/\rho)$$

$$dp/d\rho = \gamma p/\rho$$

Substituting in equation (i)

$$a = \sqrt{[(\gamma p)/\rho]}$$

The equation of state for a perfect gas (equation 1.25c) gives

$$p = \rho RT$$

Substituting in equation (ii)

$$a = \sqrt{(\gamma RT)} = \sqrt{(1.4 \times 287 \times 293.15)} = 343 \text{ m/s}$$

If it is assumed that flow of air is incompressible for $M < 0.2$ we have from equation 1.32

$$u = M\sqrt{(\gamma RT)} = 0.2\sqrt{(1.4 \times 287 \times 293.15)} = 68.6 \text{ m/s}$$

Thus the flow of air at 20 °C may be assumed incompressible (constant density) for bulk velocities less than about 70 m/s.

(ii) For liquids the isentropic bulk modulus of elasticity is given by equation 1.23.

$$K_s = \rho \left(\frac{\partial p}{\partial \rho}\right)_s$$

Substituting in equation (i)

$$a = \sqrt{(K_s/\rho)}$$

For liquids the numerical difference between the isothermal and isentropic bulk moduli is negligible. Therefore a single value of K for liquids is usually quoted

$$a = \left(\frac{2.14 \times 10^9}{10^3}\right)^{\frac{1}{2}} = 1460 \text{ m/s}$$

For all bulk velocities of liquids encountered in practice it is customary to treat the flow as incompressible (refer to chapter 9 for 'water hammer' effects).

*Problems*

1 Determine the dimensions (M, L, T) of the following quantities and suggest appropriate SI units. (i) kinetic energy (ii) power (iii) kinematic viscosity (iv) surface tension.
[$ML^2T^{-2}$, J ≡ N m, $ML^2T^{-3}$, W ≡ J $s^{-1}$, $L^2T^{-1}$, $m^2s^{-1}$, $MT^{-2}$, N $m^{-1}$]

2 Determine the conversion factors for (i) kinematic viscosity in SI units to cgs and FPS units (ii) pressure in pascals to $lbf/ft^2$.
[1 $m^2/s$ = $10^4$ $cm^2/s$ = 10.76 $ft^2/s$, 1 Pa = $2.089 \times 10^{-2}$ $lbf/ft^2$]

3 Determine the depression of the meniscus when a tube 2 mm diameter containing mercury is immersed in an open bath of mercury. Assume $\gamma$ = 480 mN/m and an angle of contact of 45°.
[5.088 mm]

4 Determine the internal pressure in a spherical droplet of water, diameter 1 μm, which is subjected to an external air pressure of 2 bar. Assume $\gamma$ = 74 mN/m.
[4.96 bar]

5 Determine the mass lifted and the work done by the large piston of the hydraulic jack in example 1.4 when the large piston is initially 800 mm below the small piston (i) by considering the potential energy change, and (ii) by evaluating $\int p \, d\upsilon$ when $p_2 = p_1 + \rho g y$
[518 kg, 50.8 J]

6 A square plate of mass 3 kg and side length 750 mm is placed on a flat plate inclined at 20° to the horizontal. An oil film exists on the plane and it is observed that the square plate slides down the plane at a velocity of 100 mm/s. If the thickness of the oil film is 0.1 mm determine the viscosity of the oil in poise.
[0.179 P]

7 A piston 100 mm diameter and 100 mm long slides in a cylinder 101 mm diameter. The space between the piston and cylinder walls is filled with oil of kinematic viscosity 280 $mm^2/s$ and density 950 $kg/m^3$. Determine the axial force required to maintain a velocity of 15 mm/s.
[0.25 N]

8 The thrust of a shaft is taken by a collar bearing provided with

a forced lubrication system which maintains a film of oil of uniform thickness 0.25 mm between the surface of the collar and the bearing. The external and internal diameters of the collar are 150 mm and 100 mm. If the kinematic viscosity of the oil film is 4 stokes, find the power absorbed in the bearing when the shaft rotates at 300 rev/min. Density of oil is 850 kg/m$^3$.

[53.6 W]

9 Discuss the significance of the following statements in terms of saturation pressure and temperature or vapour pressure. (a) A vapour lock can form in the crest of a syphon and prevent flow (b) a pressurised engine cooling system is preferable to a thermo-syphon system when driving in mountainous country (c) caravanners use bottles of butane liquified petroleum gas in France and propane LPG in Scandinavia.

10 Calculate the temperature to which the water in a pressure cooker would have to be raised before a blow-off valve set to 1 bar would operate. Atmospheric pressure is 1 bar.

[120.2 °C]

11 Show that density change due to either isothermal or adiabatic bulk compression of a liquid is given by the same expression $d\rho = (\rho/K_T)dp$. Hence determine the density of sea-water at the bottom of an ocean at which point the absolute pressure is 110 MPa. It may be assumed that the density of sea-water at sea-level is 1026 kg/m$^3$ and atmospheric pressure 100 kPa. K = 2.4 GPa.

[1075 kg/m$^3$]

12 Assuming that a fluid can be called incompressible when its change of density with pressure under isothermal conditions is less than 0.2%, determine the maximum pressure change possible for water to be called incompressible. K = 2.14 GPa.

[2.8 MPa]

13 Liquid is pumped through a rigid pipe 200 mm diameter until a blockage occurs at some unknown point. A piston is inserted in one end of the pipe and is forced without leakage through a distance of 100 mm. If the pressure in the pipe increases by 200 kN/m$^2$ determine the position of the blockage relative to the piston end of the pipe. Assume a constant modulus of rigidity for the liquid of 1.4 GN/m$^2$.

[700 m]

14 An aircraft flies at an altitude of 10 000 m where the pressure and density are 0.265 bar and 0.41 kg/m$^3$ respectively. Determine the aircraft speed for a Mach number of 1.7. Assume $\gamma$ = 1.4.

[511 m/s]

15 A valve at the end of a water pipe 25 m long is closed suddenly causing a 'water hammer' effect in which a pressure wave is propagated along the pipe. Determine the time for the wave to be propagated to the end of the pipe and reflected back to the valve. K = 2.14 GPa.

[0.0342 s]

# 2 STATICS OF FLUIDS

In a static fluid system there is no relative motion between fluid particles hence there are no shear surface forces. The only forces present are normal surface forces due to pressure and body forces due to gravity. The relationships obtained for a static fluid also apply to a system moving with constant velocity because there are no additional forces involved. If the system is subjected to constant acceleration then the relationships are modified due to inertia body forces but again there is no relative motion between fluid particles.

## 2.1 PRESSURE DISTRIBUTIONS IN STATIC FLUID SYSTEMS

The forces acting on a rectangular element, within a body of fluid which is stationary or moving with constant velocity (a = 0), are normal forces due to pressure and a body force due to gravity (figure 2.1).

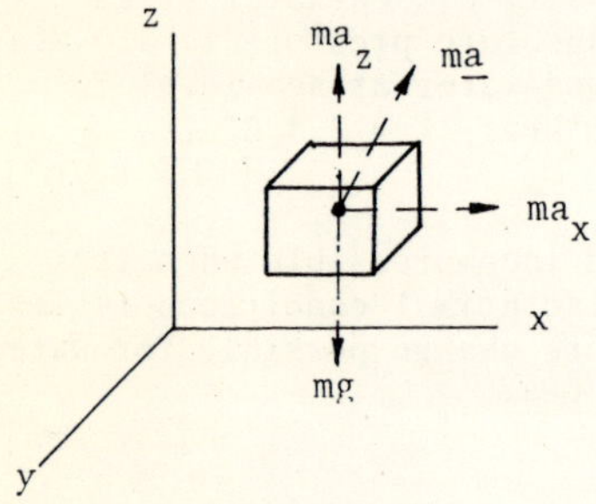


Figure 2.1

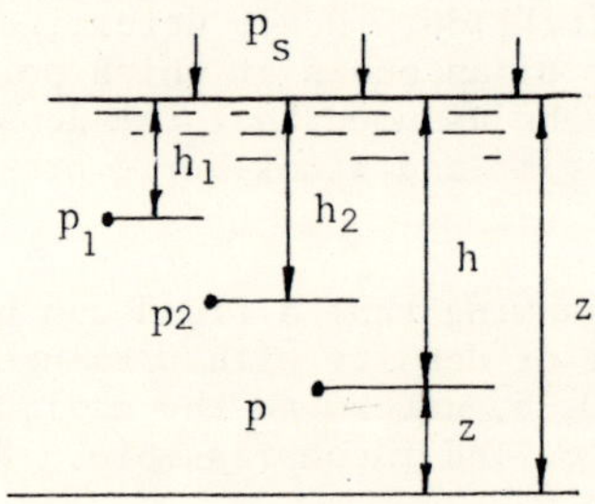


Figure 2.2

Resolving forces in the x, y and z directions and applying Pascal's law, it can be shown that

$$\frac{\partial p}{\partial x} = 0, \ \frac{\partial p}{\partial y} = 0, \ \frac{\partial p}{\partial z} + \rho g = 0 \qquad (2.1)$$

$$\text{or} \quad \text{grad } p + \underline{k}\rho g = 0 \qquad (2.2)$$

From the condition $\partial p/\partial x = 0$ and $\partial p/\partial y = 0$, it follows that the pressure must be constant at all points in a horizontal plane in a continuous fluid.

The condition $\partial p/\partial z + \rho g = 0$ may be expressed in ordinary derivations since p is not a function of x and y. Hence

$$\frac{dp}{dz} = -\rho g \qquad (2.3)$$

It follows that the pressure gradient in a fluid at rest is directly proportional to the density. Conversely, since dp/dz does not vary horizontally, the density is constant at all points in a horizontal plane.

### 2.1.1 Pressure Distribution in a Fluid of Constant Density

For a fluid of constant density (e.g. a liquid) integration of equation 2.3 yields

$$p = -\rho g z + C$$

Consider now the case of a liquid with a free surface exposed to pressure, $p_s$. When $z = z_s$, $p = p_s$. Hence

$$p = \rho g(z_s - z) + p_s \qquad (2.4a)$$

If the depth below the free surface is denoted by h

$$p = \rho g h + p_s \qquad (2.4b)$$

The difference in pressure between two points (figure 2.2) is

$$p_2 - p_1 = \rho g(h_2 - h_1) \qquad (2.5)$$

### 2.1.2 Pressure Measurement

The basic relationship between pressure, density and elevation, given by equation 2.5, may be used as a primary standard for measuring pressure

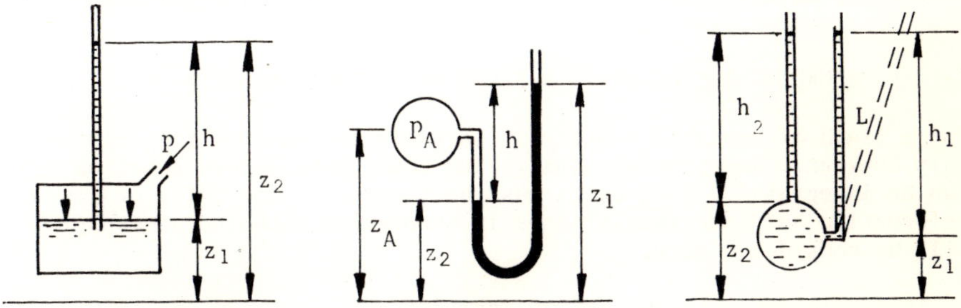


(a) absolute pressure (b) gauge pressure (c) piezometric pressure

Figure 2.3

A barometer (figure 2.3a) is used for measuring absolute pressure. It is conventionally used for measuring the absolute pressure of the atmosphere but it can also be used for any absolute pressure. Neglecting vapour pressure

$$p = \rho g h \qquad (2.6)$$

A U-tube manometer (figure 2.3b) is used for measuring gauge pressure. If $(p_g)_A$ is the gauge pressure of fluid in a vessel A

$$(p_g)_A = \rho_m gh - \rho_A g\,(z_A - z_2) \qquad (2.7)$$

If fluid A is a gas, the term $\rho_A g(z_A - z_2)$ is usually neglected.

The U-tube can also be used as a differential manometer by connecting each leg to a pressure source e.g. to measure pressure loss in a pipe.

A piezometer tube (figure 2.3c) is used for measuring liquid pressure. The length, h, of the liquid column is often referred to as the pressure head and if the tube is open to atmosphere, the corresponding pressure is the gauge pressure $p_g = \rho gh$. The sum of the pressure head and the elevation above a chosen datum (h + z) is constant in a continuous static fluid and is referred to as piezometric head. The corresponding piezometric pressure, p*, is defined

$$p^* = p_g + \rho gz \qquad (2.8)$$

The piezometric pressure can also be evaluated at any point in a fluid in motion but it will only be constant for constant area, frictionless flow.

The sensitivity of any of the above devices may be increased by inclining the tube. The vertical height of liquid, h, which measures the pressure, is unchanged but the column length, L, increases with inclination and can be more accurately read.

Pressure can also be measured with non-fluid devices such as bourdon-tube gauges, aneroid-barometers, and transducers. These devices must be calibrated, usually against a known fluid pressure e.g. in a dead-weight tester.

## 2.2 PRESSURE DISTRIBUTION IN A FLUID OF VARIABLE DENSITY

For a fluid of variable density it is necessary to know the relationship between p and ρ under equilibrium conditions before equation 2.3 can be integrated. Such cases must be treated on the basis of given information but some special ones in connection with the atmosphere will be considered here.

It is desirable to specify an *International Standard Atmosphere* for many purposes especially in connection with aeronautical engineering. From the surface of the earth to a height of about 11 000 m lies the *troposphere* in which region the temperature decreases linearly with increase in altitude i.e. $(dT/dz) = \text{const} = -\lambda$ where $\lambda \simeq 0.0065$ K/m is the temperature lapse rate. Combining equation 2.3 with $\lambda = -(dT/dz)$ and integrating between the datum z = 0, where $p = p_0$ and $T = T_0$, and a second elevation z, where p = p and T = T, we obtain

$$\frac{p}{p_0} = \left(\frac{T}{T_0}\right)^{g/\lambda R} = \left(\frac{T_0 - \lambda z}{T_0}\right)^{g/\lambda R} \qquad (2.9)$$

Above the troposphere to a height of about 20 000 m is the *strato-*

*sphere* in which region the temperature is constant at 216.7 K. Combining equation 2.3 with the equation of state $\rho = p/RT$ and integrating between stations 1 and 2, we obtain

$$\ln\left(\frac{p_2}{p_1}\right) = -\frac{g}{RT}(z_2 - z_1) \qquad (2.10)$$

Above the stratosphere, up to an altitude of 32 000 m, the temperature increases by 11.8 K and from then on there is a succession of changes which is beyond the scope of the present work. A model atmosphere of special interest in meteorology is one with an adiabatic lapse rate i.e. $p_1/p_2 = (T_1/T_2)^{(\gamma-1)/\gamma}$.

## 2.3 PRESSURE DISTRIBUTION IN AN ACCELERATING FLUID SYSTEM

When a body of fluid is accelerated as a whole without relative motion between fluid particles there are no shear stresses present. Thus, a static analysis may be made if the additional inertia forces are taken into account.

### 2.3.1 Linear Motion with Constant Acceleration

For linear acceleration consider the forces acting on a rectangular element within a body of fluid being subjected to constant acceleration $\underline{a}$ in the zx plane (figure 2.1). The acceleration $\underline{a}$ may be resolved into two components, $a_y$ and $a_z$, and Newton's second law applied. The only forces acting are normal surface forces (pressure) and body forces due to gravitational and acceleration effects. Analysis gives the pressure at a point as

$$p = -\rho[a_x x + (a_z + g)z] + C \qquad (2.11)$$

For a line of constant pressure, equation 2.11 may be differentiated to give

$$\frac{dz}{dx} = -\frac{a_x}{a_z + g} \qquad (2.12)$$

If there is a free surface it is a line of constant pressure and its slope is given by equation 2.12. Lines of constant pressure are parallel to the free surface and the resultant acceleration is normal to the free surface.

### 2.3.2 Rotational Motion with Uniform Angular Velocity

For rotational motion it is more convenient to use cylindrical coordinates. Consider the forces acting on a segmental element rotating at constant angular velocity $\omega$ (figure 2.4). The only forces acting in the radial direction are pressure forces and a body

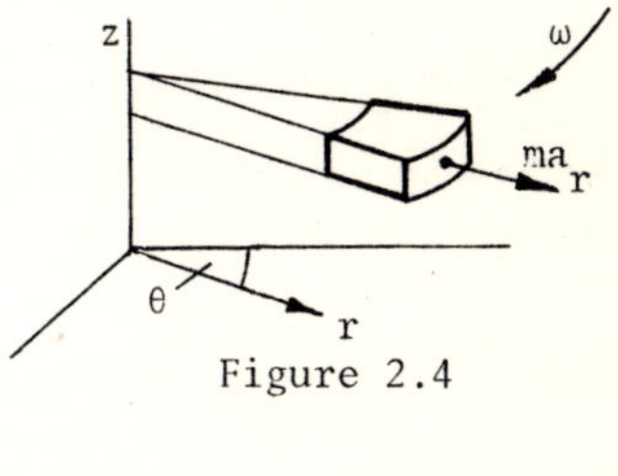

Figure 2.4

force, $ma_r$, due to centrifugal effects. Analysis gives the pressure at a point as

$$p = \rho \left(\frac{\omega^2 r^2}{2} - gz\right) + C \qquad (2.13)$$

For a line of constant pressure

$$\frac{dz}{dr} = \frac{\omega^2 r}{g} \qquad (2.14)$$

and $$z = \frac{\omega^2 r^2}{2g} + C \qquad (2.15)$$

which is the equation of a parabola symmetrical about the z-axis. The pressure at all points on a free surface is constant. Therefore, if a cylindrical body of liquid is rotated about a vertical axis, the free surface is a paraboloid.

A fluid rotating with constant angular velocity and having the characteristics described is known as a *forced vortex* because an input torque and energy are required to maintain the vortex. A more accurate name for the motion is *solid body rotation*.

## 2.4 FORCES ON SOLID SURFACES IMMERSED IN FLUIDS

The equations derived in the previous sections can be used to determine the pressure at any point in a fluid. If there is a solid surface in the fluid then the pressure, p, exerted on each element of area, dA, on the surface produces a normal pressure force, $d\underline{F}$. The total force on the surface, $\underline{F}$, may then be determined from knowledge of the pressure distribution and the geometry of the surface. Only liquids of constant density are dealt with in the following analysis because the changes in pressure in a gas are normally negligible over surfaces of practical interest.

### 2.4.1 Forces on Plane Surfaces

Consider a plane surface immersed in a stationary liquid (figure 2.5). The total area of the surface is

$$A = \int_A b \, dy$$

The first moment of area about the free surface is

$$A\bar{Y} = \int_A (b \, dy)y$$

The second moment of area about the free surface is

$$I_{ss} = \int_A (b \, dy)y^2$$

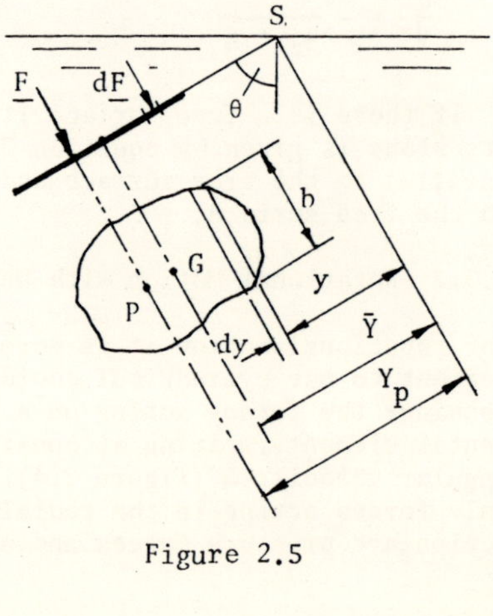

Figure 2.5

The total force on the surface is

$$\underline{F} = \int_A d\underline{F} = \int_A p(b\ dy) = \int_A \rho gh(b\ dy) = \int_A \rho gy \cos\theta(b\ dy)$$

$$\underline{F} = \rho g\bar{Y}A \cos\theta = \text{pressure at centroid} \times \text{area of surface} \quad (2.16)$$

The force, $\underline{F}$, acts at a point, P, known as the centre of pressure, distance, $Y_p$, from the liquid surface, measured in the plane of the solid surface. Taking force moments about point S

$$Y_p = \frac{I_{ss}}{A\bar{Y}} = \frac{\text{2nd moment of area}}{\text{1st moment of area}} \quad (2.17)$$

2.4.2 Forces on Irregular Surfaces

On any curved or warped surface, the forces acting upon different elements of area vary in direction (figure 2.6). However, horizontal and vertical components may be used to simplify the analysis.

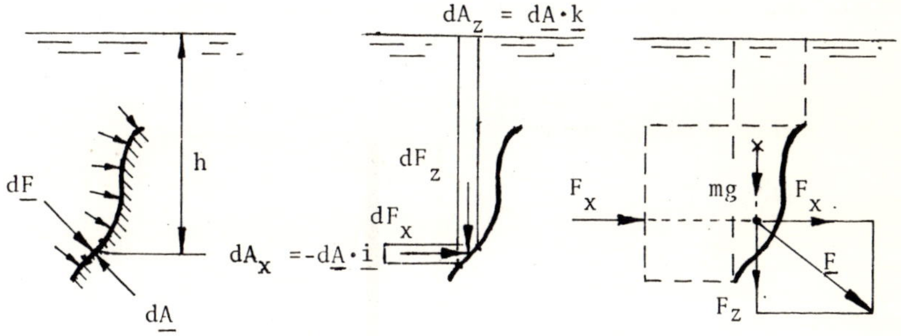


Figure 2.6

The force on the element of area $d\underline{A}$ is given by $d\underline{F} = -p\ d\underline{A}$ and the components of the force in the x and z directions are

$$dF_x = -p(-d\underline{A}\cdot\underline{i}) = p\ dA_x$$

and $$dF_z = -p(d\underline{A}\cdot\underline{k}) = -p\ dA_z$$

The total forces are

$$F_x = \int p\ dA_x = \int \rho gh\ dA_x \quad (2.18)$$

and $$F_z = -\int p\ dA_z = -\int \rho gh\ dA_z \quad (2.19)$$

Hence, the horizontal force, $F_x$, acting on any irregular area is the force acting on the projection of the area on a vertical plane and may be evaluated as in 2.4.1. The vertical force, $F_z$, is the weight of the volume of liquid above the surface, mg. The magnitude and direction of the resultant force, $\underline{F}$, may now be determined.

## 2.5 BUOYANCY FORCES

If a body is immersed in a fluid there is an upward buoyancy force,

$F_b$, exerted on the body, equal in magnitude to the weight of the volume of fluid displaced. The buoyancy force acts at the centre of buoyancy, B, which is the centroid of the displaced body of fluid.

A completely submerged body is in stable equilibrium only if its centre of gravity, G, lies below the centre of buoyancy, in which case any angular displacement produces a couple which restores the body to the equilibrium position.

A floating body can be in stable equilibrium even if its centre of gravity lies above its centre of buoyancy. This is because the centre of buoyancy can move, causing the line of action of the buoyancy force to cut the axis of symmetry at a point M. If M lies above G an anticlockwise couple restores the body to the equilibrium position but if M lies below G a clockwise couple produces overturning. The distance GM is known as the metacentric height.

*Example 2.1*

Determine the reading on pressure gauge A shown in figure 2.7. Assume $\rho_w$ = 1000 kg/m$^3$, $\rho_m$ = 13 600 kg/m$^3$

From equation 2.4, the absolute pressure recorded by the gauge is

$$p_A = \rho_w g h_w + p_a \qquad \text{(i)}$$

The pressure of the air in the container is recorded by the manometer.

$$p_{air} = \rho_m g h_m + p_a \qquad \text{(ii)}$$

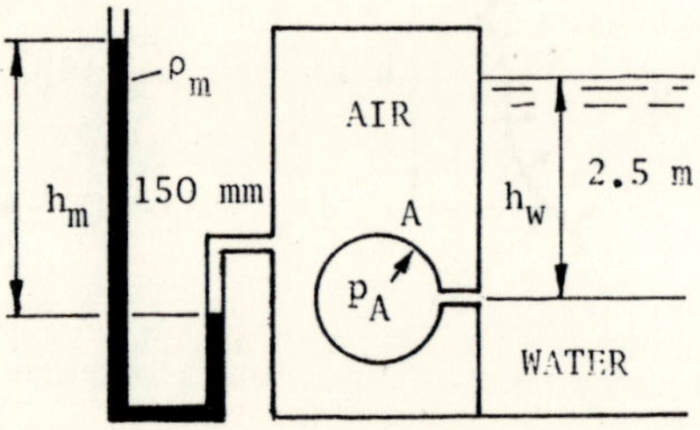


Figure 2.7

From equations (i) and (ii) the reading on pressure gauge A is

$$(p_g)_A = p_A - p_{air} = \rho_w g h_w - \rho_m g h_m$$

$$= 10^3 \times 9.81 \times 2.5 - 13.6 \times 10^3 \times 9.81 \times 0.15 = 4.5 \text{ kPa}$$

*Example 2.2*

A sensitive manometer is formed from a U-tube with a reservoir at the top of each limb (figure 2.8). The cross-sectional area of each reservoir is 20 times that of the tube. Oil of density 880 kg/m$^3$ occupies part of one limb and water of density 1000 kg/m$^3$ occupies the other. Initially, the difference in pressure, $p_A - p_B$, between the two vessels is 200 Pa. Determine the increase in the pressure difference applied across the two free surfaces in the reservoirs if the surface of separation between oil and water is caused to rise through 50 mm, (a) when vessels A and B contain petrol of density 780 kg/m$^3$, and (b) when the vessels contain air of density 1.2 kg/m$^3$.

From equation 2.4 and the fact that the pressure is constant at all points in a horizontal line xx in a continuous fluid we have initially

$$p_A + \rho g h_A + \rho_o g h_o$$
$$= p_B + \rho g h_B + \rho_w g h_w$$

Therefore

$$p_B - p_A = (\rho g h_A + \rho_o g h_o) - (\rho g h_B + \rho_w g h_w) \qquad \text{(i)}$$

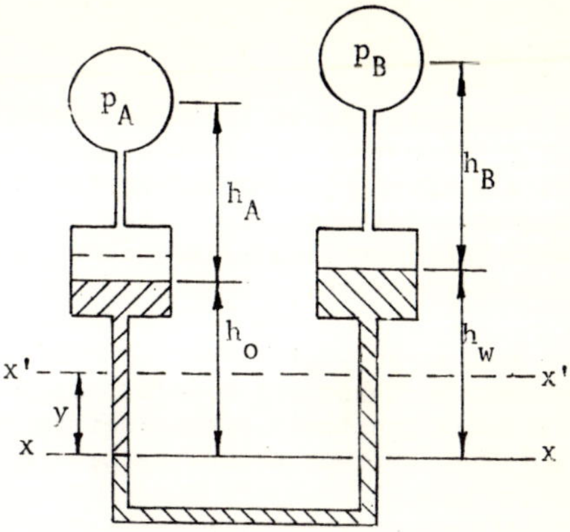


Figure 2.8

Let the pressures change to $p_A'$ and $p_B'$, and let the corresponding movement of the meniscus be y. The pressure along line x'x', must be constant. Therefore

$$p_A' + \rho g(h_A - y/20) + \rho_o g(h_o - y + y/20)$$

$$= p_B' + \rho g(h_B + y/20) + \rho_w g(h_w - y - y/20)$$

$$p_B' - p_A' = (\rho g h_A + \rho_o g h_o) - (\rho g h_B + \rho_w g h_w) - \frac{19}{20}\rho_o g y$$

$$- \frac{2}{20}\rho g y + \frac{21}{20}\rho_w g y \qquad \text{(ii)}$$

From equations (i) and (ii)

$$p_B' - p_A' = (p_B - p_A) + \left[- \frac{19}{20}\rho_o g y - \frac{2}{20}\rho g y + \frac{21}{20}\rho_w g y\right] \qquad \text{(iii)}$$

The second group of terms on the right hand side of the equation is the increase in pressure difference between the two vessels.

(a) When vessels A and B contain petrol

$$p_B' - p_A' = 2 \times 10^2 + \frac{9.81}{20}[-19 \times 880 \times 0.05 - 780 \times 2 \times 0.05 + 21 \times 10^3 \times 0.05]$$

$$= 2 \times 10^3 + \frac{9.81}{20}(-836 - 78 + 1050)$$

$$= 200 + 66.7 = 266.7 \text{ Pa}$$

The increase in pressure difference is 66.7 Pa

(b) When vessels A and B contain air

$$p_B' - p_A' = 2 \times 10^2 + \frac{9.81}{20}(-836 - 1.2 \times 2 \times 0.05 + 1050)$$

$$= 200 + 105 = 305 \text{ Pa}$$

The increase in pressure difference is 105 Pa. The contribution of the column of air is negligible.

*Example 2.3*

A model atmosphere is to be investigated in order to predict the drag on a space vehicle. The atmosphere in question follows the gas law $pv = RT(1 - p/b)$ where R is the characteristic gas constant and b is a given constant. The absolute temperature, T, decreases linearly with increase in altitude i.e. $(dT/dz) = -\lambda$ where $\lambda$ is the temperature lapse rate.

Derive an expression for T in terms of $T_o$, p, $p_0$, b, R and g where the subscript o refers to datum conditions and g is the gravitational constant. Hence determine the density of the atmosphere and the altitude where the pressure is 15 kPa.

Assume $p_0$ = 300 kPa, $T_o$ = 200 K, b = 3.2 MPa, R = 500 J/kg K, g = 15 m/s$^2$, $\lambda$ = 0.006 K/m.

Combining equation 2.3 with $dT/dz = -\lambda$ we obtain

$$dp = (\rho g/\lambda)\ dT \qquad \text{(i)}$$

We are given

$$v = 1/\rho = (RT/p)(1 - p/b) \qquad \text{(ii)}$$

From equations (i) and (ii)

$$\left(\frac{dp}{p} - \frac{dp}{b}\right) = \frac{g}{\lambda R}\ \frac{dT}{T}$$

$$\int\left(\frac{dp}{p} - \frac{dp}{b}\right) = \frac{g}{\lambda R}\int\frac{dT}{T}$$

$$\ln\left(\frac{p}{p_0}\right) - \frac{1}{b}(p - p_0) = \frac{g}{\lambda R}\ln\left(\frac{T}{T_o}\right)$$

$$\ln\left[\frac{15 \times 10^3}{3 \times 10^5}\right] - \frac{(1.5 - 30)10^4}{3.2 \times 10^6} = \frac{15}{0.006 \times 500}\ln\left(\frac{T}{200}\right)$$

$$T = 112 \text{ K}$$

$$\rho = (p/RT)(1 - p/b)^{-1}$$

$$= [(15 \times 10^3)/(500 \times 112)][1 - 15 \times 10^3/3.2 \times 10^6]^{-1} = 0.269 \text{ kg/m}^3$$

$$T = T_o - \lambda z$$

and $z = (T_o - T)/\lambda = (200 - 112)/0.006 = 14\ 667$ m

*Example 2.4*

An oil tanker has atmospheric vents at A and B and contains oil of density 850 kg/m$^3$ to a depth of 2 m (figure 2.9). Determine (i) the maximum uniform acceleration to the right for no spillage from A, and (ii) the pressure at A if the acceleration is 8 m/s$^2$ to the right when A is shut and B is open.

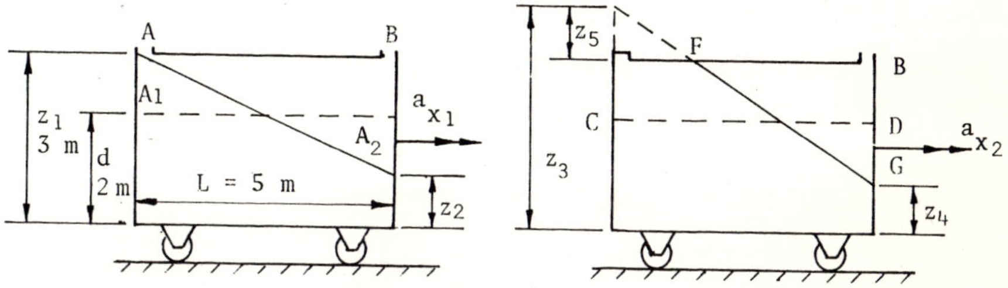


Figure 2.9

For horizontal acceleration only, equations 2.11 and 2.12 give

$$p = -\rho(a_x x + gz) + C \qquad \text{(i)}$$

$$\text{and} \quad \frac{dz}{dx} = -\frac{a_x}{g} \qquad \text{(ii)}$$

(i) When the liquid level reaches A, the geometry of the surface gives $z_1 = 3$ m and $z_2 = 1$ m. Equation (ii) gives

$$\frac{dz}{dx} = \frac{(z_2 - z_1)}{L} = -\frac{a_x}{g}$$

$$a_x = \frac{g(z_1 - z_2)}{L} = \frac{9.81(3 - 1)}{5} = 3.92 \text{ m/s}^2$$

(ii) When vent A is shut and the tanker is accelerated at 8 m/s$^2$ the slope of the surface is given by

$$\frac{dz}{dx} = \frac{z_4 - z_3}{L} = -\frac{a_x}{g}$$

$$z_3 - z_4 = \frac{La_x}{g} = \frac{5 \times 8}{9.81} = 4.08 \text{ m}$$

$$\text{Also} \quad \frac{dz}{dx} = \frac{BG}{FB} = \frac{z_4 - z_3}{L} = \frac{-4.08}{5} = -0.816$$

Area of triangle FBG = $\frac{1}{2}$FB × BG = $\frac{1}{2}(BG)^2/0.816$

For no spillage

Area FBG = Area ABCD

$\frac{1}{2}(BG)^2/0.816 = 5 \times 1$

BG = 2.856 m

$z_4 = z_1 - BG = 3 - 2.856 = 0.144$ m

From equation (i)

$$p = -\rho(a_x x + gz) + C$$

At point G, $x = 5$ m, $z = 0.144$ m, $p = p_a$. Hence $C = p_a + 35.2 \times 10^3$.

$$p = -\rho(a_x x + gz) + p_a + 35.2 \times 10^3$$

At point A, $x = 0$, $z = 3$ m, $p = p_A$. Therefore

$$p_A - p_a = -850(8 \times 0 + 9.81 \times 3) + 35.2 \times 10^3 = 10.2 \text{ kPa}$$

Note that the gauge pressure, $(p_g)_A = p_A - p_a = 10.2$ kPa, could also be obtained by applying the hydrostatic law to a head of oil $z_5 = 1.224$ m obtained by the imaginary extension of the free surface line.

*Example 2.5*

The cylindrical fuel tank of a rocket is 3 m diameter and its axis is parallel to that of the rocket. The tank is pressurised to 200 kPa and the initial depth of fuel is 3.5 m. (i) If the take-off acceleration of the rocket is 10 m/s$^2$, determine the fluid pressure at the bottom of the tank immediately after firing. (ii) If the net accelerating force on the rocket remains constant, determine the pressure at the tank bottom when 40% of the fuel has burned the total mass of the rocket is 70% of its initial value and the acceleration is 20 m/s$^2$. $\rho_{fuel} = 850$ kg/m$^3$.

For vertical acceleration only, equations 2.11 and 2.12 give

$$p = -\rho(a_z + g)z + C \quad \text{(i)}$$

$$\frac{dz}{dx} = -\frac{0}{a_z + g} = 0$$

Therefore the free surface is horizontal

(i) when $z = 3.5$ m, $p = 200$ kPa

Therefore, from equation (i)

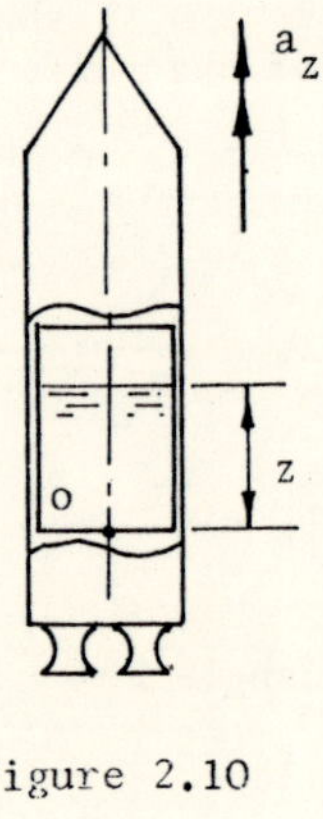


Figure 2.10

$$C = p + \rho(a_z + g)z = 200 \times 10^3 + 850(10 + 9.81)3.5 = 258.9 \text{ kPa}$$

The pressure at the bottom of the tank ($z = 0$) is

$$p_B = -\rho(a_z + g)z + 258.9 \times 10^3 = -850(10 + 9.81)0 + 258.9 \times 10^3$$

$= 258.9$ kPa

(ii) Let F be the accelerating force on the rocket and m the initial mass of the rocket. From Newton's second law

$$F = ma_{z_1} \qquad \text{(ii)}$$

When the mass reduces to 0.7m for constant F, we have

$$F = (0.7m)a_{z_2} \qquad \text{(iii)}$$

From equations (ii) and (iii)

$$a_{z_2} = \frac{a_{z_1}}{0.7} = \frac{20}{0.7} = 28.57 \text{ m/s}^2$$

When 40% of the fuel burns the level of fuel drops to $(0.6 \times 3.5) = 2.1$ m. Assuming that the tank is pressurised as in case (i) $p = 200$ kPa when $z = 2.1$ m.

$$C = 200 \times 10^3 + 850(28.57 + 9.81)2.1 = 268.5 \text{ kPa}$$

$$p_B = -850(28.57 + 9.81)0 + 268.5 \times 10^3 = 268.5 \text{ kPa}$$

*Example 2.6*

A cylindrical tank 100 mm diameter and 300 mm high is open at the top. When stationary it is partly filled with water to a depth of 150 mm. If the tank is now rotated about its vertical axis determine the maximum angular speed in rad/s for the water to be retained.

The tank is now sealed by a 100 mm diameter disc placed parallel to, and 200 mm above, the bottom of the tank. If the tank is rotated at the speed determined in (i) determine the upthrust on the disc.

From equation 2.14 the profile of the free surface is $dz/dr = \omega^2 r/g$ which is the equation of a parabola symmetrical about the z-axis. From the properties of a paraboloid, the volume shown hatched is equal to half that of the enclosing cylinder (figure 2.11a). Equating liquid volumes before and after rotation we have

$$\pi r_2^2 z_0 + \tfrac{1}{2}\pi r_2^2(z_2 - z_0) = \pi r_2^2 d$$

$$z_0 = 2d - z_2 = (2 \times 150) - 300 = 0$$

Thus the free surface just touches the bottom of the tank.

From equation 2.15 we have

$$z = \frac{\omega^2 r^2}{2g} + c \qquad \text{(i)}$$

When $z = z_o = 0$, $r = r_o = 0$. Therefore $c = 0$

At the top edge of the tank $z_2 = 300$ mm, $r_2 = 50$ mm

$$\omega = \left(\frac{2gz_2}{r^2}\right)^{\frac{1}{2}} = \left(\frac{2 \times 9.81 \times 0.3}{0.05^2}\right)^{\frac{1}{2}} = 48.5 \text{ rad/s}$$

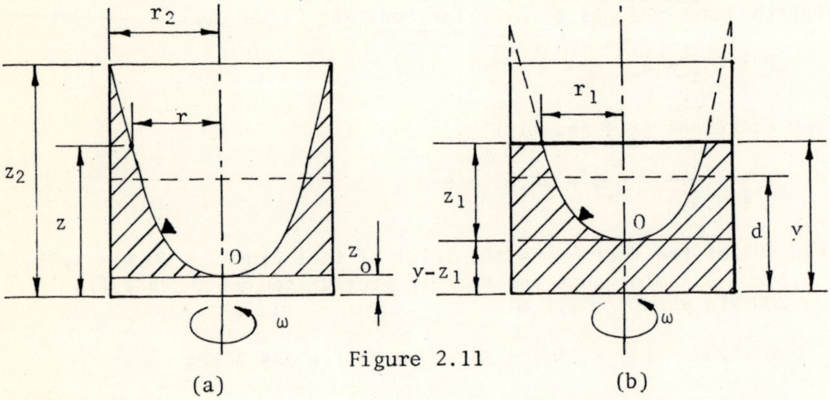


Figure 2.11

When the tank is sealed by a disc and rotated at 48.5 rad/s, the profile of the free surface is as shown in figure 2.11b. Equating liquid volumes before and after rotation, we have

$$\pi(r_2^2 - r_1^2)z_1 + \tfrac{1}{2}\pi r_1^2 z_1 = \pi r_2^2[d - (y - z_1)]$$

$$\tfrac{1}{2}r_1^2 z_1 = r_2^2(y - d) \qquad \text{(ii)}$$

Taking datum at point 0, we have from equation (i)

$$z_1 = \frac{\omega^2 r_1^2}{2g} \qquad \text{(iii)}$$

From equations (ii) and (iii)

$$\frac{\omega^2 r_1^4}{4g} = r_2^2(y - d)$$

$$r_1 = \left(\frac{4gr_2^2(y - d)}{\omega^2}\right)^{\frac{1}{4}} = \left(\frac{4 \times 9.81 \times 0.05^2(0.2 - 0.15)}{48.5^2}\right)^{\frac{1}{4}} = 38 \text{ mm}$$

$$z_1 = \frac{\omega^2 r_1^2}{2g} = \frac{48.5^2 \times 0.038^2}{2 \times 9.81} = 173 \text{ mm}$$

The pressure at any point within a rotating cylinder of liquid is given by equation 2.13

$$p = \rho\left(\frac{\omega^2 r^2}{2} - gz\right) + c$$

When $r = 0$, $z = 0$, $p = p_a$. Therefore $c = p_a$

$$p_g = p - p_a = \rho\left(\frac{\omega^2 r^2}{2} - gz\right) \qquad \text{(iv)}$$

Equation (iv) gives the gauge pressure at any radius r acting on an annular element of area $2\pi r\ dr$. The net force on the wetted surface of the disc is

$$F = \int_{r_1}^{r_2} p_g 2\pi r\ dr = 2\pi\rho\int_{r_1}^{r_2}\left(\frac{\omega^2 r^2}{2} - gz\right) r\ dr$$

Integrating

$$F = \pi(r_2{}^2 - r_1{}^2)\left(\frac{\rho\omega^2}{4}(r_2{}^2 + r_1{}^2) - \rho g z_1\right)$$

$$= \pi(0.05^2 - 0.038^2)\left(\frac{10^3 \times 48.5^2}{4}(0.05^2 + 0.038^2)\right) -$$

$$\pi(0.05^2 - 0.038^2)(10^3 \times 9.81 \times 0.173) = 2.06\ \text{N}$$

This force is equal to the hydrostatic force on the disc due to the imaginary extension of the free surface line.

*Example 2.7*

A vertical butterfly valve closes a circular hole of 1.0 m diameter in the side of a tank and is hinged about its horizontal centre line. The tank contains water whose surface is 0.5 m above the top edge of the gate. Calculate the moment about the hinge line which is required to keep the valve closed.

The position of the centre of pressure, measured in the plane of the valve, is given by equation 2.17.

$$Y_p = \frac{I_{SS}}{A\overline{Y}} \qquad \text{(i)}$$

From the parallel axes theorem

$$I_{SS} = I_{GG} + A\overline{Y}^2 \qquad \text{(ii)}$$

Therefore, from equations (i) and (ii)

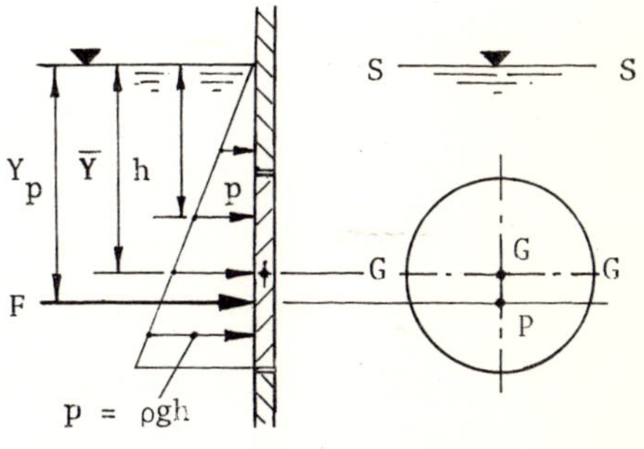


Figure 2.12

$$Y_p = Y + \frac{I_{GG}}{A\overline{Y}}$$

For a circular plate, $I_{aa} = \pi D^4/64$ and $A = \pi D^2/4$. Therefore

$$Y_p = \overline{Y} + \frac{D^2}{16\overline{Y}} \qquad \text{(iii)}$$

Note that the centre of pressure always lies below the centroid.

The hydrostatic force acting on the valve is given by equation 2.16

$$F = \rho g\overline{Y}A\cos\theta = 10^3 \times 9.81 \times 1 \times 1 \times \tfrac{1}{4}\pi^2 \times 1^2 = 7.705 \text{ kN}$$

The moment about the hinge line is

$$M = F(Y_p - \overline{Y}) = \frac{FD^2}{16\overline{Y}} = \frac{7.705 \times 10^3 \times 1^2}{16 \times 1} = 482 \text{ N m}$$

*Example 2.8*

A horizontal square duct, 200 mm by 200 mm, conveys a salt solution to evaporating pans. The end of the duct is closed by a vertical square flat valve hinged about its horizontal centre line. The duct normally conveys a salt solution of density 1200 kg/m$^3$ but, after the valve is left closed for a long period, the salt settles with the result that the density varies linearly from 1100 kg/m$^3$ at the top of the duct to 1300 kg/m$^3$ at the bottom. Working from first principles, and assuming that the pressure exerted by the solution at the top of the gate is atmospheric, determine the moment required to maintain the valve closed.

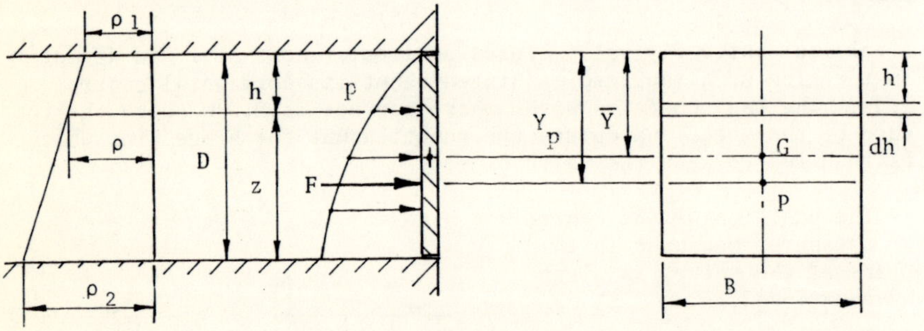


Figure 2.13

At any elevation z, the pressure gradient is given by equation 2.3

$$\frac{dp}{dz} = -\rho g \qquad \text{(i)}$$

If the depth below the top edge of the duct is denoted by h we have

$$h = D - z$$

and $dh = -dz$ (ii)

From equations (i) and (ii) the gauge pressure at any depth h is given by

$$p = \int \rho g \, dh \qquad \text{(iii)}$$

From the density distribution diagram, the density at any depth h is given by

$$\rho = \rho_1 + h(\rho_2 - \rho_1)/D \qquad \text{(iv)}$$

From equations (iii) and (iv)

$$p = \int g[\rho_1 + h(\rho_2 - \rho_1)/D]dh = g[\rho_1 h + \frac{h^2}{2D}(\rho_2 - \rho_1)]$$

The hydrostatic force F on the valve of width B is

$$F = \int pB\ dh = \int_0^D g[\rho_1 h + \frac{h^2}{2D}(\rho_2 - \rho_1)]dh = \frac{gBD^2}{6}(2\rho_1 + \rho_2) \qquad \text{(v)}$$

The moment M about the top edge of the valve is

$$M = \int pB\ dh\ h = \int_0^D gB[\rho_1 h^2 + \frac{h^3}{2D}(\rho_2 - \rho_1)]dh$$

$$= \frac{gBD^3}{24}(5\rho_1 + 3\rho_2) \qquad \text{(vi)}$$

$$Y_p = \frac{M}{F} = \frac{(gBD^3/24)(5\rho_1 + 3\rho_2)}{(gBD^2/6)(2\rho_1 + \rho_2)} = \frac{D(5\rho_1 + 3\rho_2)}{4(2\rho_1 + \rho_2)} \qquad \text{(vii)}$$

Note that for a constant density fluid, $\rho_1 = \rho_2 = \rho$, and for a rectangular surface with its top edge coincident with the liquid surface, $Y_p = 2D/3$.

The moment about the hinge point is given by

$$M_H = F(Y_p - \bar{Y}) = \frac{gBD^2}{6}(2\rho_1 + \rho_2)\left[\frac{D(5\rho_1 + 3\rho_2)}{4(2\rho_1 + \rho_2)} - \bar{Y}\right]$$

$$= \frac{9.81 \times 0.2 \times 0.2^2}{6}(2 \times 1.1 + 1.3)10^3\left[\frac{0.2(5 \times 1.1 + 3 \times 1.3)}{4(2 \times 1.1 + 1.3)} - 0.1\right]$$

$$= 1.57\ \text{N m}$$

*Example 2.9*

A sluice gate is formed from sheet metal such that the internal boundary of the section satisfies the relationship $z = 4x^2$. The gate is pivoted at point A and its centre of gravity is at G as shown in figure 2.14. When the water surface is level with the pivot A, determine the magnitude and direction of the resultant pressure force on the gate and the turning moment required to open the gate. The gate is 1 m wide and has a weight of 10 kN.

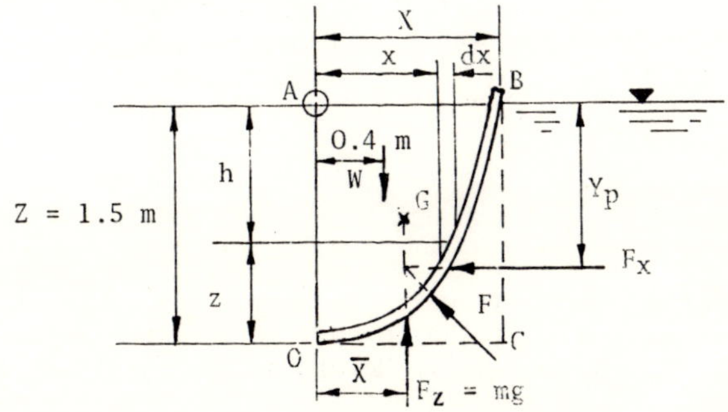


Figure 2.14

Using the method outlined in section 2.4.2, we must consider the vertical and horizontal forces acting on the gate. From equation 2.19, the vertical force, $F_z$, is equal to the weight of the water displaced over the gate width L. The force $F_z$ acts at the centroid of the area A of the section, distance X from the z axis. Considering the moment on an element of area $dA = (1.5 - 4x^2)dx$ and integrating, we obtain

$$\overline{X} = \frac{\int x \, dA}{\int dA} = \frac{\int_0^X x(1.5 - 4x^2)dx}{\int_0^X (1.5 - 4x^2)dx}$$

The given relationship is $z = 4x^2$. Therefore, when $Z = 1.5$ m, $X = 0.612$ m.

$$\overline{X} = \frac{\int_0^{0.612} x(1.5 - 4x^2)dx}{\int_0^{0.612} (1.5 - 4x^2)dx} = \frac{0.14}{0.612} = 0.23 \text{ m}$$

From equation 2.19, the vertical force, $F_z$, is given by

$$F_z = mg = \rho gLA = 10^3 \times 9.81 \times 1 \times 0.612 = 6 \text{ kN}$$

From equations 2.16 and 2.18, the horizontal force, $F_x$, is the force acting on the projection of the curved surface on a vertical plane BC.

$$F_x = \rho g\overline{Y} \, (ZL) = 10^3 \times 9.81 \times \tfrac{1}{2} \times 1.5 \times 1.5 \times 1 = 11.04 \text{ kN}$$

The magnitude and direction of the resultant force are given by

$$F = \sqrt{(F_z^2 + F_x^2)} = \sqrt{(6^2 + 11.04^2)} = 12.57 \text{ kN}$$

$$\alpha = \tan^{-1} (F_z/F_x) = \tan^{-1} (6/11.05) = 28.5°$$

The anticlockwise turning moment, M, required to open the gate is given by

$$M + F_z\overline{X} = F_xY_p + W \times S$$

Note from example 2.9 that for a vertical rectangular surface, with its upper edge coincident with the water surface, $Y_p = 2Z/3$.

$$M = 11.04 \times 10^3 \times \frac{2}{3} \times 1.5 + 10^4 \times 0.4 - 6 \times 10^3 \times 0.612$$

$$= 11.37 \text{ kN}$$

*Example 2.10*

A floating platform for drilling oil wells in deep water is constructed as a square floor supported at the corners by vertical cylinders to provide buoyancy (figure 2.15). Determine the condition for neutral stability about an axis parallel to AA.

Let the original depth of immersion of the cylinders be h and the height of the centre of gravity, G, of the platform be L above the water-line. Consider the incremental volumes of the vertical cylinders that are moved below or above the water-line as the platform tips. For simplicity, consider these incremental volumes to be in the shape of right circular discs whose top and bottom surfaces are parallel.

When the platform tips through a small angle $\theta = AA'/R$ the weight of the volume of water displaced by the cylinders remains unchanged. Therefore the buoyancy force remains at

$$F_b = 4\rho g \tfrac{1}{4}\pi D^2 h$$

The moment due to the movement of the centre of buoyancy from B to B' is given by

$$M_t = BB' \times F_b$$

$$= \rho g \pi D^2 h BB' \qquad \text{(i)}$$

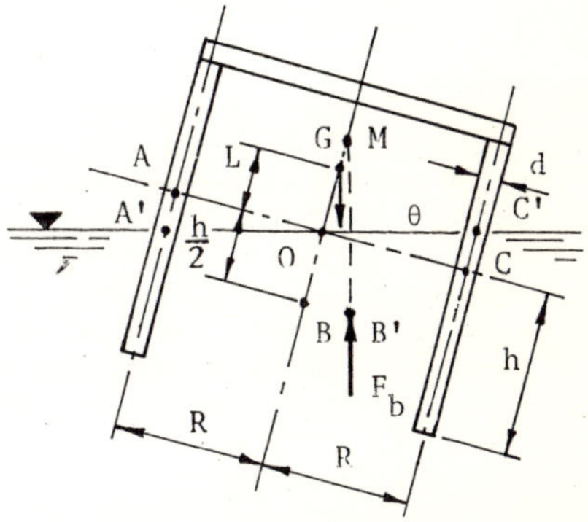


Figure 2.15

The weight of the volume of water displaced by the length of cylinder AA' is equal to that due to the length of cylinder CC'. Therefore, the total moment due to the altered displacement is also given by

$$M_t = 4\rho g \tfrac{1}{4}\pi D^2 AA' \times R = \rho g \pi D^2 R^2 \theta \qquad \text{(ii)}$$

Equating (i) and (ii)

$$BB' = \frac{R^2\theta}{h}$$

and $$BM = \frac{BB'}{\theta} = \frac{R^2}{h} \qquad \text{(iii)}$$

The metacentric height is GM

$$GM = BM - BG = \frac{R^2}{h} - (L + \frac{h}{2}) \qquad \text{(iv)}$$

For neutral stability the metacentric height M coincides with the centre of gravity G i.e. GM = 0. From equation (iv), the condition for neutral stability is therefore

$$L = \frac{R^2}{h} - \frac{h}{2}$$

Note that if GM is negative (i.e. if M lies below G) the platform is unstable.

A more general expression for BM may be deduced from equation (iii)

$$BM = \frac{R^2}{h} = \frac{4 \times \frac{1}{4}\pi D^2 R^2}{4 \times \frac{1}{4}\pi D^2 h} = \frac{I_{ss}}{V} \qquad \text{(v)}$$

Where $I_{ss}$ is the second moment of area of the water-line plane about OO and V is the volume of liquid displaced. (A more rigorous proof is obtained by considering a small element of area a in the water-line plane, distance x from OO, and integrating.)

*Example 2.11*

The level of water in a container is to be controlled by the arrangement shown in figure 2.16. A plane gate, 0.5 m square, is hinged about its upper horizontal edge, O. The arm OS is rigidly attached to the gate at O and its other end S is connected to the float F by means of the wire SA. In the condition shown, the surface level in the container is 2 m above the hinge line O and the length of the wire has been adjusted so that it is straight but free from tension. The float is a vertical circular cylinder, 1.5 m long, of mass 785 kg. It is required to be just sufficiently large to open the gate when the surface level rises through 0.2 m. Determine the diameter of the float and check if the float is stable for all conditions. The centre of gravity of the float is at the mid-point of the axis.

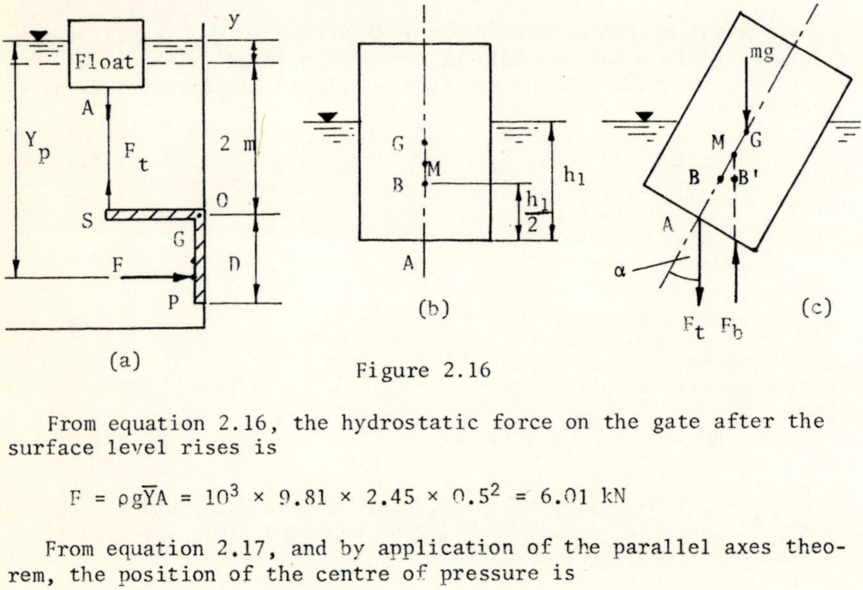


Figure 2.16

From equation 2.16, the hydrostatic force on the gate after the surface level rises is

$$F = \rho g\bar{Y}A = 10^3 \times 9.81 \times 2.45 \times 0.5^2 = 6.01 \text{ kN}$$

From equation 2.17, and by application of the parallel axes theorem, the position of the centre of pressure is

$$Y_p = \frac{I_{ss}}{A\bar{Y}} = \frac{BD\bar{Y}^2 + BD^3/12}{BD\bar{Y}} = 2.45 + \frac{0.5^2}{12 \times 2.45} = 2.46 \text{ m}$$

The moment about hinge O due to the hydrostatic force is

$$M = F(Y_p - 2.2) = 6.01 \times 10^3(2.46 - 2.2) = 1.556 \text{ kN m} \quad \text{(i)}$$

The moment about hinge 0 is also equal to that produced by the increase in buoyancy force due to the displaced water (buoyancy due to displaced air is neglected).

$$M = F_t OS = \rho g \tfrac{1}{4}\pi D^2 y OS = 10^3 \times 9.81 \times \tfrac{1}{4}\pi D^2 \times 0.2 \times 1$$

$$= 1.54 D^2 \text{ kN m} \qquad \text{(ii)}$$

Equating (i) and (ii)

$$D = \left(\frac{1.556}{1.54}\right)^{\frac{1}{2}} \simeq 1 \text{ m}$$

Initially, the wire is free from tension and exerts no force on the float (figure 2.16b). The buoyancy force, $F_b$, is equal to the weight, mg, of the float. Therefore

$$h_1 = \frac{mg}{\rho g \frac{1}{4}\pi D^2} = \frac{785 \times 4}{10^3 \times \pi \times 1^2} = 1 \text{ m}$$

$$BG = \frac{L}{2} - \frac{h_1}{2} = 0.75 - 0.5 = 0.25 \text{ m}$$

From example 2.10 (equation v) we have

$$BM = \frac{I_{ss}}{V_1} = \frac{\pi D^4/64}{(\pi D^2/4)h_1} = \frac{1^2}{16 \times 1} = 0.0625 \text{ m}$$

$$GM = BG - BM = 0.25 - 0.0625 = 0.1875 \text{ m}$$

Thus, the metacentric height, GM, is negative (M below G) and overturning would occur for any inclination of the axis. The float is unstable.

Now consider the float with the retaining wire attached (figure 2.16c). The buoyancy force, $F_b$, is equal to the sum of the weight of the float, mg, and the wire tension, $F_t$.

$$F_b = mg + F_t = 785 \times 9.81 + 1.556 \times 10^3 = 9.257 \text{ kN}$$

$$h_2 = h_1 + 0.2 = 1 + 0.2 = 1.2 \text{ m}$$

$$BG = AG - \frac{h_2}{2} = 0.75 - 0.6 = 0.15 \text{ m}$$

$$BM = \frac{I_{ss}}{V_2} = \frac{\pi D^4/64}{(\pi D^2/4)h_2} = \frac{1^2}{16 \times 1.2} = 0.0521 \text{ m}$$

$$GM = BG - BM = 0.15 - 0.0521 = 0.0979 \text{ m}$$

Consider moments about G when the float is displaced through an angle $\alpha$ from the vertical (figure 2.16c). For stability, the restoring moment, $M_t$, due to wire tension must be greater than the overturning moment, $M_b$, due to the buoyancy force.

$$M_t = F_t \times AG \sin\alpha = 1.556 \times 10^3 \times 0.75 \sin\alpha = 1.167 \sin\alpha \text{ kN}$$

$$M_b = F_b \times GM \sin\alpha = 9.257 \times 10^3 \times 0.0979 \sin\alpha = 0.906 \sin\alpha \text{ kN}$$

The float is stable since $M_t > M_b$.

*Problems*

1 A U-tube manometer is used to measure the pressure of kerosine in a pipeline A (figure 2.3b). The manometer fluid is mercury and the recorded head, h, is 500 mm. If the mercury-kerosine interface is 750 mm above the pipe centre line determine the gauge pressure of the kerosine. If conditions in the pipeline now change so that the mercury-air interface falls 80 mm below the mercury interface with no change in manometer position, determine the absolute pressure in the pipeline. Atmospheric pressure is 1 bar. $\rho_k$ = 820 kg/m$^3$, $\rho_m$ = 13 600 kg/m$^3$.

[73 kPa, 97.65 kPa]

2 A mercury manometer is used to measure the pressure difference between two pressure vessels. For the arrangement and readings shown in figure 2.17, determine the piezometric pressure difference $p_A^* - p_B^*$ when the vessels A and B contain (i) water and (ii) air of density 1.5 kg/m$^3$.

[49.38 kPa, 53.36 kPa]

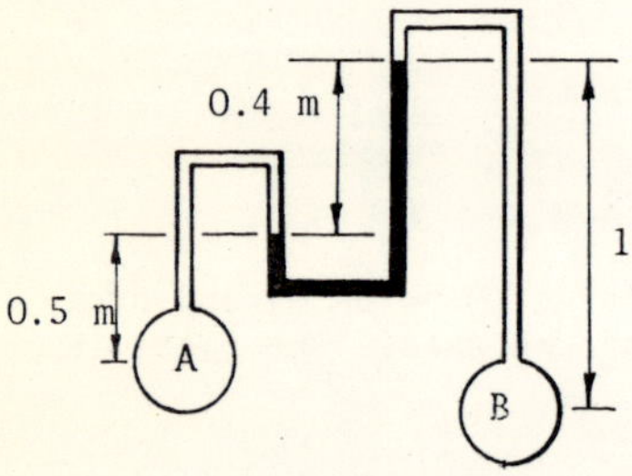


Figure 2.17

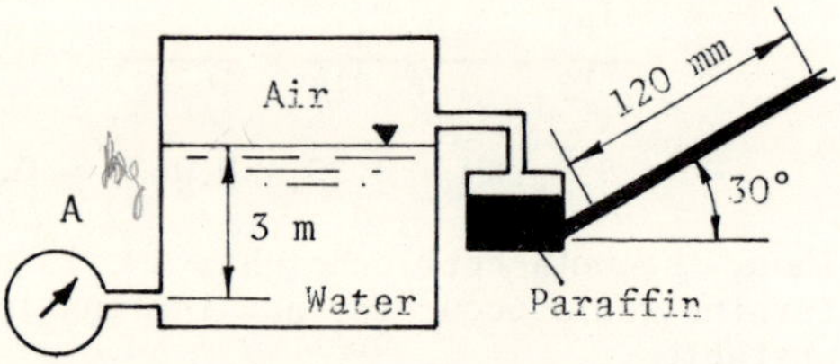


Figure 2.18

3 Determine the reading on pressure gauge A in figure 2.18. $\rho_w$ = 1000 kg/m$^3$, $\rho_p$ = 780 kg/m$^3$, $\rho_a$ = 1.2 kg/m$^3$.

[29.89 kPa]

4 In a chemical reactor 65 m tall the density $\rho$ of a fluid mixture in kg/m$^3$ varies with distance z in metres above the bottom of the reactor according to the relationship $\rho = 1 - (z/150) + (z/300)^2$. Assuming the fluid mixture to be stationary determine the pressure difference between the bottom and top of the reactor.

[509 Pa]

5 Show that, for an adiabatic atmosphere $p/\rho^\gamma$ = const

$$g(z - z_0) = RT_1 \frac{\gamma}{\gamma - 1}\left\{1 - \left(\frac{p}{p_0}\right)^{(\gamma-1)/\gamma}\right\}$$

where subscript o refers to the reference state.

6 Determine the air density at an altitude of 15000 m in the International Standard Atmosphere. The stratosphere may be assumed to commence at 11000 m at which point the pressure is 0.227 bar.

[0.1942 kg/m$^3$]

7 In the lower part of the earth's atmosphere the temperature T may be considered to vary with altitude z according to the linear relation $T = T_0 - \lambda z$. Prove that the variation of pressure is given by the relation

$$\frac{p}{p_0} = \left(1 - \frac{\lambda z}{T_0}\right)^{g/\lambda R}$$

where g is the gravitational constant ($9.81\ m/s^2$), R is the characteristic gas constant for air (287 J/kg K) and $dT/dz = -\lambda$ is the temperature lapse rate (0.0065 K/m). Assuming that $dp = -\rho g\ dz$, determine the density of air at the bottom of a mine shaft 800 m below sea level and compare your calculated value with that obtained from the I.S.A. tables. Assume sea level conditions of $p_0 = 1.0132$ bar and $T_0 = 288.15$ K.

[1.3244 kg/m$^3$]

8 The petrol tank in a car is basically a rectangular box 750 mm long, 400 mm wide and 300 mm high. The tank is mounted horizontally in the car with the long side parallel to the axis of the rear axle. A fuel level sensor is positioned 50 mm above the bottom face and 150 mm from both the front and nearside faces of the tank. When the petrol level drops below the sensor a fuel warning light comes on. Assuming that when the car is stationary the level in the tank is 60 mm above the bottom, determine the linear acceleration of the car when the light just comes on. Assume $\rho_p = 700\ kg/m^3$.

[1.962 m/s$^2$]

9 Derive an expression for the variation of pressure in a fluid in relative equilibrium. A thin-walled, open-topped tank in the form of a cube of 2 m side is initially full of oil of relative density 0.88. It is accelerated uniformly at 5 m/s$^2$ up a long, straight slope at $\tan^{-1}(\frac{1}{4})$ to the horizontal, the base of the tank remaining parallel to the slope, and the two side faces remaining parallel to the direction of motion. Calculate (a) the volume of oil remaining in the tank when no more spilling occurs, and (b) the pressure at the lowest corners of the tank.

[4.9 m$^3$, 16.75 kPa, 3.77 kPa]

10 A cylindrical vessel 600 mm and 300 mm high is open at the top except for a lip 50 mm wide all round and normal to the side. It contains water to a height of 200 mm above the bottom and is rotated with axis vertical. Calculate the rev/min at which the water reaches the inside edge of the lip and the total pressure force against the underside of the lip.

[90.8 rev/min, 53.2 N]

11 A novel form of angular speed measurement is shown in figure 2.19 where a float connected to a displacement transducer moves with the water level at the cylinder axis. Derive (i) a relationship between the float movement and angular velocity $\omega$, and (ii) the maximum value of $\omega$ the instrument can record and the corresponding initial depth of water.

[$h = 255 \times 10^{-6}\omega^2$ m, 20 rad/s]

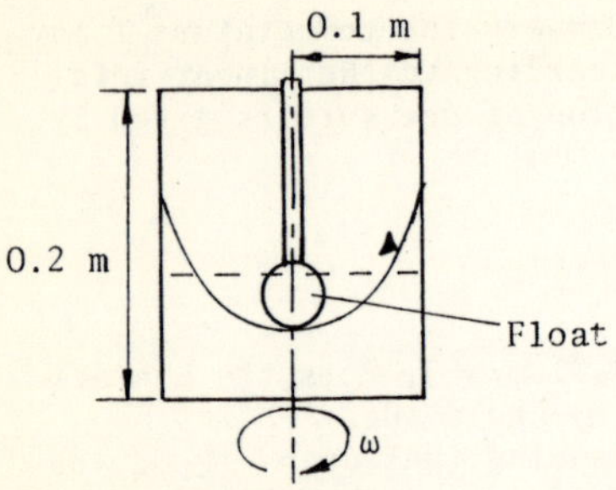


Figure 2.19

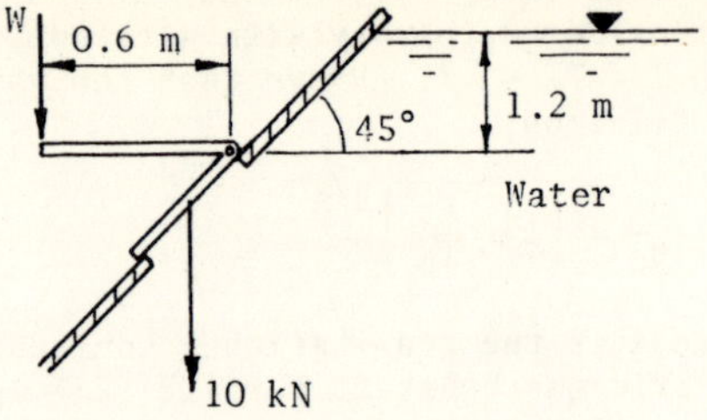


Figure 2.20

12 A circular opening in the sloping side of a reservoir containing water is 0.8 m diameter. It is closed by means of a circular gate which is hinged about its top edge. The gate is kept closed against water pressure partly by its own weight of 10 kN and partly by a load of W acting on a horizontal lever arm of length 0.6 m (figure 2.20). Calculate the value of W such that the gate will begin to open when the water level is 1.2 m above the top of the gate.

[391 N]

13 A rectangular tank length L, width B and depth D is used for storing liquid. Determine the magnitude and point of action of the force on (a) the face of width B and depth D and (b) the bottom face of length L and width B when the tank is half full of liquid of density $\rho_1$, and half full of liquid of lower density $\rho_2$, the liquids being clearly separated. Would the position and magnitude of the forces vary if the two liquids are thoroughly mixed to give a uniform density of $(\rho_1 + \rho_2)/2$?

[$BD^2g(3\rho_2 + \rho_1)/8$, $(D/6)(11\rho_2 + 5\rho_1)/(3\rho_2 + \rho_1)$, $(gDBL/2)(\rho_1 + \rho_2)$, (a)Yes, (b)No]

14 A circular opening of radius R is situated in the vertical side of a tank containing liquid of density $\rho$. The opening is sealed by a pair of semi-circular doors with the split-line along the vertical diameter of the opening. The hinge-line of each door is vertical and a distance $(1 + x)R$ from the split-line. The liquid free surface is a distance 4R above the top edge of the opening. Show that the moment about the hinge required to keep each door closed is given by

$$5\rho g R^4[(1 + x)\pi/2 - 2/3]$$

15 Figure 2.21 shows a vertical section through a tank of rectangular plan, fitted with a gate hinged at O. The gate section shown is a circular arc of radius R, subtending an angle of 90°. The tank has an open top and is filled with liquid of density $\rho$ to a depth H. The gate is kept closed by the application at P of a horizontal force F. Determine (a) the magnitude of the horizontal force F required to keep the gate closed, and (b) the horizontal and vertical components of the reaction at the hinge O, for values H = 5 m, R = 1 m. Neglect the weight of the gate, and consider a gate width of 1 m.
$\rho = 1000$ kg/m$^3$ [44.1 kN, $R_x = 0$, $R_y = 46.9$ kN]

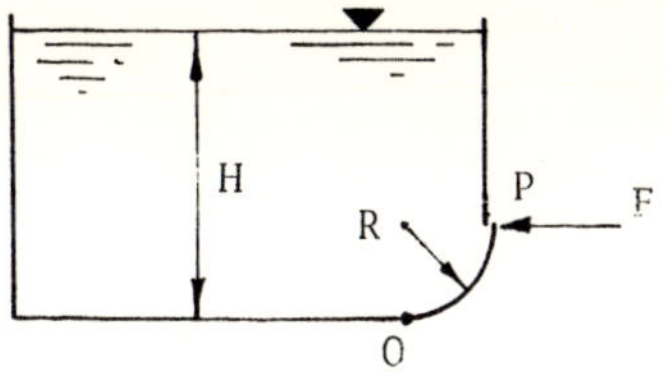


Figure 2.21

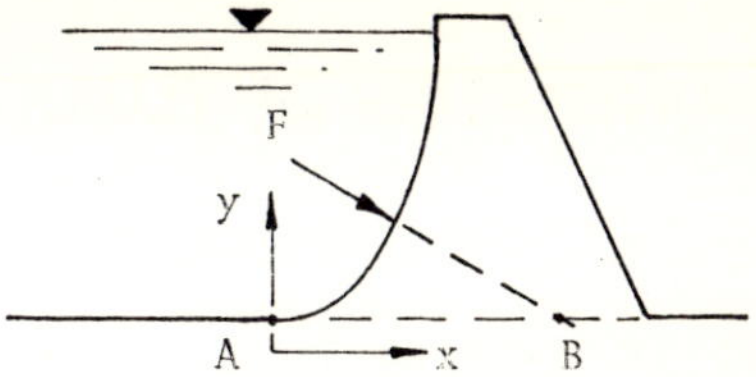


Figure 2.22

16 The face of a dam shown in figure 2.22 is curved according to the relation $y = x^2/24$, where y and x are in metres. The height of the free surface above the horizontal plane through A is 15 m. Calculate the resultant force due to the fresh water acting on unit breadth of the dam and determine the position of the point B at which the line of action of the force cuts the horizontal plane through A.

[2.164 MN, AB = 10.08 m]

17 A fuel tank of triangular cross-section floats in a fresh water lake such that the base of the tank is horizontal and its top corner is in the free surface. The tank is 6 m long and has sides of 2 m. Determine (i) the magnitude and position of the force acting on each triangular end of the tank, (ii) the magnitude and position of the forces acting on the other sides, and (iii) the upthrust exerted on the tank by the water.

[22.6 kN, 1.3 m, 120 kN, 1.732 m, 102 kN]

18 A hydrometer uses the principle of buoyancy to determine the specific gravities, S, of liquids. It is constructed as shown in figure 2.23 with a vertical constant-area stem and a weighted bulb to ensure stable equilibrium. The hydrometer sinks to different depths in different liquids and the specific gravity is recorded by graduations on the stem. If the hydrometer is calibrated in water show that the distance y obtained when the hydrometer is immersed in a liquid with $S \neq 1$ is given by $y = V(S - 1)/AS$ where A is the cross-sectional area of the stem and V is the submerged volume for S = 1.

19 A conical space capsule is designed so that when it returns to earth it floats on the sea as shown in figure 2.24. Determine the value of OG for neutral equilibrium. $\rho_{sw} = 1025$ kg/m$^3$.

[5 m]

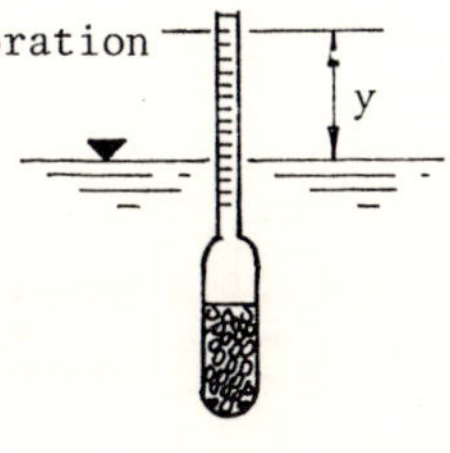


Figure 2.23

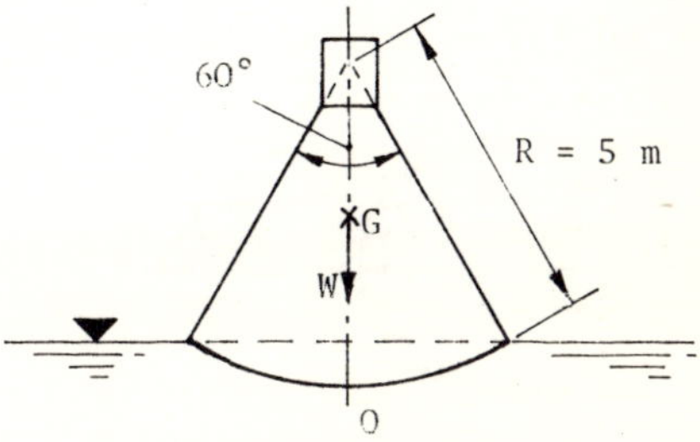


Figure 2.24

# 3 KINEMATICS OF FLUID FLOW

In this chapter the motion of fluids will be discussed in terms of displacements, velocities and accelerations without reference to the forces that cause the motion.

## 3.1 FLOW FIELDS

A *flow field* is a region in which the flow is defined in terms of space and time coordinates. *Steady* flow is constant with respect to time whereas *unsteady* flow varies. Turbulent flow is usually treated as steady provided the time-average flow is constant.

In this text the *Eulerian* method of analysis is used. Thus, the motion is described in terms of the velocity, acceleration, etc., at a particular point in the flow field instead of considering the motion of an individual fluid particle (*Lagrangian* method). The position in the field and the velocity, u, are specified by either cartesian (figure 3.1a) or cylindrical (figure 3.1b) coordinates.

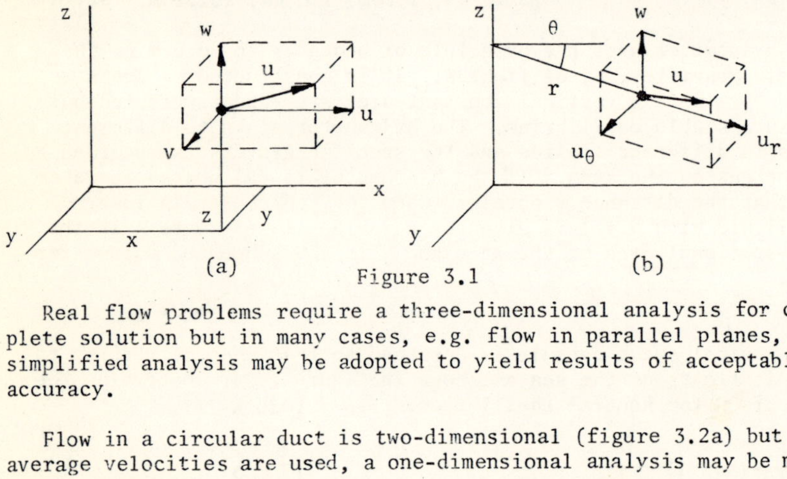


Figure 3.1

Real flow problems require a three-dimensional analysis for complete solution but in many cases, e.g. flow in parallel planes, a simplified analysis may be adopted to yield results of acceptable accuracy.

Flow in a circular duct is two-dimensional (figure 3.2a) but if average velocities are used, a one-dimensional analysis may be made.

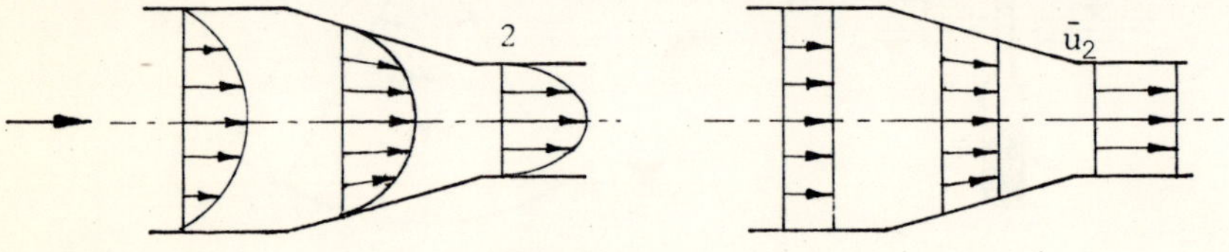


(a) Figure 3.2 (b)

Flow over an aircraft wing of finite span is truly three-dimensional (figure 3.3a) but the wing may be considered to be an aerofoil section of infinite length and a two-dimensional analysis made (figure 3.3b).

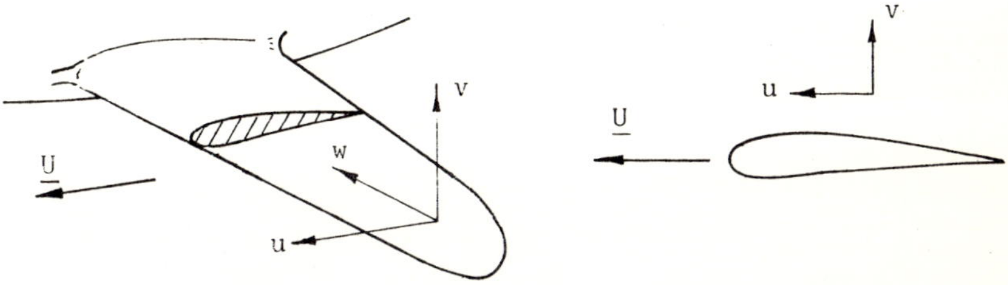


Figure 3.3

When considering moving bodies, it is necessary to select a frame of reference. Velocities are measured relative to earth or relative to the moving body.

Figure 3.4 shows an aerofoil moving with velocity $\underline{U}$, relative to earth, through a fluid which is stationary except where disturbed locally by the aerofoil. Point P is fixed relative to the aerofoil and also moves with velocity $\underline{U}$. Fluid passing through point P has velocity $\underline{u}$, relative to earth, and velocity $\underline{u}_R$, relative to the aerofoil. Velocity $\underline{u}_R$ is given by the vector difference

$$\underline{u}_R = \underline{u} - \underline{U}$$

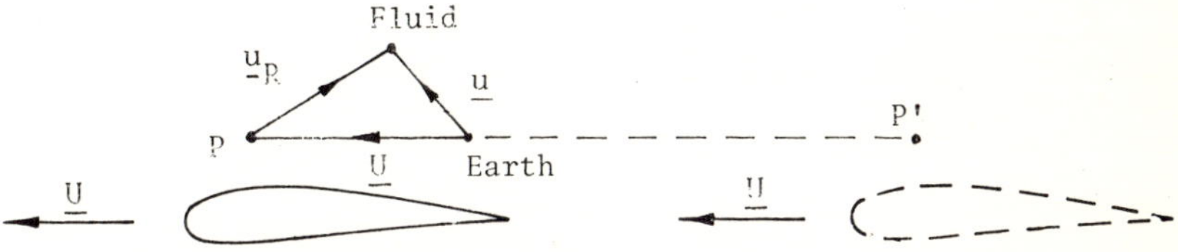


Figure 3.4

The point P is fixed relative to the aerofoil therefore a stationary observer would see a succession of P points and a flow pattern based on $\underline{u}$ which is unsteady. An observer moving with the aerofoil would see a steady flow pattern based on $\underline{u}_R$. It is possible to superimpose a velocity $-\underline{U}$ on the whole field to cause the fluid to flow over a stationary aerofoil and give a steady flow pattern. This principle is used in wind-tunnel testing.

## 3.2 FLOW LINES

The nature of fluid flow in a particular situation can be shown pictorially by the use of streamlines (figure 3.5a). A *streamline* is a line which shows the direction of the velocity vector at any instant.

At every point on the line the velocity is tangential to the line. Thus, there can be no flow across a streamline. A number of adjacent streamlines may form an imaginary flow passage known as a *streamtube* (figure 3.5b) and if the cross-sectional area is small enough for the velocity to be considered constant the flow passage is called a *stream filament*.

A *pathline* is the line of motion traced out by an individual fluid particle in a finite time.

A *streakline* is a line of fluid particles, all of which passed through the same point in the flow field at a previous time. In experimental work, streaklines are obtained by injecting dye, smoke or particles into the moving fluid and observing the subsequent flow pattern.

Streamlines, pathlines and streaklines are identical for steady flow.

For the aerofoil of figure 3.4, streamlines can be drawn for steady flow based on $\underline{u}_R$ and pathlines for unsteady flow based on $\underline{u}$. The resulting patterns are shown in figure 3.5.

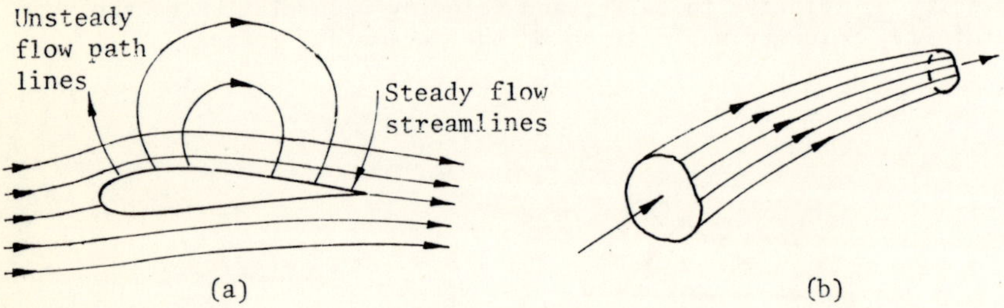


(a) (b)

Figure 3.5

3.3 ACCELERATION

In general, the velocity, $\underline{u}$, of a fluid particle is a function of both position and time. Mathematically we may state

$$\underline{u} = \underline{u}(s,t)$$

and
$$d\underline{u} = \frac{\partial \underline{u}}{\partial s} ds + \frac{\partial \underline{u}}{\partial t} dt$$

Acceleration
$$\underline{a} = \frac{D\underline{u}}{Dt} = \underline{u}\frac{\partial \underline{u}}{\partial s} + \frac{\partial \underline{u}}{\partial t} \tag{3.1}$$

SUBSTANTIAL ACCELERATION = CONVECTIVE ACCELERATION + LOCAL ACCELERATION

For steady flow the local acceleration $\partial \underline{u}/\partial t = 0$.

Now consider two-dimensional flow along a curved streamline. The components of acceleration are tangential ($a_s$) and normal ($a_n$) to the streamline. Referring to figure 3.6 we obtain

$$a_s = u\frac{\partial u_s}{\partial s} + \frac{\partial u_s}{\partial t} = u\frac{\partial u}{\partial s} + \frac{\partial u}{\partial t} \quad (3.2a)$$

$$a_n = u\frac{\partial u_n}{\partial s} + \frac{\partial u_n}{\partial t} = \frac{u^2}{r} + \frac{\partial u_n}{\partial t} \quad (3.2b)$$

For steady flow $a_n = u^2/r$ as obtained in section 2.3.2.

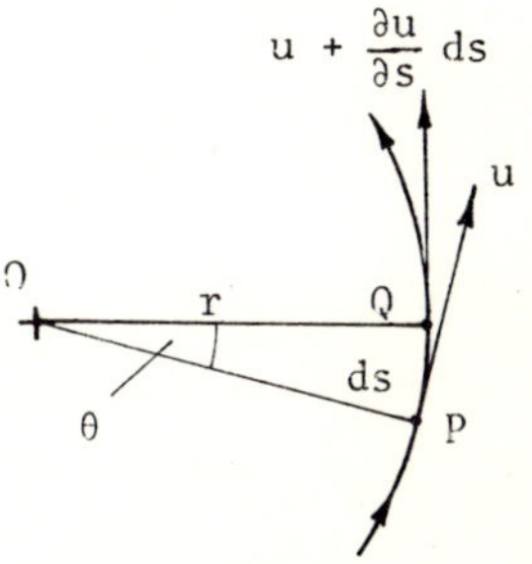


Figure 3.6

## 3.4 CONSERVATION OF MASS AND THE CONTINUITY EQUATION

The principle of conservation of mass states that in the absence of energy-mass conversion matter can neither be created nor destroyed. Figure 3.7 applies this principle to fluid flow through a control volume V having control surface S. The volumetric flow rate through the surface element $d\underline{A}$ is obtained using the *normal* component of the velocity which gives $\underline{u}\cdot d\underline{A} = u \cos\theta \, dA$ and the mass flow rate $\rho\underline{u}\cdot d\underline{A}$.

The total flow rate over the whole surface is

$$\iint_S \rho\underline{u}\cdot d\underline{A}$$

The rate of mass decrease within V with respect to time is

$$-\frac{\partial}{\partial t}\iiint_V \rho \; dV$$

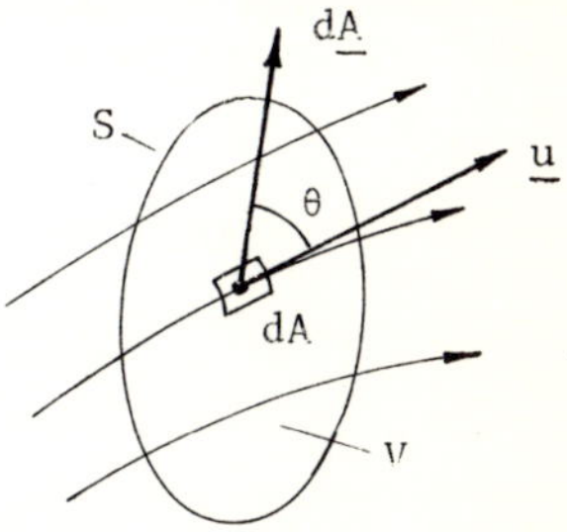


Figure 3.7

Conservation of mass gives the *continuity* equation

$$\iint_S \rho\underline{u}\cdot d\underline{A} + \frac{\partial}{\partial t}\iiint_V \rho \; dV = 0 \quad (3.3)$$

For any steady flow or *incompressible* ($\rho$ = const) unsteady flow the time dependent term disappears.

$$\iint_S \rho\underline{u}\cdot d\underline{A} = 0 \quad (3.4)$$

### 3.4.1 One-dimensional Flow

For the simple case of steady flow in which the velocity and density are assumed uniform across finite areas at entry and exit, the continuity equation simplifies to a one-dimensional situation (figure 3.8). Then with mean velocity $\bar{u}$ perpendicular to A, we obtain

$$\rho_2 \bar{u}_2 A_2 - \rho_1 \bar{u}_1 A_1 = 0$$

and $$\dot{m} = \rho_1 \bar{u}_1 A_1 = \rho_2 \bar{u}_2 A_2 \qquad (3.5a)$$

where $\dot{m}$ is the mass flow rate.

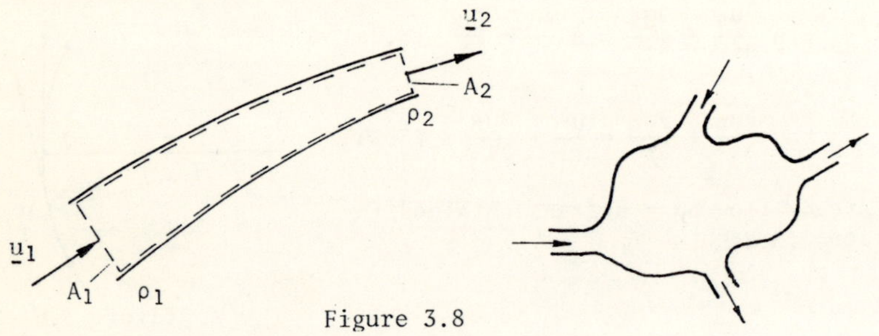


Figure 3.8

For multi-stream flow through control volume

$$\dot{m} = \Sigma(\rho\bar{u}A)_{IN} = \Sigma(\rho\bar{u}A)_{OUT} \qquad (3.5b)$$

3.4.2 Conservation of Mass at a Point

In order to obtain a relationship for conservation of mass at a point, equation 3.3 is applied to a cubic element which is then allowed to shrink to a point. This analysis gives a solution in the form

$$\frac{\partial}{\partial x}(\rho u) + \frac{\partial}{\partial y}(\rho v) + \frac{\partial}{\partial z}(\rho w) = \text{div } \rho \underline{u} = -\frac{\partial \rho}{\partial t} \qquad (3.6)$$

For two-dimensional, incompressible flow this reduces to

$$\frac{\partial u}{\partial x} + \frac{\partial v}{\partial y} = 0 \qquad (3.7a)$$

In polar coordinates

$$\frac{\partial}{\partial r}(u_r r) + \frac{\partial u_\theta}{\partial \theta} = 0 \qquad (3.7b)$$

3.5 THE STREAM FUNCTION

For incompressible, steady flow a *stream function*, $\Psi$, may be introduced to describe the flow pattern algebraically. For two-dimensional flow, streamlines are given by the function $\Psi = \Psi(x, y)$. Each streamline has a particular constant value $\Psi = c$ and the algebraic difference $\Psi_2 - \Psi_1$ between two streamlines $\Psi_1$ and $\Psi_2$ (say) represents the constant volumetric flow rate between those two streamlines with the sign convention for $\Psi$ chosen such that $\Psi$ increases to the left looking downstream. The velocity vectors may be found by differentiating the stream function.

$$u = \frac{\partial \Psi}{\partial y} \qquad (3.8a)$$

$$v = -\frac{\partial \Psi}{\partial x} \qquad (3.8b)$$

$$u_r = \frac{1}{r}\frac{\partial \Psi}{\partial \theta} \qquad (3.9a)$$

$$u_\theta = -\frac{\partial \Psi}{\partial r} \qquad (3.9b)$$

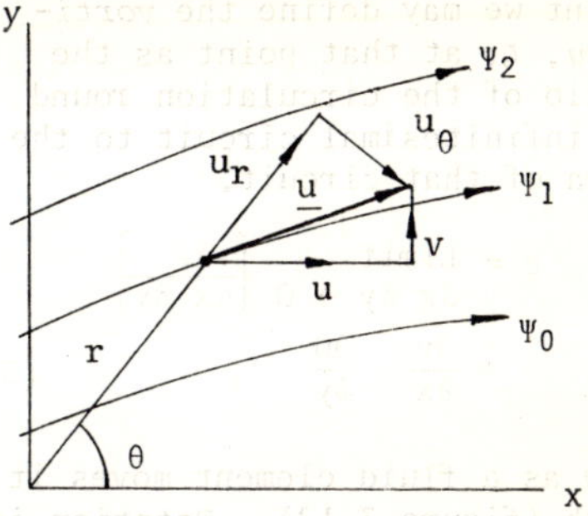

Figure 3.9

The existence of Ψ implies continuity and vice versa. Every physically possible two-dimensional, steady flow must have a stream function. Thus, Ψ exists for possible real flows as well as ideal flows.

## 3.6 CIRCULATION, VORTICITY AND ROTATION

In section 3.4 the *normal* component of velocity, $\underline{u}\cos\theta$, on a control surface was used to obtain the fluid flow. The *tangential* component of velocity, $\underline{u}\sin\theta$, will now be used to obtain the fluid swirl. The swirl is measured by the *circulation*, Γ, defined as the *line integral* of the tangential component of velocity around a closed curve fixed in the flow. Figure 3.10 shows the two-dimensional case. The angle of intersection, α, between the curve and the streamline is used in preference to θ. By convention, the line integral is evaluated in an anti-clockwise direction because circulation is measured positive clockwise along a positive x, y or z axis.

$$d\Gamma = (\underline{u}\cos\alpha)ds$$

$$= (\underline{u}\sin\theta)ds$$

$$\Gamma = \oint_c (\underline{u}\cos\alpha)ds \qquad (3.10a)$$

$$\text{or} \quad \Gamma = \oint_c \underline{u}\cdot d\underline{s} \qquad (3.10b)$$

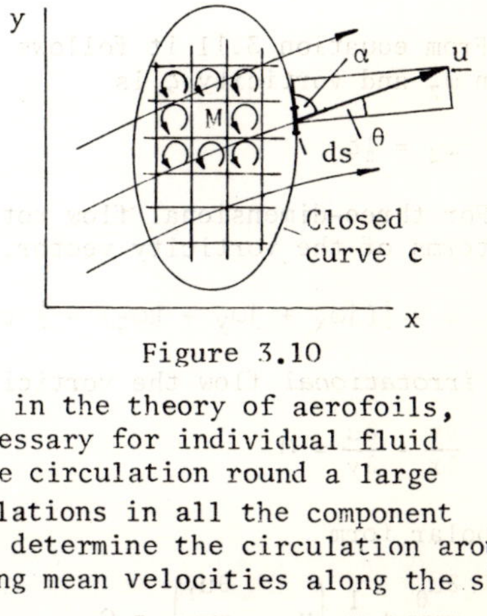

Figure 3.10

Circulation is an important concept in the theory of aerofoils, fans, etc. Note that it is not necessary for individual fluid elements to travel the circuit. The circulation round a large circuit equals the sum of the circulations in all the component small circuits, e.g. M. We may now determine the circulation around a small rectangular element, M, using mean velocities along the sides (figure 3.11).

$$\delta\Gamma_M = \oint_c \underline{u}\cdot d\underline{s} = \delta x\,\delta y\left(\frac{\partial v}{\partial x} - \frac{\partial u}{\partial y}\right)$$

$$\frac{\delta\Gamma_M}{\delta x\,\delta y} = \frac{\partial v}{\partial x} - \frac{\partial u}{\partial y}$$

As the element shrinks to a point we may define the *vorticity*, $\zeta$, at that point as the ratio of the circulation round an infinitesimal circuit to the area of that circuit.

$$\zeta = \lim_{\delta x\, \delta y \to 0} \left(\frac{\delta\Gamma}{\delta x\, \delta y}\right)$$

$$= \frac{\partial v}{\partial x} - \frac{\partial u}{\partial y} \qquad (3.11)$$

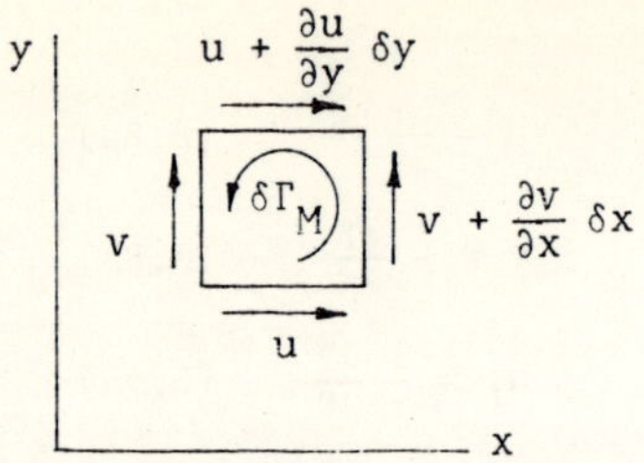

Figure 3.11

Now as a fluid element moves it may undergo distortion, rotation or both (figure 3.12). *Rotation* is defined as the average angular velocity, $\omega_z$, of any two mutually perpendicular line elements in the plane of flow, e.g. aa and bb.

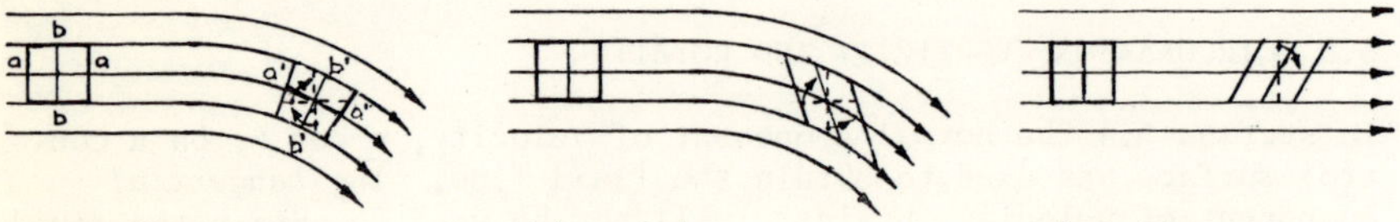

Rotation
No distortion

Distortion
No rotation

Rotation
and distortion

Figure 3.12

It can be shown that

$$\textit{rotation}\ \omega_z = \tfrac{1}{2}\left(\frac{\partial v}{\partial x} - \frac{\partial u}{\partial y}\right) \qquad (3.12a)$$

From equation 3.11 it follows that the relationship between rotation $\omega_z$ and vorticity $\zeta$ is

$$\omega_z = \tfrac{1}{2}\zeta \qquad (3.12b)$$

For three-dimensional flow rotation is defined in vector form in terms of the vorticity vector, curl $\underline{u}$.

$$\omega = \tfrac{1}{2}(i\omega_x + j\omega_y + k\omega_z) = \tfrac{1}{2}\ \text{curl}\ \underline{u} \qquad (3.13)$$

For *irrotational* flow the vorticity is everywhere zero. Therefore

$$\frac{\partial v}{\partial x} - \frac{\partial u}{\partial y} = 0 \qquad (3.14a)$$

In polar form

$$\frac{\partial u_\theta}{\partial r} + \frac{1}{r}\left(u_\theta - \frac{\partial u_r}{\partial \theta}\right) = 0 \qquad (3.14b)$$

It should be noted that the concept of rotational or irrotational flow applies to individual fluid elements and not to the motion of the fluid as a whole which can be in a curved path (figure 3.13).

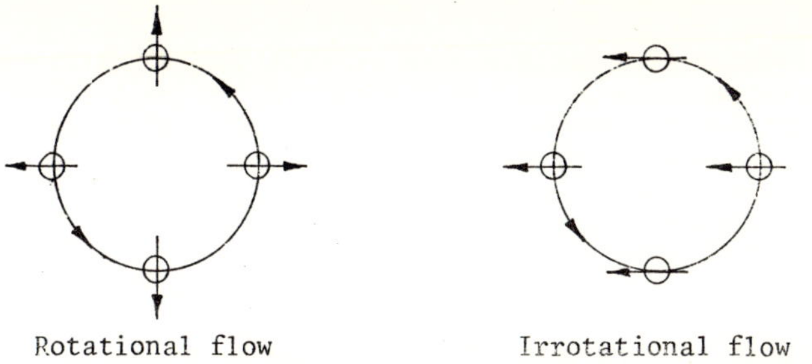


Figure 3.13

Rotation is due to the action of shear (viscous) forces. Thus the flow of an *ideal* or frictionless fluid must be irrotational if no rotation is present initially.

## 3.7 VELOCITY POTENTIAL FUNCTION

For irrotational flow a *velocity potential function*, $\phi$, is introduced whose derivative with respect to the flow direction gives the velocity in that direction. For two-dimensional flow

$$u = \frac{\partial \phi}{\partial x}, \quad v = \frac{\partial \phi}{\partial y} \tag{3.15}$$

In polar form

$$u_\theta = \frac{1}{r}\frac{\partial \phi}{\partial \theta}, \quad u_r = \frac{\partial \phi}{\partial r} \tag{3.16}$$

For three-dimensional flow, velocity is defined in vector form in terms of the gradient of the velocity potential.

$$\underline{u} = \text{grad } \phi = \nabla \phi \tag{3.17}$$

The stream function, $\Psi$, and the velocity potential function, $\phi$, may be combined to form the Laplace equation.

$$\frac{\partial^2 \Psi}{\partial x^2} + \frac{\partial^2 \Psi}{\partial y^2} = 0 \quad \text{or} \quad \nabla^2 \Psi = 0 \tag{3.18}$$

$$\frac{\partial^2 \phi}{\partial x^2} + \frac{\partial^2 \phi}{\partial y^2} = 0 \quad \text{or} \quad \nabla^2 \phi = 0 \tag{3.19}$$

Any function of $\Psi$ or $\phi$ that satisfies the Laplace equation must represent a possible incompressible, irrotational flow. The Laplace equation is linear therefore the principle of superposition applies and the combined flow, $\Psi$, of two flows $\Psi_1$ and $\Psi_2$ is given by $\Psi = \Psi_1 + \Psi_2$. Similarly, $\phi = \phi_1 + \phi_2$. Lines of constant $\Psi$ and $\phi$ for a given flow are orthogonal therefore the flow may be represented by a mesh of $\Psi$ and $\phi$ lines.

It must be stressed at this stage that stream function is valid for any physically possible flow, whether rotational or irrotational. Velocity potential is valid only for irrotational flow.

## 3.8 SOLUTION OF FLOW PATTERNS

Solutions to particular flow patterns can be obtained by analytical, graphical, numerical methods and by analogies. In the analytical method the flow pattern is obtained by the combination of Ψ-functions for simple flows. Some basic patterns are given below.

(i) Uniform flow

$$\Psi = uy - vx$$

$$= \underline{u}r \sin(\theta - \alpha) \qquad (3.20)$$

$$\phi = ux + vy$$

$$= \underline{u}r \cos(\theta - \alpha) \qquad (3.21)$$

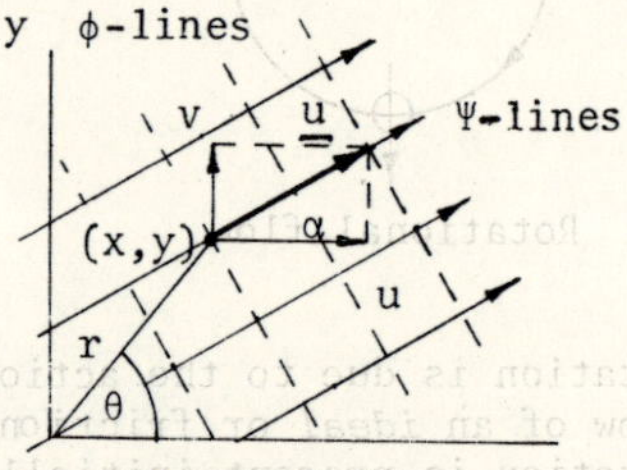

Figure 3.14

(ii) Source (and sink)

The strength, Q, of a source is defined as the total flow per unit time per unit depth.

$$\Psi = \frac{Q}{2\pi}\,\theta = m\theta \qquad (3.22)$$

$$\phi = \frac{Q}{2\pi}\ln r = m \ln r \qquad (3.23)$$

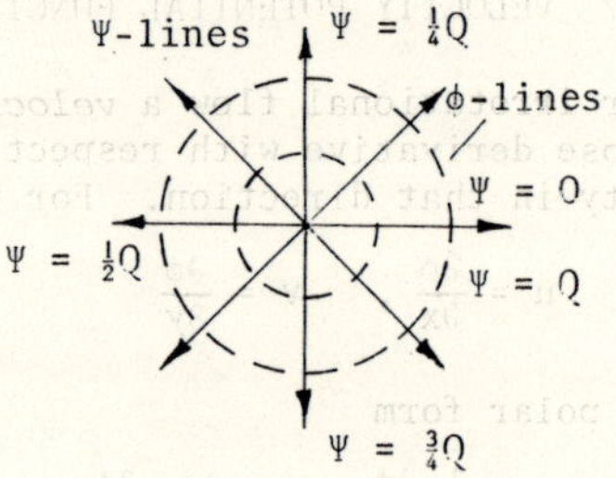

Figure 3.15

(iii) Irrotational vortex

For an irrotational vortex $u_\theta r = C$
Strength of vortex $K = 2\pi u_\theta r = 2\pi C$
(+ve ↺)

$$\Psi = -\frac{K}{2\pi}\ln r \qquad (3.24)$$

$$\phi = \frac{K}{2\pi\theta} \qquad (3.25)$$

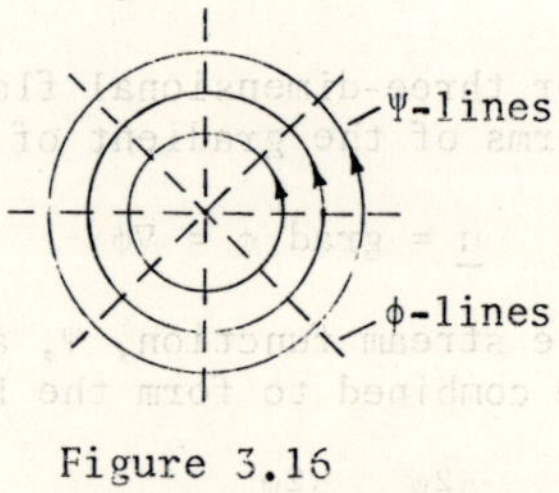

Figure 3.16

(iv) Doublet

For a doublet of strength μ

$$\Psi = -\frac{\mu \sin\theta}{r} \qquad (3.26)$$

$$\phi = \frac{\mu \cos\theta}{r} \qquad (3.27)$$

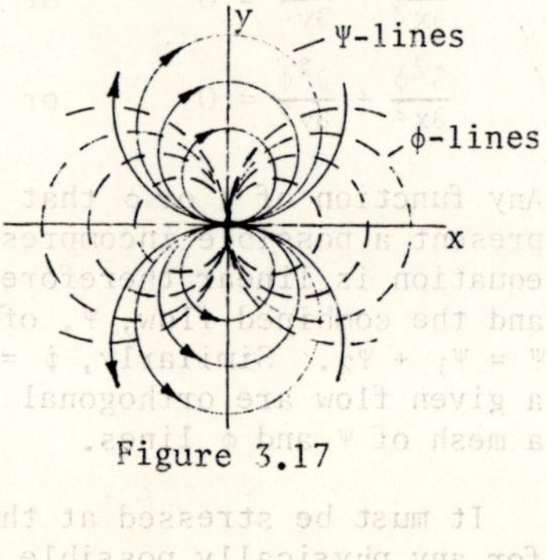

Figure 3.17

*Example 3.1*

For a cylinder of radius a, moving at steady velocity, $\underline{u}$, through a stationary fluid, the steady flow streamline pattern is shown in figure 3.18. The velocity components relative to the cylinder are $(u_r)_R = U(1 - a^2/r^2) \cos\theta$ and $(u_\theta)_R = -U(1 + a^2/r^2) \sin\theta$.

Determine (i) the unsteady flow pattern, and (ii) the steady and unsteady velocities at the point $\theta = 45°$, $r = 35$ mm, for a cylinder of diameter 40 mm moving at 10 m/s through a stationary fluid.

The steady flow pattern shown is based on velocities, $\underline{u}_R$, relative to the cylinder, and would be seen by an observer moving with the cylinder. The unsteady flow pattern is based on velocities, $\underline{u}$, relative to earth and would be seen by an observer at rest. It may be deduced by superimposing a velocity, $-\underline{U}$, on the steady flow pattern. The uniform flow part of the velocity components is nullified and the given expressions for $(u_r)_R$ and $(u_\theta)_R$ reduce to

$$u_r = -(Ua^2/r^2) \cos\theta$$

and $u_\theta = -(Ua^2/r^2) \sin\theta$

Hence it can be shown that

$$\Psi = -Ua^2 \sin\theta/r$$

and $\underline{u} = \surd(u_r{}^2 + u_\theta{}^2) = Ua^2/r^2$

This gives the steady flow pattern shown in figure 3.18b. It is of interest to note that this pattern is that of a doublet outside the cylinder. Substituting values, we have

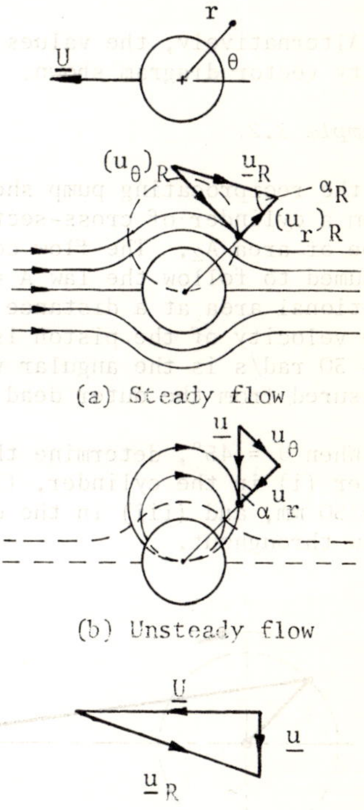


Figure 3.18

$$(u_r)_R = U\left(1 - \frac{a^2}{r^2}\right)\cos\theta = 10\left(1 - \frac{20^2}{35^2}\right)\cos 45 = 4.76 \text{ m/s}$$

$$(u_\theta)_R = -U\left(1 + \frac{a^2}{r^2}\right)\sin\theta = -10\left(1 + \frac{20^2}{35^2}\right)\sin 45 = -9.38 \text{ m/s}$$

$$\underline{u}_R = \surd(4.76^2 + 9.38^2) = 10.52 \text{ m/s}$$

$\alpha_R = \tan^{-1}(9.38/4.76) = 63.1°$

$$u_r = -\frac{Ua^2 \cos\theta}{r^2} = -\frac{10 \times 20^2 \times \cos 45}{35^2} = -2.31 \text{ m/s}$$

$$u_\theta = -\frac{Ua^2 \sin\theta}{r^2} = -\frac{10 \times 20^2 \times \sin 45}{35^2} = -2.31 \text{ m/s}$$

$$\underline{u} = \sqrt{(2.31^2 + 2.31^2)} = 3.27 \text{ m/s}$$

$$\alpha = \tan^{-1}(2.31/2.31) = 45°$$

Alternatively, the values $\underline{u}$ and $\alpha$ could be obtained from the velocity vector diagram shown.

*Example 3.2*

In the reciprocating pump shown in figure 3.19, water is delivered from a cylinder of cross-sectional area $A_1$, through a valve, to a pipe of area $A_2$. The flow contraction, of length 100 mm, may be assumed to follow the law $A = (5 - 40x)\ 10^{-3}\ \text{m}^2$ where A is the cross-sectional area at a distance x m from commencement of contraction. The velocity of the piston is $v = 0.05\omega(\sin\theta + 0.1 \sin 2\theta)$ where $\omega = 30$ rad/s is the angular velocity of the crank and $\theta$ is the angle measured from the outer dead centre position.

When $\theta = 45°$, determine the instantaneous acceleration of the water (i) in the cylinder, (ii) at a point in the contraction where x = 50 mm, and (iii) in the delivery pipe. Assume one-dimensional flow throughout.

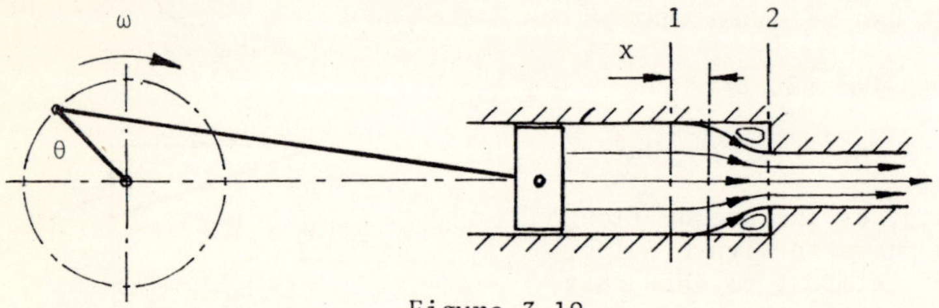


Figure 3.19

The mean velocity, $\bar{u}$ of the water in the cylinder is equal to the piston velocity, v. Therefore

$$\bar{u} = 0.05\omega(\sin\theta + 0.1 \sin 2\theta) = 0.05 \times 30(\sin 45 + 0.1 \sin 90)$$

$$= 1.21 \text{ m/s}$$

and $$\frac{\partial \bar{u}_1}{\partial t} = \frac{\partial \theta}{\partial t}\frac{\partial \bar{u}_1}{\partial \theta} = 0.05\omega^2(\cos\theta + 0.2 \cos 2\theta) \qquad \text{(i)}$$

At any station in the flow where the area is A and the velocity is $\bar{u}$, we have

$$\frac{\partial \bar{u}}{\partial t} = \frac{A_1}{A}\frac{\partial \bar{u}_1}{\partial t} = 0.05\omega^2(\cos\theta + 0.2\cos 2\theta)\frac{A_1}{A} \qquad \text{(ii)}$$

From the continuity equation 3.5a

$$\dot{V} = \bar{u}_1 A_1 = \bar{u}_1(5 - 40x_1)\ 10^{-3} = 1.21 \times 5 \times 10^{-3} = 6.05 \times 10^{-3}\ \text{m}^3/\text{s}$$

At any station in the flow contraction

$$\bar{u} = \frac{\dot{V}}{A} = 6.05\ (5 - 40x)^{-1}$$

$$\frac{\partial \bar{u}}{\partial x} = 6.05 \times 40\ (5 - 40x)^{-2}$$

and $$\bar{u}\frac{\partial \bar{u}}{\partial x} = 6.05^2 \times 40\ (5 - 40x)^{-3} = 1.464 \times 10^3\ (5 - 40x)^{-3} \qquad \text{(iii)}$$

From equation 3.1, the acceleration at any point is

$$\frac{d\bar{u}}{dt} = \bar{u}\frac{\partial \bar{u}}{\partial x} + \frac{\partial \bar{u}}{\partial t}$$

In the cylinder $\partial\bar{u}/\partial x = 0$. Therefore from equation (i)

$$\frac{d\bar{u}}{dt} = \frac{\partial \bar{u}_1}{\partial t} = 0.05 \times 30^2(\cos 45 + 0.2\cos 90) = 31.82\ \text{m/s}^2$$

In the flow contraction, where x = 50 mm, equations (ii) and (iii) give

$$\frac{d\bar{u}}{dt} = \bar{u}\frac{\partial \bar{u}}{\partial x} + \frac{A_1}{A}\frac{\partial \bar{u}_1}{\partial t} = 1.464 \times 10^3\ (5 - 40 \times 0.05)^{-3} + \frac{5 \times 31.82}{(5 - 40 \times 0.05)} = 107.3\ \text{m/s}^2$$

In the delivery pipe $\partial\bar{u}/\partial x = 0$. Therefore from equation (ii)

$$\frac{d\bar{u}}{dt} = \frac{A_1}{A_2}\frac{\partial \bar{u}_1}{\partial t} = \frac{5 \times 31.82}{(5 - 40 \times 0.1)} = 159\ \text{m/s}^2$$

*Example 3.3*

Air enters the 1.5 m diameter circular inlet duct of a stationary jet engine at 50 m/s with temperature 285 K and pressure 100 kN/m$^2$. Fuel is consumed by the engine at the rate of 1.2 kg/s. If the exhaust gases leave the 0.7 m diameter exhaust jet pipe at the design pressure of 101 kN/m$^2$ and temperature 800 K determine the gas exit velocity. $R_{air}$ = 287 J/kg K, $R_{gas}$ = 290 J/kg K.

Assuming one-dimensional flow and applying the continuity equation 3.5b to flow through the jet engine (figure 3.20) we have

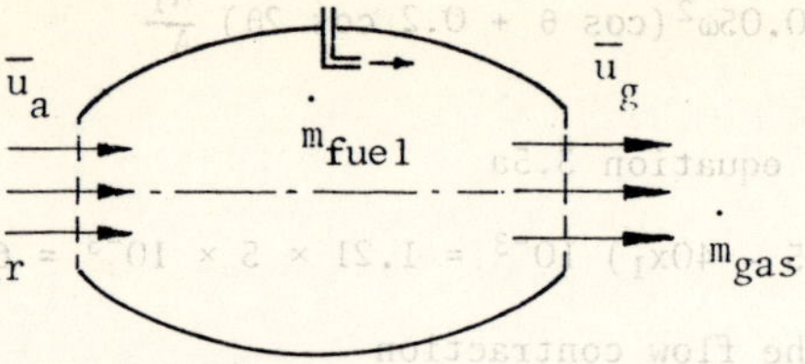


Figure 3.20

$$\dot{m}_{air} + \dot{m}_{fuel} = \dot{m}_{gas}$$

or $$(\rho A\bar{u})_{air} + \dot{m}_{fuel} = (\rho A\bar{u})_{gas} \qquad \text{(i)}$$

Substituting $\rho = p/RT$ in equation (i), we obtain

$$\left(\frac{pA\bar{u}}{RT}\right)_{air} + \dot{m}_{fuel} = \left(\frac{pA\bar{u}}{RT}\right)_{gas}$$

$$\frac{10^5 \times \pi \times 1.5^2 \times 50}{287 \times 4 \times 285} + 1.2 = \frac{101 \times 10^3 \times \pi \times 0.7^2 \times \bar{u}_g}{290 \times 4 \times 800}$$

$$\bar{u}_g = 650 \text{ m/s}$$

*Example 3.4*

Derive the equation of continuity for a two-dimensional, incompressible steady flow and determine if the flow field, represented by velocity components $u = x^2y + y^2$ and $v = x^2 - y^2x$, satisfies continuity.

Write down the velocity components, u and v, in terms of partial derivatives of the stream function $\Psi$ and hence determine (i) the stream function for the flow, and (ii) the velocity at the point $x = 2$, $y = 3$.

Consider the flow through a rectangular element of unit depth (figure 3.21). The average velocities are shown. For an incompressible fluid, volumetric flow in equals that out. Therefore

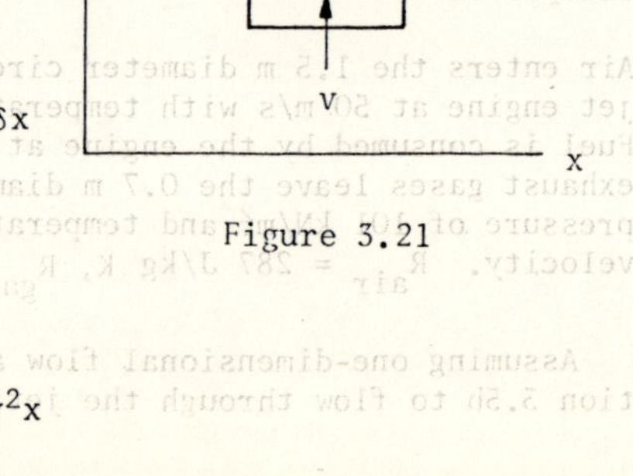


Figure 3.21

$$u\,\delta y + v\,\delta x = \left(u + \frac{\partial u}{\partial x}\,\delta x\right)\delta y + \left(v + \frac{\delta v}{\delta y}\,\delta y\right)\delta x$$

and $$\frac{\partial u}{\partial x} + \frac{\partial v}{\partial y} = 0 \qquad \text{(i)}$$

The given velocity components are

$$u = x^2y + y^2 \qquad v = x^2 - y^2x$$

$$\frac{\partial u}{\partial x} = 2xy \qquad \frac{\partial v}{\partial y} = -2xy$$

Substituting in continuity equation (i) we have

$$2xy - 2xy = 0$$

Therefore, the given flow field satisfies continuity.

(i) The velocity component, u, expressed in terms of $\Psi$ and (x, y) is

$$u = \frac{\partial \Psi}{\partial y} = x^2y + y^2$$

Hence

$$\Psi = \frac{x^2y^2}{2} + \frac{y^3}{3} + f(x) + C_1 \qquad \text{(ii)}$$

and $$\frac{\partial \Psi}{\partial x} = xy^2 + f'(x) \qquad \text{(iii)}$$

Now $$v = -\frac{\partial \Psi}{\partial x} = x^2 - y^2x \qquad \text{(iv)}$$

Therefore from equations (iii) and (iv)

$$-x^2 + y^2x = xy^2 + f'(x)$$

$$f'(x) = -x^2$$

and $$f(x) = -\frac{x^3}{3} + C_2$$

Substituting for f(x) in equation (ii)

$$\Psi = \frac{x^2y^2}{2} + \frac{y^3}{3} - \frac{x^3}{3} + C_3$$

(ii) At the point x = 2, y = 3 the velocity components are

$$u = x^2y + y^2 = 2^2 \times 3 + 3^2 = 21$$

$$v = x^2 - y^2x = 2^2 - 3^2 \times 2 = -14$$

$$\underline{u} = \sqrt{(u^2 + v^2)} = \sqrt{[21^2 + (-14)^2]}$$

$$= 25.24$$

$$\theta = \tan (14/21) = 33.7$$

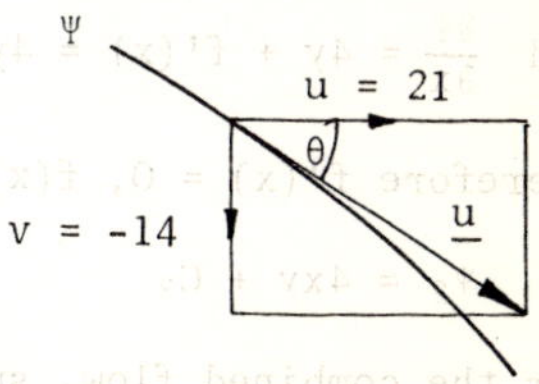


Figure 3.22

*Example 3.5*

A flow pattern is obtained by the superposition of two flow patterns 1 and 2 defined as follows.

$$\Psi_1 = y^3/3 - x^2y + 2xy \text{ and } \phi_2 = 2x^2 - 2y^2$$

Show that both of the flow patterns are real and irrotational and obtain the stream function for the combined flow pattern. Determine the circulation for the combined flow pattern about a closed rectangular path defined by the cartesian co-ordinates (0, 0), (0, 1), (2, 1) and (2, 0).

For real flow, equation 3.7 must be satisfied and for irrotational flow equation 3.14 must be satisfied. Alternatively, the conditions for real, irrotational flow are satisfied by the Laplace equations $\nabla^2\Psi = 0$ and $\nabla^2\phi = 0$. For the given flow patterns

$$\Psi_1 = y^3/3 - x^2y + 2xy \qquad \phi_2 = 2x^2 - 2y^2$$

$$\frac{\partial^2\Psi}{\partial x^2} = -2y \qquad \frac{\partial^2\phi}{\partial x^2} = 4$$

$$\frac{\partial^2\Psi}{\partial y^2} = 2y \qquad \frac{\partial^2\phi}{\partial y^2} = -4$$

$$\frac{\partial^2\Psi}{\partial x^2} + \frac{\partial^2\Psi}{\partial y^2} = 0 \qquad \frac{\partial^2\phi}{\partial x^2} + \frac{\partial^2\phi}{\partial y^2} = 0$$

Therefore both flow patterns are real and irrotational.

Now $\phi_2 = 2x^2 - 2y^2$

$$\frac{\partial\phi}{\partial x} = 4x = u = \frac{\partial\Psi}{\partial y}$$

and $$\frac{\partial\phi}{\partial y} = -4y = v = -\frac{\partial\Psi}{\partial x}$$

Now $u = \dfrac{\partial\Psi}{\partial y} = 4x$

Hence

$$\Psi_2 = 4xy + f(x) + C_1$$

and $$\frac{\partial\Psi}{\partial x} = 4y + f'(x) = 4y$$

Therefore $f'(x) = 0$, $f(x) = 0$ and the stream function for flow 2 is

$$\Psi_2 = 4xy + C_3$$

For the combined flow, superposition of flows 1 and 2 gives

$$\Psi = \Psi_1 + \Psi_2 = y^3/3 - x^2y + 6xy + C$$

The velocity components for the combined flow are

$$u = \frac{\partial\Psi}{\partial y} = y^2 - x^2 + 6x$$

and $$v = -\frac{\partial\Psi}{\partial x} = 2xy - 6y$$

$$\Gamma = \int_a^b v\ dy + \int_b^c u\ dx + \int_c^d v\ dy + \int_d^a u\ dx$$

$$= \int_0^1 (-2y)\ dy + \int_2^0 (1 - x^2 + 6x)\ dx$$

$$+ \int_1^0 (-6y)\ dy + \int_0^2 (-x^2 + 6x)\ dx$$

$$= 0$$

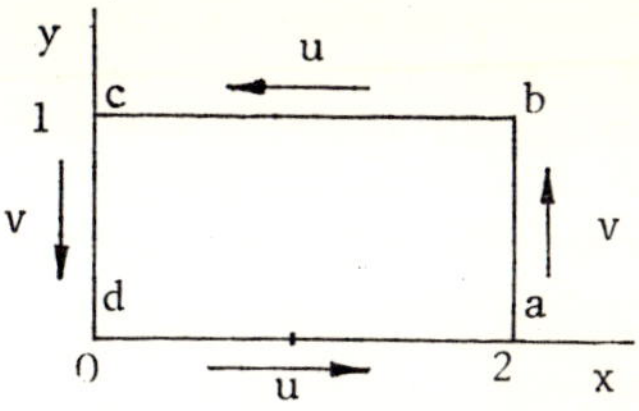

Figure 3.23

*Example 3.6*

Given that the equation for vorticity, expressed in polar co-ordinates is

$$\zeta = \frac{\partial u_\theta}{\partial r} + \frac{1}{r}\left(u_\theta - \frac{\partial u_r}{\partial \theta}\right)$$

determine expressions for the stream function, $\Psi$ and the velocity potential, $\phi$, for an irrotational circular vortex in terms of the strength, K, of the vortex. Hence, show that the circulation, $\Gamma$, is equal to K for a streamline of an irrotational vortex and equal to zero for any closed curve which does not enclose the axis of the vortex.

For an irrotational vortex

$$\frac{\partial u_\theta}{\partial r} + \frac{1}{r}\left(u_\theta - \frac{\partial u_r}{\partial \theta}\right) = 0$$

For a circular vortex the streamlines are concentric and $u_r = 0$. Hence

$$\frac{\partial u_\theta}{\partial r} + \frac{u_\theta}{r} = 0$$

and $u_\theta r = C$

The strength K of the vortex is defined $K = 2\pi C = 2\pi u_\theta r$.

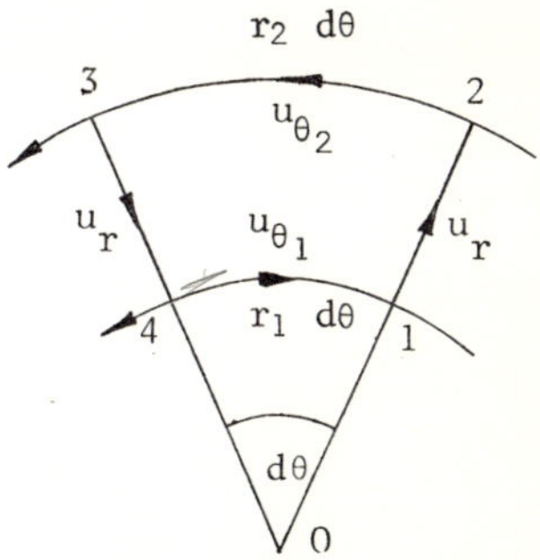

Irrotational vortex

Figure 3.24

Now $u_r = 0 = \dfrac{\partial \phi}{\partial r} = \dfrac{1}{r}\dfrac{\partial \Psi}{\partial \theta}$

and $u_\theta = \dfrac{1}{r}\dfrac{\partial \phi}{\partial \theta} = -\dfrac{\partial \Psi}{\partial r}$

Hence

$$\phi = u_\theta r\theta = \frac{K\theta}{2\pi}$$

and $\Psi = -\int u_\theta\ dr = -\int \dfrac{K}{2\pi}\dfrac{dr}{r} = -\dfrac{K}{2\pi}\ln r$

From equation 3.10 the circulation around a streamline is

$$\Gamma = u_\theta \oint ds = u_\theta 2\pi r = K \qquad \text{Q.E.D.}$$

The circulation around the path 1-2-3-4-1 (figure 3.24) is

$$\Gamma = \int_1^2 0dr + \int_2^3 u_{\theta_2} r_2\, d\theta + \int_3^4 0dr + \int_4^1 u_{\theta_2} r_1\, d\theta$$

$$= 0 + C\theta + 0 - C\theta = 0 \qquad \text{Q.E.D.}$$

These results and example 3.5 confirm that the circulation around a closed curve in an irrotational field is zero except where the curve encloses a singular point. In this example, the origin or core of the vortex is the singularity. At a singular point, the velocity or acceleration is theoretically infinity or zero and vorticity exists at the point.

*Example 3.7*

Starting with the velocity components in terms of the stream function, derive an expression for the superposition of a source and parallel flow of an ideal, incompressible fluid. Hence, determine expressions for the position of the stagnation point and the stagnation streamline. Sketch, in good proportion, the resulting streamline pattern.

For uniform flow parallel to the x axis, equations 3.8 give the velocity components as

$$v = -\frac{\partial \Psi}{\partial x} = 0$$

and $$u = \frac{\partial \Psi}{\partial y} = U$$

Therefore

$$\Psi = Uy + C$$

Let $\Psi = 0$ when $y = 0$. Then $c = 0$, and

$$\Psi_1 = Uy$$

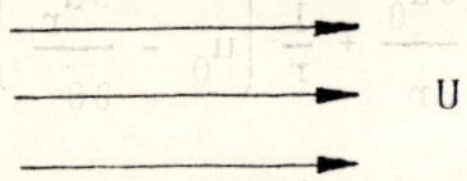


(a) uniform flow

For a source of strength, Q, defined as the total flow per unit time per unit depth, equations 3.9 give

$$u_\theta = -\frac{\partial \Psi}{\partial r} = 0$$

and $$u_r = \frac{1}{r}\frac{\partial \Psi}{\partial \theta} = \frac{Q}{2\pi r}$$

Therefore

$$\frac{\partial \Psi}{\partial \theta} = \frac{Q}{2\pi} = m$$

and $$\Psi = m\theta + C$$

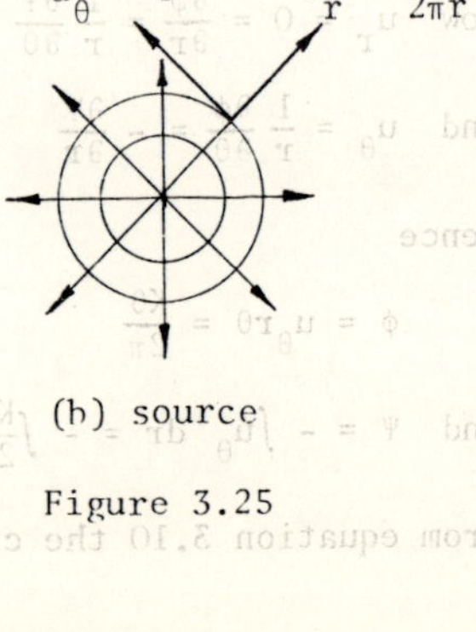


(b) source

Figure 3.25

Let $\Psi = 0$ when $\theta = 0$. Then $c = 0$, and

$$\Psi_2 = m\theta$$

For a combination of a source and uniform flow (-ve right to left)

$$\Psi = \Psi_1 + \Psi_2 = -Uy + m\theta$$

or $\quad \Psi = -Ur \sin\theta + m\theta \qquad \text{(i)}$

The velocity components for the combined flow are

$$u_r = \frac{1}{r}\frac{\partial\Psi}{\partial\theta} = \frac{1}{r}(-Ur\cos\theta + m) \qquad \text{(ii)}$$

and $\quad u_\theta = -\dfrac{\partial\Psi}{\partial r} = U\sin\theta \qquad \text{(iii)}$

A *stagnation point*, O, occurs where the fluid is brought to rest and the velocity components $u_r$ and $u_\theta$ are both zero. Therefore, from equations (ii) and (iii), the stagnation point occurs at $\theta = 0$ and $r = m/U$.

The streamline $\Psi = 0$ gives the stagnation streamline and the body shape. Equation (i) gives

$$0 = -Ur\sin\theta + m\theta$$

From this equation the stagnation streamline is the x axis ($\theta = 0$) and a body of shape $r = m\theta/U\sin\theta$.

The combined flow may be sketched by superimposing a source of strength $Q = 16$ (say) with uniform flow as shown in figure 3.26.

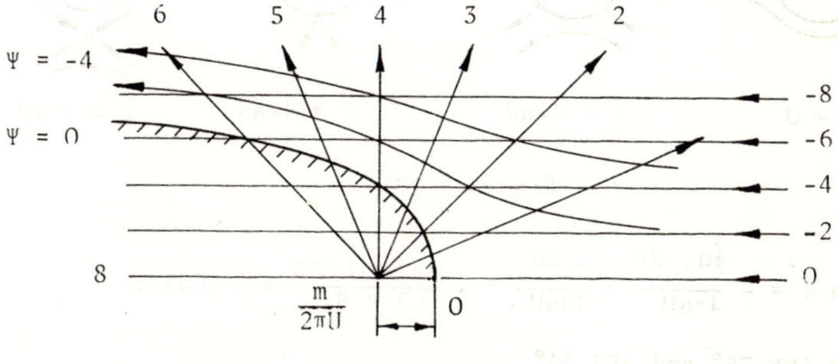


Figure 3.26

*Example 3.8*

The superposition of a circulation of strength, $-\Gamma$, on uniform flow of velocity, U, normal to a circular cylinder of radius, a, produces a streamline pattern represented by the stream function

$$\Psi = U\left(r - \frac{a^2}{r}\right)\sin\theta + \frac{\Gamma}{2\pi}\ln r$$

Sketch flow patterns for different values of $\Gamma$, clearly marking the position of the stagnation points. Hence determine the position of the stagnation points when a cylinder of diameter 150 mm is rotated at $8\pi$ rad/s in an airstream of velocity 8 m/s perpendicular to the cylinder axis. Assume $\rho = 1.2$ kg/m$^3$.

The given stream function is formed by the combination of (i) a uniform flow, (ii) a doublet and (iii) an irrotational vortex having circulation $-\Gamma$. Note that $\mu = Ua^2$.

$$\Psi = U\left(r - \frac{a^2}{r}\right)\sin\theta + \frac{\Gamma}{2\pi}\ln r$$

and $$\frac{\partial\Psi}{\partial r} = U\left(1 + \frac{a^2}{r^2}\right)\sin\theta + \frac{\Gamma}{2\pi r}$$

At a point on the cylinder surface $r = a$. Therefore

$$u_\theta = -\frac{\partial\Psi}{\partial r} = -2U\sin\theta - \frac{\Gamma}{2\pi a}$$

At a stagnation point $u_\theta = 0$, $u_r = 0$. Therefore

$$\sin\theta = -\frac{\Gamma}{4\pi aU}$$

The positions of the stagnation points for different values of $\Gamma$ are shown in figure 3.27. Note that the condition $\Gamma = 0$ corresponds to flow normal to a cylinder. (Combination of uniform flow and doublet.)

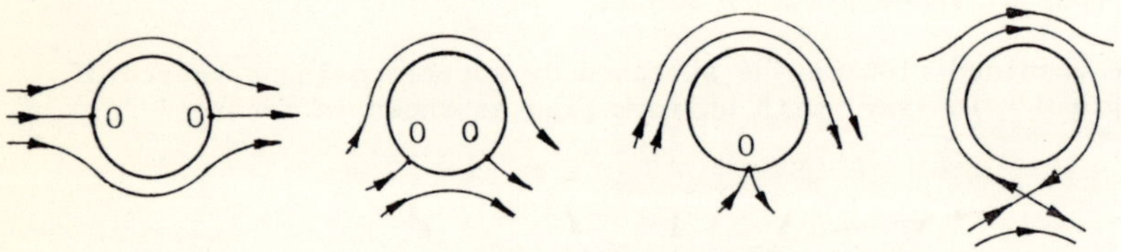


Figure 3.27

$$\sin\theta = -\frac{\oint u_\theta\,ds}{4\pi aU} = \frac{2\pi a u_\theta}{4\pi aU} = -\frac{8\pi \times 0.075}{2 \times 8} = -0.1178$$

$$\theta = 186.76° \text{ and } 353.24°$$

Problems

1 For a cylinder of radius a, rotating with angular velocity in a uniform fluid flow of velocity U, perpendicular to the cylinder axis, the velocity components relative to the cylinder axis are $(u_r)_R = -U(1 - a^2/r^2)\cos\theta$ and $(u_\theta)_R = -U(1 + a^2/r^2)\sin\theta - \omega a^2/r$. Determine the steady and unsteady flow velocities at the point $\theta = 30°$, $r = 40$ mm for a cylinder of radius 25 mm rotating at

300 rev/min in a uniform air-stream of 15 m/s.

[$u_R$ = 13.49 m/s, 24°, $\underline{u}$ = 6.12 m/s, 64°]

2 A venturi meter, inserted in a pipe of diameter 100 mm, consists of a converging conical section of included angle 60°, a parallel throat section of diameter 50 mm and a diverging conical section of included angle 10°. Determine the acceleration at the mid-points of the convergent, throat and divergent sections at the instant when the flow is 0.01 m /s. Assume (i) steady-flow conditions and (ii) unsteady-flow with an acceleration in the 100 mm pipe of 40 $m/s^2$. Assume one-dimensional flow.

[157.4 $m/s^2$, 228.5, 0, 160 $m/s^2$, -23.9 $m/s^2$, 47.2 $m/s^2$]

3 Water flows at 5 m/s through a 200 mm diameter pipe. The pipe then branches into two pipes of diameters 100 mm and 150 mm, respectively. The mass flow rate through the 100 mm pipe is found to be 50 kg/s. Determine the velocity of flow in the 150 mm pipe. Assume $\rho$ = 1000 $kg/m^3$.

[6.06 m/s]

4 A two-dimensional field is described by

$$u = y^2 - x^2 + y$$

$$v = 2xy + x$$

Show that this field satisfies continuity and is also irrotational. Find the stream function.

[$\Psi = y^3/3 - x^2y + y^2/2 - x^2/2 + C$]

5 Draw the streamlines for the field represented by $\Psi = xy$ for values of $\Psi$ = 0, 1, 2, 3. Find the corresponding potential, $\phi$, if one exists and draw the equipotential lines. Show that the $\Psi$ lines and $\phi$ lines are orthogonal. What actual flow pattern might this represent?

[$\phi = x^2/2 - y^2/2 + C$, Flow at 90° corner]

6 (a) Working from first principles show that for steady incompressible flow

$$\frac{\partial^2\phi}{\partial x^2} + \frac{\partial^2\phi}{\partial y^2} = 0$$

where $\phi$ is the velocity potential.

(b) Show that the velocity potential $\phi = 2(x^2 - y^2)$ represents a steady motion and find a corresponding expression for the stream function. Show approximately a few of these streamlines. What would be the magnitude and direction of flow at the coordinate (4, 2)?

[$\Psi = 4xy$, 17.9, -27.1°]

7 Two flow patterns are superimposed. One flow pattern is described by the stream function $\Psi = 3x^2 - 3y^2 + 2x$ and the other is described by the velocity potential $\phi = x^2 + x - y^2$. Show that the individual

flows are real and irrotational and determine the stream function for the combined flows. What will be the velocity at the point $x = 3$, $y = 4$ in the combined flow?

[$\Psi = 3x^2 - 3y^2 + 2x + 2xy + y + C$, 34.1]

8 The x and y components of velocity in an ideal, two-dimensional flow are given by $u = 3x^2 - 3y^2$ and $v = 6xy$. (i) Show that the flow is both real and irrotational, and (ii) determine the circulation around a rectangle which is formed by straight lines joining the points (1, 1), (1, -1), (0, -1) and (0, 1) lying in an x-y plane.

[$\Gamma = 0$]

9 A vortex pair consists of two vortices of equal strength, K, but opposite sign. Obtain expressions for the stream and velocity potential functions. Hence, show that, in the unsteady-flow pattern, both vortex centres move in a direction perpendicular to the line joining their centres, distance, d, apart, with velocity $v = K/2\pi d$. Sketch the unsteady and steady flow patterns.

10 Sketch, in good proportion, the streamline pattern for the superposition of a source and sink of equal strengths, a finite distance apart, and a uniform flow.

11 A cylinder of liquid rotates with an angular velocity of 10 rad/s. Determine (i) the circulation for a curve enclosed by radii 0.75 m and 1.2 m and included angle 30°, and (ii) the circulation for a curve enclosed by a circular streamline of radius 0.75 m.

[4.595 $m^2/s$, 35.35 $m^2/s$]

12 Assuming two-dimensional flow, the stream function for the vaneless diffuser of a centrifugal pump is given by the combination of a source and an irrotational vortex. Hence

$$\Psi = q\theta/2\pi + (\Gamma/2\pi)\ln r$$

where q is the flow per unit depth ($m^2/s$) and $\Gamma$ is the circulation ($m^2/s$) equal to the strength of the vortex $K = 2\pi u_\theta r$.

(i) Determine the expression for the velocity potential, $\phi$, and, by obtaining equations for $\Psi = 0$ and $\phi = 0$, sketch the flow net
(ii) The two-dimensional vaneless diffuser of a centrifugal pump has entry and exit diameters of 500 mm and 800 mm, respectively. Water enters at a velocity of 10 m/s and an angle of 60° with a radial line. Obtain a numerical expression for the stream function in terms of $\theta$ and r and calculate the magnitude and direction of the exit velocity.

[$\phi = \frac{q}{2\pi}\ln r - \frac{\Gamma\theta}{2\pi}$, $\Psi = 1.25\theta + 2.165\ln r$, 6.247 m/s, 60°]

# 4 DYNAMICS OF FLUID FLOW

In this chapter the forces and energies involved in fluid flow will be considered. The conservation laws of mass, momentum and energy form the basis of solution for all problems in fluid dynamics.

## 4.1 CONSERVATION OF MOMENTUM

The law of conservation of momentum is embodied in Newton's second law of motion which states that the rate of change of momentum of a given mass is equal to the net force acting on the mass i.e.

$$\Sigma \underline{F} = \frac{d}{dt}(m\underline{u}) \tag{4.1}$$

This law is not a true conservation law in the sense that momentum is conserved under all conditions but if $\Sigma \underline{F} = 0$ then momentum must be conserved.

## 4.2 EQUATIONS OF MOTION

The sum of the surface and body forces acting on a fluid element may be equated to the rate of change of momentum to obtain the *equations of motion*.

If shear or viscous forces are neglected we obtain the *Euler equations of motion* for an *ideal* fluid.

$$-\frac{1}{\rho}\frac{\partial p}{\partial x} + X = u\frac{\partial u}{\partial x} + v\frac{\partial u}{\partial y} + w\frac{\partial u}{\partial z} + \frac{\partial u}{\partial t} \tag{4.2a}$$

$$-\frac{1}{\rho}\frac{\partial p}{\partial y} + Y = u\frac{\partial v}{\partial x} + v\frac{\partial v}{\partial y} + w\frac{\partial v}{\partial z} + \frac{\partial v}{\partial t} \tag{4.2b}$$

$$-\frac{1}{\rho}\frac{\partial p}{\partial z} + Z = u\frac{\partial w}{\partial x} + v\frac{\partial w}{\partial y} + w\frac{\partial w}{\partial z} + \frac{\partial w}{\partial t} \tag{4.2c}$$

where X, Y and Z are the body forces per unit mass.

These three equations may be expressed in vector notation as

$$-\frac{1}{\rho}\,\text{grad}\,p + \underline{S} = \frac{D\underline{u}}{Dt} \tag{4.3}$$

where $\underline{S} = \underline{i}X + \underline{j}Y + \underline{k}Z$ and $\underline{u} = \underline{i}u + \underline{j}v + \underline{k}w$

The Euler equations of motion are valid at any point in a flow field for the unsteady, three-dimensional, compressible flow of an ideal fluid.

If shear or viscous forces are included we obtain the *Navier-Stokes equations of motion* for a *real* fluid.

$$-\frac{1}{\rho}\text{ grad } p + \underline{S} + \frac{1}{3}\frac{\mu}{\rho}\text{ grad div }\underline{u} + \frac{\mu}{\rho}\nabla^2\ \underline{u} = \frac{D\underline{u}}{Dt} \qquad (4.4)$$

The equations of motion are not amenable to complete solution and for integration over a finite region simplifying assumptions have to be made.

### 4.2.1 Irrotational Flow and Flow along a Streamline

Throughout the whole of an irrotational flow field or along a streamline in any ideal flow the Euler equations 4.2 reduce to a single integrable equation

$$-\frac{1}{\rho}\,dp + X\,dx + Y\,dy + Z\,dz = d\left(\frac{u^2}{2}\right) - \frac{1}{\rho}\frac{\partial p}{\partial t}\,dt \qquad (4.5)$$

### 4.2.2 Steady-state Euler Equation of Motion

For steady flow ($\partial p/\partial t = 0$) situations in which the sole body force is that due to gravity (weight), $Z = -g$ and $X = Y = 0$. Therefore

$$\frac{dp}{\rho} + \underline{u}\,d\underline{u} + g\,dz = 0 \qquad (4.6)$$

This is the *steady-state Euler equation of motion* which is strictly valid for the flow of an ideal fluid but it may be applied to real fluid flow if there are no significant velocity gradients due to eddies or solid surfaces. Integration is possible for incompressible flow where $\rho$ is constant or for compressible flow where $\rho$ is a known function of p (e.g. isentropic flow $p/\rho^\gamma$ = const).

## 4.3 THE BERNOULLI EQUATION

For constant density fluid, equation 4.6 may be integrated between two stations 1 and 2 in the flow to yield

$$\frac{p_1}{\rho} + \tfrac{1}{2}\underline{u}_1^{\,2} + gz_1 = \frac{p_2}{\rho} + \tfrac{1}{2}\underline{u}_2^{\,2} + gz_2 = \text{constant} = e_o \qquad (4.7a)$$

This is the *Bernoulli equation* which strictly applies to the steady frictionless flow of a constant density (incompressible) fluid, along a streamline, or the whole field of irrotational flow. Each of the terms has units of specific energy N m/kg (= $m^2/s^2$), therefore the Bernoulli equation is also an energy equation. It expresses conservation of 'mechanical' energy since each term is based on force quantities. Alternative forms of the Bernoulli equation are

$$\frac{p_1}{\rho g} + \frac{\underline{u}_1^{\,2}}{2g} + z_1 = \frac{p_2}{\rho g} + \frac{\underline{u}_2^{\,2}}{2g} + z_2 = H_o \qquad (4.7b)$$

where $H_o$ is the *Bernoulli* or *total head*

and $p_1 + \frac{1}{2}\rho \underline{u}_1^2 + \rho gz = p_2 + \frac{1}{2}\rho \underline{u}_2^2 + \rho gz_2 = p_o$ (4.7c)

where $p_o$ is the *stagnation* or *total head* pressure.

The Bernoulli equations 4.7 may be used for total flow through a finite flow area. In this case a mean velocity $\bar{u}$ may be used although to be strictly accurate a kinetic energy correction factor should also be applied. Equation 4.7a becomes

$$\frac{p_1}{\rho} + \frac{1}{2}\bar{u}_1^2 + gz_1 = \frac{p_2}{\rho} + \frac{1}{2}\bar{u}_2^2 + gz_2 = e_o \qquad (4.7d)$$

### 4.3.1 Pressure Variation Across Curved Streamlines

It is usually assumed in flow problems that the pressure normal to the flow is constant. This assumption is not always valid and it is necessary here to consider the effect of flow curvature on the pressure variation across the flow.

If r is the radius to the centre of curvature of a streamline in a curved flow it can be shown that

$$\frac{dH_o}{dr} = \frac{u_\theta}{g}\left(\frac{du_\theta}{dr} + \frac{u_\theta}{r}\right) = \frac{u_\theta \zeta}{g} \qquad (4.8)$$

and $$\frac{dp^*}{dr} = \frac{\rho u_\theta^2}{r} \qquad (4.9)$$

Equation 4.9 may be used to predict the variation in piezometric pressure across a curved flow if the velocity variation is known. For vortex motion, where there is a single centre of curvature, equations 4.8 and 4.9 can be integrated.

(i) Free or Irrotational Vortex

For an irrotational vortex, defined in section 3.8, the vorticity, $\zeta$, is zero. Therefore

$$\frac{dH_o}{dr} = \frac{u_\theta}{g}\left(\frac{du_\theta}{dr} + \frac{u_\theta}{r}\right) = 0$$

Thus $u_\theta r$ = constant = C (4.10)

and $H_o$ = constant (4.11)

This is a special case of circular flow with the Bernoulli head, $H_o$, constant over the whole region of flow. It follows that for two points 1 and 2 in an irrotational vortex

$$H_{o1} = H_{o2}$$

$$\text{and} \quad p_2^* - p_1^* = \frac{\rho C^2}{2}\left(\frac{1}{r_1^2} - \frac{1}{r_2^2}\right) \qquad (4.12)$$

(ii) Forced or Rotational Vortex

For a rotational vortex, defined in section 2.3.2, all the fluid particles move with constant angular velocity $\omega = u_\theta/r = du_\theta/dr$. The vorticity is finite at all points in the flow but there is no relative motion between particles and no frictional effects. Therefore the Bernoulli equation may be applied

$$\frac{dH_o}{dr} = \frac{u_\theta}{g}\left(\frac{du_\theta}{dr} + \frac{u_\theta}{r}\right) = \frac{2\omega^2 r}{g} \qquad (4.13)$$

Integrating between two points 1 and 2

$$H_{o2} - H_{o1} = \omega^2(r_2^2 - r_1^2)/g \qquad (4.14)$$

$$\text{and} \quad p_2^* - p_1^* = \rho\omega^2(r_2^2 - r_1^2)/2 \qquad (4.15)$$

In the theoretical irrotational vortex, the relationship $u_\theta r = C$ implies that $u_\theta \to \infty$ as $r \to 0$. In a real fluid, viscosity causes a central core to rotate as a solid body and $u_\theta$ increases linearly from zero at the axis to a finite value some distance from the core, when there is a transition from the rotational to the irrotational velocity distribution. This is a closer approximation than the simple irrotational vortex to the motion of air in tornadoes and behind wing tips, and of water in whirlpools and at propellers and impellers.

### 4.3.2 Applications of the Bernoulli Equation

No flow is ideal but there are many practical applications where the Bernoulli equation may be used as a basis of analysis.

Consider the terms in the Bernoulli equation. The potential energy term $gz$ increases linearly with elevation and the kinetic energy term $u^2/2$ increases with decreasing area ($\dot{V} = uA = \text{const}$). Thus in ideal, incompressible flow the pressure of the fluid decreases with increase in elevation or decrease in flow area. The two effects are independent.

Venturi meters, orifice plates and nozzles (figure 4.1a) are flow meters which depend on the measurement of the variable pressure drop between an upstream station 1 and a throat station 2. The areas are constant.

A rotameter (figure 4.1c) depends on the measurement of a variable flow area by means of the position of a float in a tapered glass tube. The pressure drop is almost constant.

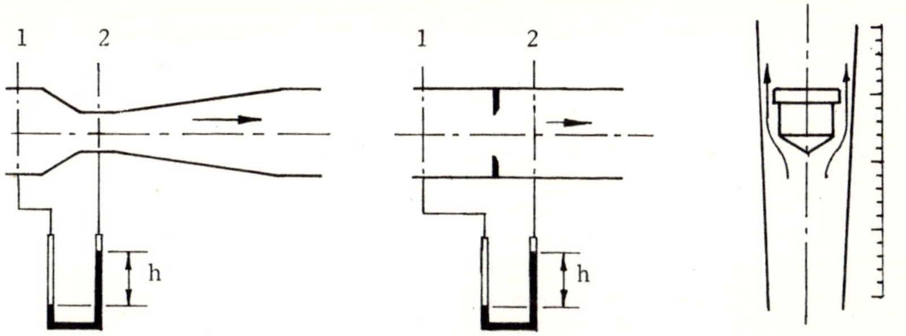

(a) Venturi meter (b) Orifice plate (c) Rotameter

Figure 4.1

The pressure p in the Bernoulli equation is the static pressure which determines the internal state of the fluid. It is measured by means of a small hole in the boundary wall normal to the fluid flow.

The *stagnation* or *total head* pressure, $p_0$, is that produced when a fluid is brought to rest (strictly under frictionless, adiabatic conditions). It is measured by means of a *pitot* or *total-head tube* which is simply a circular tube with its open end facing the flow. The fluid is brought to rest at a *stagnation point*, 0, on the nose and the stagnation pressure, $p_0$, at this point is recorded on a pressure gauge or manometer. Applying the Bernoulli equation between stations 1 and 2

$$p_0 = p_1 + \tfrac{1}{2}\rho u_1^2 \tag{4.16}$$

The *stagnation pressure*, $p_0$, is the sum of the *static pressure*, p, and the *dynamic pressure* $\rho u^2/2$.

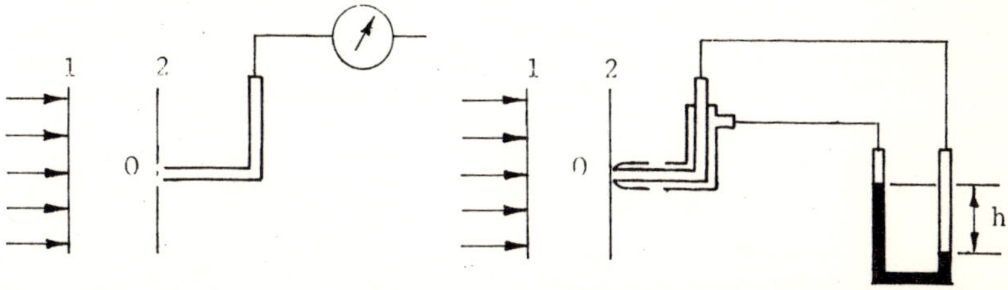

(a) Pitot tube (b) Pitot-static tube

Figure 4.2

A pitot-static tube consists of a pitot tube surrounded by a concentric tube having a circumferential ring of small holes to measure static pressure. If the two tubes are connected across a manometer the *dynamic pressure* is measured directly and for a fluid of known density the velocity can be calculated. Equation 4.16 gives

$$u = \sqrt{[2(p_o - p_1)/\rho]} \qquad (4.17)$$

It should be noted that equation 4.17 is strictly valid for incompressible flow. If the flow is compressible (e.g. gas flow) equation 4.17 becomes increasingly inaccurate with increasing velocity (increasing Mach number).

Free-surface flow (e.g. channels, streams) is normally measured by means of a weir. The Bernoulli equation can be used to obtain a relationship and other techniques are available (see chapters 5 and 8).

## 4.4 THE MOMENTUM THEOREM

The Euler equations may be integrated over an irrotational field to give the pressure forces on the boundaries of a finite system. This solution is restricted to ideal fluid flow. For the flow of a real fluid, Newton's second law may be applied in a more general form to give the *momentum theorem*.

Consider the fluid which is continuously entering and leaving a stationary control volume V bounded by control surface S (figure 4.3).

From Newton's second law

$$\Sigma \underline{F} = \frac{d}{dt}\int_V \rho \underline{u}\, dV + \int_S \underline{u}(\rho\underline{u}\cdot d\underline{A}) \qquad (4.18)$$

$$= \frac{d}{dt}\int_V \rho \underline{u}\, dV + \int_S \underline{u}\, d\dot{m}$$

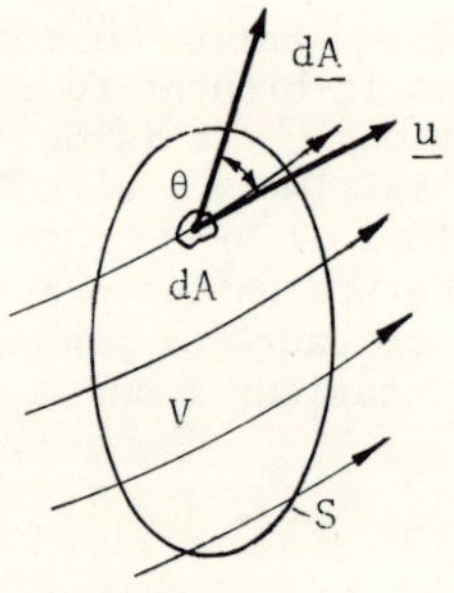


Figure 4.3

The first term on the right hand side is the time rate of change of momentum within the control volume and the second term is the time rate of net efflux of momentum across the control surface.

For steady flow, the first term disappears and, if the flow is now restricted to single entry and exit stations 1 and 2 (figure 4.4a) with uniform properties at each station, we obtain

$$\Sigma \underline{F} = \dot{m}\,(\underline{u}_2 - \underline{u}_1) \qquad (4.19)$$

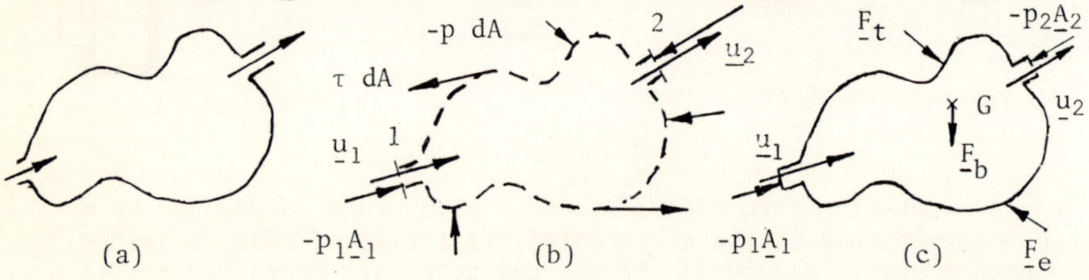


Figure 4.4

$\Sigma\underline{F}$ is the summation of all the forces acting on the control volume i.e. surface forces (pressure and shear) and body forces. If the control surface is chosen as the fluid boundary, S consists of the inside surface of the duct and the entry and exit planes, as shown in figure 4.4b. The pressure and shear forces exerted by the duct on the fluid internally are difficult to evaluate and may be represented by a single resultant force $\underline{F}_{ps}$. The body force is $\underline{F}_b$. Therefore

$$\Sigma\underline{F} = -p_1\underline{A}_1 - p_2\underline{A}_2 + \underline{F}_b + \underline{F}_{ps} = \dot{m}(\underline{u}_2 - \underline{u}_1) \qquad (4.20a)$$

or $$\underline{F}_{ps} = \dot{m}(\underline{u}_2 - \underline{u}_1) + p_1\underline{A}_1 + p_2\underline{A}_2 - \underline{F}_b \qquad (4.20b)$$

It is sometimes desirable to draw the control surface outside the duct and areas $A_1$ and $A_2$. In this case, there must be an external force $\underline{F}_e$ acting on the duct to hold it in place (figure 4.4c). In addition, there is an external force $\underline{F}_t$ acting on the duct due to pressure and drag forces.

For a stationary control volume, $\underline{F}_t$ consists solely of forces due to atmospheric pressure, $p_a$, which cancel, except on areas $A_1$ and $A_2$. Hence

$$\underline{F}_e = \dot{m}\ (\underline{u}_2 - \underline{u}_1) + (p_1 - p_a)\ \underline{A}_1 + (p_2 - p_a)\ \underline{A}_2 - \underline{F}_b \qquad (4.21)$$

Note that $F_e$ is the external force exerted on the fluid inside the control volume. By Newton's third law, the force exerted by the fluid on the duct is $\underline{F}_d = -\underline{F}_e$.

For a moving control volume, equation 4.18 applies but the velocity associated with mass flow rate refers to velocity relative to S and will be written $\underline{u}_{RS}$. Hence

$$\Sigma\underline{F} = \frac{d}{dt}\int_V \rho_V\underline{u}_V\ dV + \int_S \underline{u}_S(\rho_S\underline{u}_{RS}\cdot d\underline{A}) \qquad (4.22)$$

This equation is valid for a control volume having linear velocity $\underline{U}$ relative to earth. The velocities $\underline{u}_S$ and $\underline{u}_V$ are also relative to earth. Therefore $\underline{u}_S = \underline{U} + \underline{u}_{RS}$ and $\underline{u}_V = \underline{U} + \underline{u}_{RV}$ where $\underline{u}_{RV}$ is velocity relative to V. Equation 4.22 may be expanded and modified to give

$$\Sigma\underline{F} = M_V\frac{d\underline{U}}{dt} + \frac{d}{dt}\int_V \rho_V\underline{u}_{RV}\ dV + \int_S \underline{u}_{RS}(\rho_S\underline{u}_{RS}\cdot d\underline{A}) \qquad (4.23)$$

The first term on the right hand side is the product of the mass of the whole control volume (solid and fluid) and the acceleration of this mass. The other terms are similar to those in equation 4.18 for a stationary control volume but with velocities *relative* to volume and surface, respectively.

## 4.5 ANGULAR MOMENTUM

The linear momentum equation 4.18 may be modified for *angular momentum* or *moment of momentum* by introducing a radius vector $\underline{r}$ from the origin to the point of application of the force. The forces acting on a system produce a net torque $\underline{T}$.

$$\underline{T} = \Sigma(\underline{F} \times \underline{r}) = \frac{d}{dt} \int_V (\underline{u} \times \underline{r})\rho \, dV + \int_S (\underline{u} \times \underline{r})(\rho \underline{u}_R \cdot d\underline{A}) \qquad (4.24)$$

The most important application of this relationship is steady flow through turbomachinery e.g. pumps, compressors, turbines.

Consider figure 4.5 which shows flow through the rotor of a pump. All velocities are in the plane of rotation. The velocity of a point on the rotor periphery relative to the casing is $\underline{U} = \omega r$. The velocity of the fluid relative to the casing is $\underline{u}$ and that relative to the rotor is $\underline{u}_R$.

The relationship between the velocities is given by

$$\underline{u} = \underline{U} + \underline{u}_R \qquad (4.25)$$

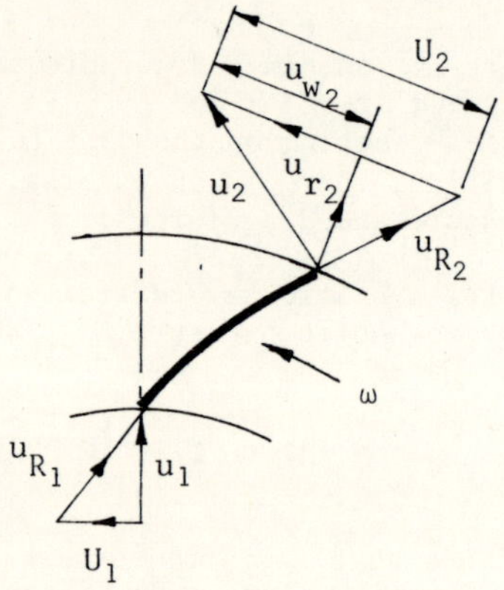


Figure 4.5

The velocity $\underline{u}$ can, in general, be resolved into three perpendicular components: one axial $u_a$, one radial $u_r$ and one tangential $u_w$.

Applying the momentum equation we may state

(1) The change in magnitude of $u_a$ produces an axial thrust $F_a$. From equation 4.19

$$F_a + \Sigma F_{pressure} = \dot{m}(u_{a_2} - u_{a_1}) \qquad (4.26)$$

(2) The change in magnitude of $u_r$ results in radial journal loads.

(3) The change in magnitude of $u_w$ results in the torque $\underline{T}$ on the rotor. Rotation is about the rotor axis rather than a point, and flow is steady. Therefore equation 4.24 reduces to the scalar equation

$$T = \int_S u_w r(\rho \underline{u}_R \cdot d\underline{A}) = \int_S u_w r \, d\dot{m} \qquad (4.27)$$

With uniform properties and single mean values for the fluid velocities at rotor entry station 1 and exit station 2 we obtain

$$T = \dot{m}\,(u_{w_2} r_2 - u_{w_1} r_1) \qquad (4.28)$$

The rate of energy transfer between rotor and fluid is

$$\dot{E} = T\omega = \dot{m}(u_{w_2} r_2 - u_{w_1} r_1)\omega = \dot{m}(u_{w_2} U_2 - u_{w_1} U_1) \qquad (4.29a)$$

For a turbine

$$\dot{E} = \dot{m}(u_{w_1} U_1 - u_{w_2} U_2) \qquad (4.29b)$$

Equation 4.29b is sometimes called the *Euler turbine equation.*

## 4.6 CONSERVATION OF ENERGY

The principle of conservation of energy was stated in section 1.6.2. This principle is embodied in the first law of thermodynamics (equation 1.15) which may be written as a time rate equation for the fluid which is continuously entering and leaving a stationary control volume (figure 4.3).

$$\frac{dQ}{dt} = \frac{dE}{dt} + \frac{dW}{dt}$$

or
$$\dot{Q} = \frac{d}{dt}\int_V e\rho \, dV + \int_S e_s(\rho \underline{u}\cdot d\underline{A}) + \dot{W} \qquad (4.30)$$

The first term on the right hand side is the time rate of change of energy within the control volume and the second term is the time rate of net flow of energy across the control surface. $\dot{W}$ consists of the time rate of flow work $\int_S (p/\rho)(\rho\underline{u}\cdot d\underline{A})$ on the control surface and the rate of shaft work $\dot{W}_x$. The control surface is chosen so that there is no fluid shear work. Hence

$$\dot{Q} = \frac{d}{dt}\int_V e\rho \, dV + \int_S e_s(\rho\underline{u}\cdot d\underline{A}) + \int_S (p/\rho)(\rho\underline{u}\cdot d\underline{A}) + \dot{W}_x \qquad (4.31)$$

This is the *unsteady-flow energy equation (U.S.F.E.E.)*. For steady flow, time rate terms disappear and equation 4.31 reduces to

$$\dot{Q} = \int_S (e_s + p/\rho)(\rho\underline{u}\cdot d\underline{A}) + \dot{W}_x$$

where $e_s = e + \frac{1}{2}u^2 + gz$ is the stored specific energy of the fluid.

$$\dot{Q} = \int_S (e + p/\rho + \tfrac{1}{2}u^2 + gz)(\rho\underline{u}\cdot d\underline{A}) + \dot{W}_x \qquad (4.32a)$$

This is the *steady-flow energy equation (S.F.E.E.)*.

If the flow is now restricted to single entry and single exit with uniform properties at each station, $\int \rho\underline{u}\cdot d\underline{A} = \dot{m}$ and we obtain the form of the S.F.E.E. most commonly used in thermodynamics.

$$\dot{Q} + \dot{m}(e_1 + p_1/\rho_1 + \tfrac{1}{2}\bar{u}_1{}^2 + gz_1) = \dot{W}_x + \dot{m}(e_2 + p_2/\rho_2 + \tfrac{1}{2}\bar{u}_2{}^2 + gz_2) \qquad (4.32b)$$

For unit mass

$$q + e_1 + p_1/\rho_1 + \tfrac{1}{2}\bar{u}_1{}^2 + gz_1 = w_x + e_2 + p_2/\rho_2 + \tfrac{1}{2}\bar{u}_2{}^2 + gz_2 \tag{4.32c}$$

In differential form putting $\upsilon = 1/\rho$

$$dq = dw_x + de + p\,d\upsilon + \upsilon\,dp + u\,du + g\,dz \tag{4.32d}$$

Note that enthalpy $h = e + p/\rho$ may be substituted into equations 4.32.

## 4.7 RELATIONSHIP BETWEEN THE EULER, BERNOULLI AND STEADY-FLOW ENERGY EQUATIONS

It was shown in section 4.3 that the terms in the Bernoulli equation represent 'mechanical' forms of energy. These terms also appear in the steady-flow energy equation together with terms representing 'thermal' forms of energy. It is now necessary to consider how these equations are related for duct flow and turbomachinery.

### 4.7.1 Incompressible Flow in Ducts

The steady-flow energy equation 4.32d applied to duct flow ($dw_x = 0$) gives

$$dq = de + p\,d\upsilon + \upsilon\,dp + u\,du + g\,dz \tag{4.33}$$

For any process, $T\,ds = de + p\,d\upsilon$ (equation 1.17). Also $\upsilon = 1/\rho$. Hence

$$\frac{dp}{\rho} + u\,du + g\,dz = dq - T\,ds \tag{4.34}$$

The Euler equation 4.6 states that the left hand side of this equation is zero for ideal or reversible flow in which case $dq_R = T\,ds$, as predicted by the second law of thermodynamics (equation 1.16b). If the flow is adiabatic and reversible $dq = 0$ and $ds = 0$. If, however, the flow is adiabatic and irreversible $dq = 0$ but $T\,ds$ is finite and positive. Thus, $T\,ds$ represents a dissipation or loss of mechanical energy into thermal energy due to frictional effects. For incompressible, adiabatic, irreversible flow in ducts, equation 4.34 can be integrated between stations 1 and 2 to give

$$\frac{p_1}{\rho} + \tfrac{1}{2}\bar{u}_1{}^2 + gz_1 = \frac{p_2}{\rho} + \tfrac{1}{2}\bar{u}_2{}^2 + gz_2 + \int_1^2 T\,ds \tag{4.35}$$

This equation is sometimes called the Bernoulli equation with losses because $\int T\,ds = e_L$ represents the dissipation or loss of mechanical energy. For flow in ducts, $e_L$ may be predicted from knowledge of the nature of flow (see chapter 7).

The steady-flow energy equation 4.32c for the same conditions gives

$$\frac{p_1}{\rho} + \tfrac{1}{2}\bar{u}_1^2 + gz_1 = \frac{p_2}{\rho} + \tfrac{1}{2}\bar{u}_2^2 + gz_2 + (e_2 - e_1) \qquad (4.36)$$

Therefore, from equations 4.35 and 4.36

$$\int_1^2 T\, ds = e_L = (e_2 - e_1) \qquad (4.37)$$

Thus, for incompressible, adiabatic irreversible flow in ducts the loss in mechanical energy due to dissipation appears as an increase in internal energy.

For ideal or reversible flow $\int T\, ds = e_L = (e_2 - e_1) = 0$ and both equations 4.35 and 4.36 reduce to the Bernoulli equation 4.7d. Thus the Bernoulli equation can be obtained from the Euler equation, derived from momentum concepts, and the steady-flow energy equation, based on energy concepts. A single statement of the Bernoulli equation satisfies both concepts. This is not true for compressible flow where density changes are significant and the momentum and energy equations must be considered separately.

### 4.7.2 Incompressible-flow Turbomachinery

In the case of adiabatic, incompressible flow with work transfer (e.g. hydraulic turbines or pumps or fans) the steady-flow energy equation 4.32c gives

$$w_x = \left(\frac{p_1}{\rho} + \tfrac{1}{2}\bar{u}_1^2 + gz_1\right) - \left(\frac{p_2}{\rho} + \tfrac{1}{2}\bar{u}_2^2 + gz_2\right) - (e_2 - e_1) \qquad (4.38)$$

For adiabatic, irreversible flow, equation 4.37 gives $(e_2 - e_1) = e_L$. Therefore the internal energy change is due to the dissipation of mechanical energy due to friction and eddies and serves no useful purpose. The useful energy transfer is that associated with the change of mechanical energy, $e_m$, of the fluid defined

$$e_m = \left(\frac{p_1}{\rho} + \tfrac{1}{2}\bar{u}_1^2 + gz_1\right) - \left(\frac{p_2}{\rho} + \tfrac{1}{2}\bar{u}_2^2 + gz_2\right) \qquad (4.39)$$

Therefore, from equations 4.38 and 4.39

$$w_x = e_m - (e_2 - e_1) \qquad (4.40)$$

The efficiency of energy conversion between fluid and rotor for an incompressible-flow machine is the *hydraulic efficiency*, $\eta_H$, defined as follows for a pump

$$(\eta_H)_p = \frac{e_m}{w_x} \qquad (4.41)$$

For a turbine, the available mechanical energy, $e_m'$, is referred to a datum of zero energy at the level of the turbine discharge or tail water level

$$(\eta_H)_T = \frac{w_x}{e_m'} \qquad (4.42)$$

*Example 4.1*

A siphon has a uniform circular bore of 75 mm and consists of a bent pipe with its crest 2 m above water level and its open end discharging into the atmosphere at a level 4 m below water level. Find the velocity of flow, the discharge and the absolute pressure at crest level, if the atmospheric pressure is equivalent to 10 m of water. Neglect losses due to friction. Determine also the maximum permissible crest height for water at 20 °C.

The Bernoulli equation 4.7d gives

$$\frac{p}{\rho g} + \frac{\bar{u}^2}{2g} + z = H_o \qquad (i)$$

For frictionless flow, the total head, $H_o$, is constant and may be represented by a *total head line* (or *energy line*) parallel to the datum line (figure 4.6). A *piezometric* or *hydraulic grade line* may also be drawn based on $(p/\rho g + z)$. The piezometric line is useful because the vertical distance between it and a point on the pipe centre line represents the pressure $p/\rho g$ in the pipe. Thus the possible onset of cavitation may be predicted.

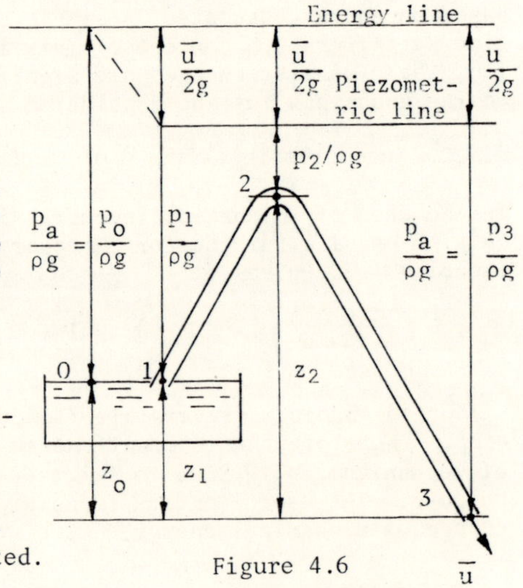


Figure 4.6

Note that absolute pressures are used in figure 4.6 but energy and piezometric lines are frequently based on gauge pressures.

For free discharge to atmosphere, $p_3 = p_a$. Applying the Bernoulli equation (i) between points 0 and 3 ($\bar{u}$ = constant in pipe)

$$\frac{p_a}{\rho g} + 0 + z_1 = \frac{p_a}{\rho g} + \frac{\bar{u}^2}{2g} + z_3$$

$$\bar{u} = \surd[2g(z_1 - z_3)] = \surd(2 \times 9.81 \times 4) = 8.85 \text{ m/s}$$

Applying the continuity equation

$$\dot{V} = \bar{u}A = \tfrac{1}{4}\pi \times 0.075^2 \times 8.85 = 0.039\ \text{m}^3/\text{s}$$

Applying equation (i) between points 0 and 2

$$\frac{p_a}{\rho g} + 0 + z_1 = \frac{p_2}{\rho g} + \frac{\bar{u}^2}{2g} + z_2$$

$$p_2 = \rho g(z_1 - z_2) - \tfrac{1}{2}\rho\bar{u}^2 + p_a \qquad \text{(ii)}$$

$$= 10^3 \times 9.81(-2) - \tfrac{1}{2} \times 10^3 \times 8.85 + 10^3 \times 9.81 \times 10$$

$$= 39.38\ \text{kPa}$$

The height of the crest above the water surface is restricted because the absolute pressure $p_2$ falls with increasing crest height and can drop below the saturation pressure, causing a vapour lock. For a saturation pressure 2.337 kPa at 20 °C, the maximum height $(z_2 - z_1)$ is given by equation (ii)

$$z_2 - z_1 = \frac{p_a - p_2}{\rho g} - \frac{\bar{u}^2}{2g} = 10 - \frac{2337}{10^3 \times 9.81} - \frac{8.85^2}{2 \times 9.81} = 9.7\ \text{m}$$

*Example 4.2*

Figure 4.7 shows a venturi mounted with a vertical axis and arranged as a suction device. The throat area is 0.00025 m$^2$ and the outlet area is 4 times the throat area. The suction pipe joins the throat at a point 0.1 m above the outlet and the surface of the sump is 1 m below this point. The venturi discharges to atmosphere and the free surface of the sump is also exposed to atmospheric pressure. Neglecting viscous effects, calculate the lowest steady volume rate of flow through the venturi at which flow will occur up the suction pipe.

Applying the Bernoulli equation 4.7d, in head units, between venturi throat and exit we have

$$\frac{p_2}{\rho g} + \frac{\bar{u}_2^2}{2g} + z_2 = \frac{p_3}{\rho g} + \frac{\bar{u}_3^2}{2g} + z_3 \qquad \text{(i)}$$

From continuity

$$\dot{V} = \bar{u}_2 A_2 = \bar{u}_3 A_3 \qquad \text{(ii)}$$

From equations (i) and (ii)

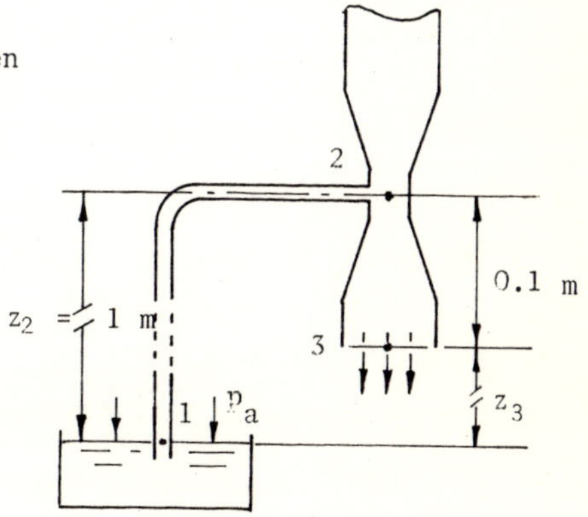


Figure 4.7

$$\dot{V} = A_2\left[1 - \left(\frac{A_2}{A_3}\right)^2\right]^{-\frac{1}{2}}\left[\frac{2(p_3 - p_2)}{\rho} + 2g(z_3 - z_2)\right]^{\frac{1}{2}} \qquad \text{(iii)}$$

Flow will commence in the suction pipe at the instant that the pressure difference across the ends of the pipe is fractionally greater than the hydrostatic pressure difference. This occurs for any flow greater than the condition $p_1{}^* = p_2{}^*$. Therefore for minimum flow

$$p_1 + \rho g z_1 = p_2 + \rho g z_2$$

or $\quad p_1 - p_2 = \rho g(z_2 - z_1) \qquad$ (iv)

From equations (iii) and (iv) and $p_1 = p_3 = p_a$ we obtain

$$\dot{V} = A_2\left[1 - \left(\frac{A_2}{A_3}\right)^2\right]^{-\frac{1}{2}} [2g(z_3 - z_1)]^{\frac{1}{2}}$$

$$= 0.00025\left(1 - \frac{1}{4^2}\right)^{-\frac{1}{2}} [2 \times 9.81(1 - 0.1)]^{\frac{1}{2}} = 0.00109 \text{ m}^3/\text{s}$$

It should be noted that the pressure at a throat can drop below the saturation pressure and cause cavitation. In this problem the throat pressure, $p_2$, can be evaluated from equation (i) or equation (iv). Cavitation does not occur at ambient conditions.

*Example 4.3*

(a) Show that, for a given flow rate through a pressure differential flow meter, the difference in pressure level registered by a manometer connected to the inlet and throat sections is independent of the angle of inclination of the flow meter.

(b) A 75 mm diameter orifice-plate is situated in an inclined pipe 150 mm diameter as shown in figure 4.8. The throat (D/2) pressure tapping is 200 mm above the inlet (D) tapping and $C_d = 0.6$. Petrol of density 780 kg/m$^3$ flows through the meter at the rate of 0.04 m$^3$/s. By application of the Bernoulli equation determine (i) the pressure difference in kPa between inlet and throat and (ii) the difference of level which would be registered by a vertical mercury manometer, the tubes above the manometer being full of petrol.

(a) Applying the Bernoulli equation 4.7d between pressure tapping stations 1 and 2 we have

$$\frac{p_1}{\rho} + \bar{u}_1{}^2 + gz_1 = \frac{p_2}{\rho} + \bar{u}_2{}^2 + gz_2$$

and $\quad (\bar{u}_2{}^2 - \bar{u}_1{}^2)^{\frac{1}{2}} = \left(\dfrac{2(p_1 - p_2)}{\rho} + 2g(z_1 - z_2)\right)^{\frac{1}{2}} \qquad$ (i)

From the continuity equation 3.5a

$$\dot{m} = \rho_1 A_1 \bar{u}_1 = \rho_2 A_2 \bar{u}_2 \qquad \text{(ii)}$$

and $$\bar{u}_1 = \frac{A_2}{A_1}\bar{u}_2 \qquad \text{(iii)}$$

Combining equations (i), (ii) and (iii)

$$\dot{m}_{\text{theoretical}} = \rho A_2 \left[1 - \left(\frac{A_2}{A_1}\right)^2\right]^{-\frac{1}{2}} \left[\frac{2(p_1 - p_2)}{\rho} + 2g(z_1 - z_2)\right]^{\frac{1}{2}}$$

This equation is based on frictionless flow. In practice, frictional effects are accounted for by a discharge coefficient, $C_d$, and the actual mass flow rate is given by

$$\dot{m} = C_d \rho A_2 \left[1 - \left(\frac{A_2}{A_1}\right)^2\right]^{-\frac{1}{2}} \left[\frac{2(p_1 - p_2)}{\rho} + 2g(z_1 - z_2)\right]^{\frac{1}{2}} \qquad \text{(v)}$$

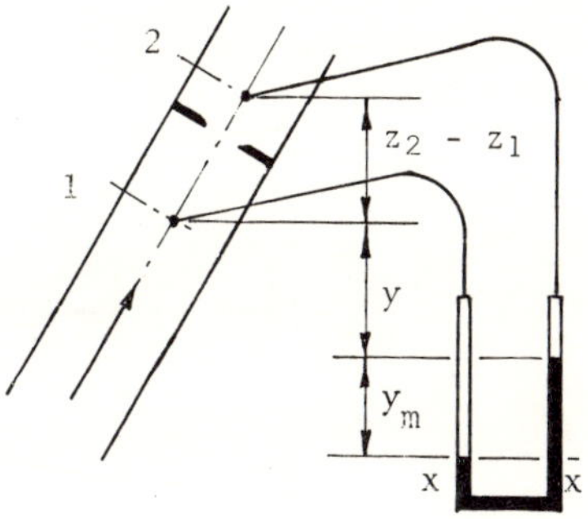


Figure 4.8

Consider now the reading on the manometer. Equating pressures on line xx, we have

$$p_1 + \rho g(y_m + y) = p_2 + \rho g[(z_2 - z_1) + y] + \rho_m g y_m \qquad \text{(vi)}$$

$$\frac{p_1 - p_2}{\rho} + g(z_1 - z_2) = y_m g \left(\frac{\rho_m}{\rho} - 1\right)$$

Substituting equation (vi) in (v)

$$\dot{m} = C_d \rho A_2 \left[1 - \left(\frac{A_2}{A_1}\right)^2\right]^{-\frac{1}{2}} \left[2 y_m g \left(\frac{\rho_m}{\rho} - 1\right)\right]^{\frac{1}{2}} \qquad \text{(vii)}$$

This expression for mass flow rate depends only on the manometer reading and is independent of the angle of inclination of the flow meter.

(b) Substituting values in equation (vii) we have

$$0.04 = 0.6 \times \tfrac{1}{4}\pi \times 0.075^2 \left[1 - \frac{1}{2^4}\right]^{-\frac{1}{2}} \left[2 \times y_m \times 9.81 \left(\frac{13.6}{0.78} - 1\right)\right]^{\frac{1}{2}}$$

$$y_m = 662 \text{ mm}$$

From equation (vi)

$$p_1 - p_2 = \rho y_m g\left(\frac{\rho_m}{\rho} - 1\right) + \rho g(z_2 - z_1)$$

$$= 780 \times 0.662 \times 9.81\left(\frac{13.6}{0.78} - 1\right) + 780 \times 9.81 \times 0.2$$

$$= 84.8 \text{ kPa}$$

*Example 4.4*

Air flows through a converging conical nozzle. The nozzle is of length 305 mm and its diameter changes linearly from 102 mm at the entry to 51 mm at the throat. At a section half-way along the nozzle a pitot-static tube is mounted on the centre-line. The pitot-static tube shows a dynamic pressure of 614 Pa. Assuming frictionless flow calculate the velocity of flow at this section and also the longitudinal gradient of static pressure. $\rho_{air} = 1.42 \text{ kg/m}^3$.

The velocity recorded by the pitot-static tube is given by equation 4.17

$$u = \sqrt{[2(p_o - p)/\rho]}$$

$$= \sqrt{[2 \times 614/1.42]}$$

$$= 29.4 \text{ m/s}$$

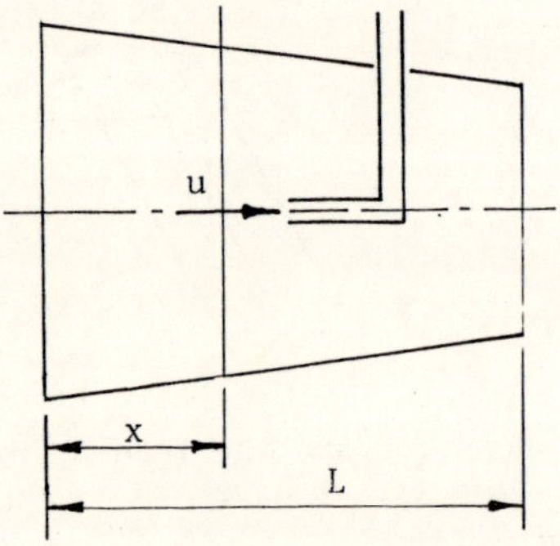

Figure 4.9

For frictionless flow the Euler equation 4.6 may be applied. For gas flow, $dz \simeq 0$, hence

$$\frac{dp}{\rho} + u\, du = 0 \qquad \text{(i)}$$

The continuity equation for incompressible flow may be expressed in differential form as follows

$$Au = \dot{V} \qquad \text{(ii)}$$

$$\ln A + \ln u = \ln \dot{V}$$

$$\frac{dA}{A} + \frac{du}{u} = 0 \qquad \text{(iii)}$$

Combining equations (i) and (iii) we have

$$dp = \rho u^2 \frac{dA}{A} \qquad \text{(iv)}$$

From the geometry of the nozzle, the cross-sectional area, A, at distance x from the nozzle entrance is

$$A = \tfrac{1}{4}\pi D^2 = \tfrac{1}{4}\pi(0.102 - 0.167x)^2 \qquad \text{(v)}$$

$$dA = -\tfrac{1}{4}\pi(0.102 - 0.167x)0.334\ dx \qquad \text{(vi)}$$

Combining equations (iv) and (vi) we have

$$\frac{dp}{dx} = \frac{-0.334\rho u^2}{(0.102 - 0.167x)} = \frac{-0.334 \times 1.42 \times 29.4^2}{(0.102 - 0.167 \times 0.152)} = -5.35\ \text{kPa/m}$$

*Example 4.5*

Water is pumped at the rate of 0.2 $m^3/s$ through the system shown in figure 4.10. Neglecting all losses calculate the power required by the pump.

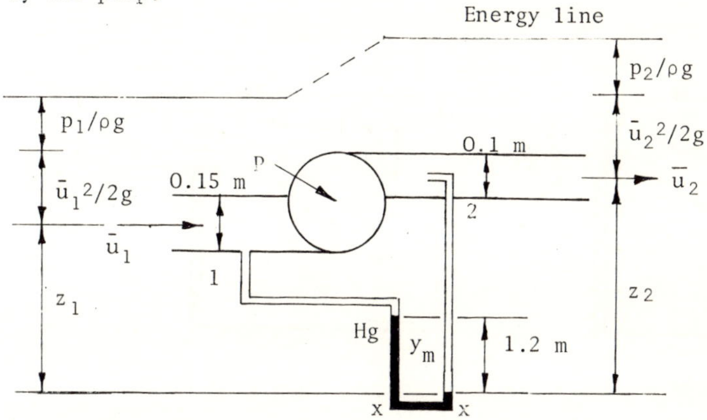


Figure 4.10

For frictionless, adiabatic flow the steady-flow energy equation 4.38 reduces to

$$w_x = \left(\frac{p_1}{\rho} + \tfrac{1}{2}\bar{u}_1^{\,2} + gz_1\right) - \left(\frac{p_2}{\rho} + \tfrac{1}{2}\bar{u}_2^{\,2} + gz_2\right) \qquad \text{(i)}$$

Multiplying throughout by the mass flow rate, $\dot{m} = \rho\dot{V}$, to give the power, P, required by the pump and rearranging, we have

$$P = \dot{m}w_x = \dot{V}[p_1 + \tfrac{1}{2}\rho\bar{u}_1^{\,2} + \rho gz_1] - \dot{V}[p_2 + \tfrac{1}{2}\rho\bar{u}_2^{\,2} + \rho gz_2] \qquad \text{(ii)}$$

At point 1 the static pressure, $p_1$, is recorded and at point 2 the pitot tube records the stagnation pressure $p_{o\,2} = p_2 + \tfrac{1}{2}\rho u_2^{\,2}$(equation 4.16). Equating pressures in line xx of manometer

$$p_1 + \rho g(z_1 - y_m) + \rho_m gy_m = p_2 + \tfrac{1}{2}\bar{u}_2^{\,2} + \rho gz_2$$

$$\text{or} \quad p_1 + \rho gz_1 - (p_2 + \tfrac{1}{2}\bar{u}_2^{\,2} + \rho gz_2) = gy_m(\rho - \rho_m) \qquad \text{(iii)}$$

Substituting equation (iii) in (ii) we have

$$P = \dot{V}[\tfrac{1}{2}\rho\bar{u}_1^{\,2} + gy_m(\rho - \rho_m)] = \dot{V}\left[\tfrac{1}{2}\rho\frac{\dot{V}^2}{A_1^{\,2}} + gy_m(\rho - \rho_m)\right]$$

$$= 0.2[\tfrac{1}{2} \times 10^3 \times 0.2^2 \times 4^2/(\pi^2 \times 0.15\ ) + 9.81 \times 1.2(1 - 13.6)10^3]$$

$P = -16.85$ kW

*Example 4.6*

The two-dimensional flow-rate of water over a sharp-crested weir is shown in figure 4.11a. Sketch the streamline pattern and the pressure distribution.

Assuming the idealised flow condition of figure 4.11b, determine the discharge over the weir. The weir is 3 m wide and the discharge coefficient is 0.61.

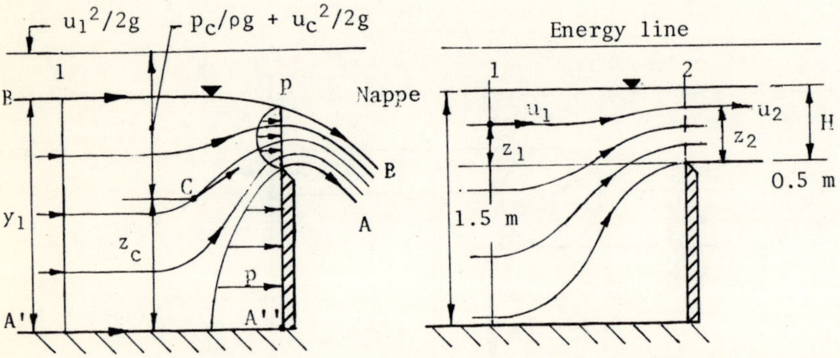


(a) Actual flow (b) Idealised flow

Figure 4.11

(i) The streamline pattern for flow over the weir is shown. At section 1 a one-dimensional situation exists with a hydrostatic pressure distribution. The gauge pressure at A' is $\rho g y_1$. At any other point, C, in the flow field, $(p_c/\rho g + u_c^2/2g)$ may be determined from the point and energy line but $u_c$ must be known before $p_c$ can be calculated. In the plane of the weir plate, the pressure distribution can be predicted qualitatively. At the base, a stagnation point occurs at A'' with pressure $(\rho g y_1 + \rho u_1^2/2)$ and at the free streamlines AA and BB the pressure is constant, with zero gauge pressure. The curvature of the streamlines in the vicinity of the plate produces non-uniform positive pressure distributions as shown. Downstream from the weir, the pressure within the nappe becomes uniform and zero (gauge) as the flow approaches the one-dimensional situation.

(ii) In the idealised flow it is assumed that the free surface remains horizontal, there are no viscous or surface tension effects and the pressure throughout the nappe is atmospheric. At section 1 the velocity is uniform ($u_1$ = const) and the hydrostatic law applies. Hence $p_1/\rho g + z_1 = H$. At section 2, $p_2 = 0$, and the velocity, $u_2$, at any point is found by applying the Bernoulli equation along a streamline

$$H + u_1^2/2g = 0 + u_2^2/2g + z_2 \qquad \text{(i)}$$

$$u_2 = \sqrt{[2g(H - z_2 + u_1^2/2g)]}$$

The discharge through an elementary strip of width B is $u_2B\ dz_2$ and the total discharge is given by

$$\dot{V} = B\sqrt{(2g)} \int_0^H (H - z_2 + u_1^2/2g)^{\frac{1}{2}}\ dz_2$$

$$= \frac{2}{3} B\sqrt{(2g)} \left[\left(H + \frac{u_1^2}{2g}\right)^{1.5} - \left(\frac{u_1^2}{2g}\right)^{1.5}\right] \qquad \text{(ii)}$$

In practice, a discharge coefficient, $C_d$, must be introduced

$$\dot{V} = \frac{2}{3} BC_d \sqrt{(2g)} \left[\left(H + \frac{u_1^2}{2g}\right)^{1.5} - \left(\frac{u_1^2}{2g}\right)^{1.5}\right] \qquad \text{(iii)}$$

As a first approximation, the velocity of approach, $u_1$, is neglected. Therefore

$$\dot{V} = \frac{2}{3} BC_d \sqrt{(2g)}H^{1.5} = \frac{2}{3} \times 3 \times 0.61 \times \sqrt{(2 \times 9.81)} \times 0.5^{1.5}$$

$$= 1.91\ m^3/s$$

The velocity of approach may now be estimated

$$u_1 = \frac{\dot{V}}{A} = \frac{1.91}{3 \times 1.5} = 0.425\ m/s$$

Substituting values in equation (iii) we obtain

$$\dot{V} = \frac{2}{3} \times 3 \times 0.61 \times \sqrt{(2 \times 9.81)}\left[\left[0.5 + \left(\frac{0.425^2}{2 \times 9.81}\right)\right]^{1.5} - \left(\frac{0.425^2}{2 \times 9.81}\right)^{1.5}\right]$$

$$= 1.96\ m^3/s$$

*Example 4.7*

An impeller, 300 mm diameter, rotates at 10 rad/s about a vertical axis in a concentric cylindrical casing containing water. Inside the impeller, the motion is that of a forced vortex but between the impeller tip and the side of the casing the motion is that of a free vortex. If the pressure head at the centre of the impeller is 600 mm and that at the side of the casing is 800 mm, determine the radius of the casing.

The impeller rotates a central core of water, of radius $r_1$, with constant angular velocity, $\omega$. From equation 4.15

$$p_2^* - p_1^* = \rho\omega^2(r_2^2 - r_1^2)/2$$

$$\text{or} \quad \left(\frac{p_2}{\rho g} + z_2\right) - \left(\frac{p_1}{\rho g} + z_1\right) = \frac{\omega^2}{2g}(r_2{}^2 - r_1{}^2) \qquad \text{(i)}$$

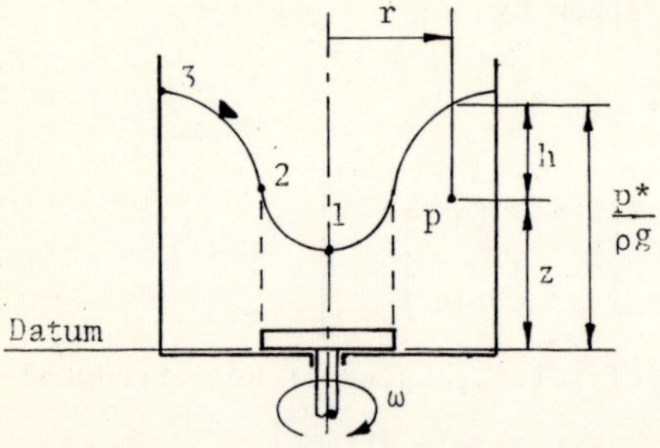


Figure 4.12

If a horizontal reference plane is taken then the pressure at any point in the plane is related to a point vertically above it in the free surface by $h = p/\rho g$. Taking a datum line through the impeller ($z_1 = z_2 = 0$) and substituting values in equation (i) we have

$$h_2 = \frac{p_2}{\rho g} = 0.6 + \frac{10^2(0.15^2 - 0)}{2 \times 9.81} = 0.715 \text{ m}$$

Outside the impeller, a free vortex exists for which $u_\theta r = C$. Assuming free vortex conditions between radii $r_2$ and $r_3$, equation 4.12 gives

$$p_3{}^* - p_2{}^* = \frac{\rho C^2}{2}\left(\frac{1}{r_2{}^2} - \frac{1}{r_3{}^2}\right)$$

$$\text{or} \quad \left(\frac{p_3}{\rho g} + z_3\right) - \left(\frac{p_2}{\rho g} + z_2\right) = \frac{C^2}{2g}\left(\frac{1}{r_2{}^2} - \frac{1}{r_3{}^2}\right) \qquad \text{(ii)}$$

Assuming a sharp transition between the forced and free vortices, we may use the $p_2/\rho g = 0.715$ m value obtained earlier. Also $C = u_{\theta 2} r_2 = 10 \times 0.15^2 = 0.225 \text{ m}^2/\text{s}$. Substituting values in equation (ii) we have

$$0.8 - 0.715 = \frac{0.225^2}{2 \times 9.81}\left(\frac{1}{0.15^2} - \frac{1}{r_3{}^2}\right)$$

$$r_3 = 295 \text{ mm}$$

*Example 4.8*

A square plate of uniform thickness is hinged about its upper edge, which is horizontal. The length of each side is 0.4 m and the plate weighs 100 N. A horizontal jet of water, 0.02 m diameter, impinges on the plate at a point 0.2 m below the upper edge so that when the plate is vertical the jet strikes it normally at its centre point. The jet has a velocity of 15 m/s. Determine the force which must be applied at the lower edge in order to keep the plate vertical. If the plate is now allowed to swing freely, find the inclination to the vertical which the plate assumes under the action of the jet.

In both cases, assume that after impact the jet flows over the surface of the plate.

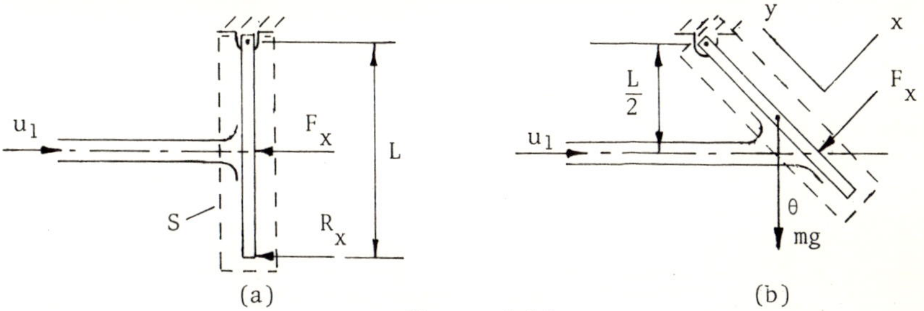


Figure 4.13

Consider a control volume enclosing the plate, as shown in figure 4.13a. The free jet and the surroundings are at atmospheric pressure, therefore there is no net pressure force on the control volume. Let a retaining force, $R_x$, be applied to the bottom of the plate to keep the plate vertical and let $F_x$ be the force exerted by the plate on the fluid at the point of contact of the jet. Taking moments about the hinge point

$$F_x L/2 = R_x L$$

$$F_x = 2R_x \qquad \text{(i)}$$

Applying the momentum equation 4.19 and continuity equation 3.5a and adopting the sign convention for the x-y coordinate system shown

$$\Sigma F = \dot{m}(\underline{u}_2 - \underline{u}_1)$$

$$-F_x = \rho_1 A_1 u_1 (0 - u_{x_1}) = 10^3 \times \tfrac{1}{4}\pi \times 0.02^2 \times 15(0 - 15)$$

$$F_x = 70.5 \text{ N} \leftarrow$$

$$R_x = F_x/2 = 70.5/2 = 35.25 \text{ N}$$

Note that, by Newton's third law, the force exerted by the fluid on the plate is equal and opposite to $F_x$ i.e. 70.5 N →.

If the plate is now allowed to swing freely to an equilibrium position the moment produced by the impinging jet is balanced by the moment produced by the weight of the plate (figure 4.13b). Let $F_x$ be the force, normal to the plate, exerted by the plate on the fluid. Taking moments about the hinge point

$$\frac{F_x L}{2 \cos \theta} = \frac{mgL \sin \theta}{2}$$

$$F_x = mg \sin \theta \cos \theta \qquad \text{(ii)}$$

Applying the momentum and continuity equations and adopting the coordinate system shown (x axis normal to the plate)

$$-F_x = \rho_1 A_1 u_1 (0 - u_{x_1}) = \rho_1 A_1 u_1 (0 - u_1 \cos\theta) \qquad \text{(iii)}$$

From equations (ii) and (iii)

$$\sin\theta = \frac{\rho_1 A_1 u_1^2}{mg} = \frac{10^3 \times \frac{1}{4}\pi \times 0.02^2 \times 15^2}{100} = 0.707$$

$$\theta = 45°$$

Note that the jet is assumed to flow parallel to the surface of the plate after impact and $F_y = 0$.

*Example 4.9*

(a) A horizontal jet of water flows smoothly on to a stationary curved vane which turns it through 135°. At the initial point of contact with the vane, the jet is 40 mm diameter and the velocity is 40 m/s. As a result of friction, the velocity of the water leaving the vane is 38 m/s. Neglecting gravity effects, determine the net force on the plate.

(b) If the vane now moves in the direction of the initial jet with a velocity of 15 m/s, determine the net force on the vane in the direction of motion.

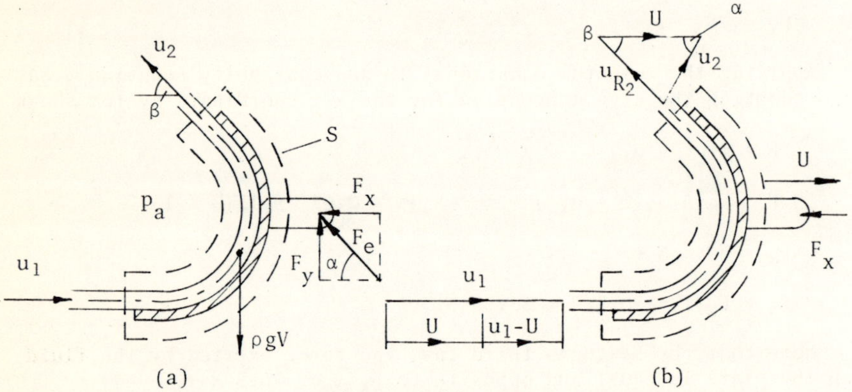


Figure 4.14

(a) Consider a control volume enclosing the vane, as shown in figure 4.14. The volume of fluid within the control volume is V. The free jet and the surroundings are at atmospheric pressure, therefore there is no net pressure force on the control volume. Let a resultant external force, $F_e$, act on the vane. Applying the momentum equation 4.19 in the x direction

$$\Sigma F_x = \dot{m}(u_{x_2} - u_{x_1})$$

where $\Sigma F_x$ is the sum of the forces acting on the fluid. In this case, the only force is the x-component of $F_e$. Therefore

$$-F_x = \dot{m}(u_{x_2} - u_{x_1}) = \dot{m}(-u_2 \cos \beta - u_1)$$

$$F_x = 10^3 \times \tfrac{1}{4}\pi \times 0.04^2 \times 40\ (38 \cos 45 + 40) = 3.36 \text{ kN}$$

Applying the momentum equation in the y direction

$$\Sigma F_y = F_y - \rho g V = \dot{m}(u_{y_2} - u_{y_1})$$

Neglecting gravity effects, $\rho g V = 0$, and we have

$$F_y = \dot{m}(u_2 \sin \beta - 0) = 10^3 \times \tfrac{1}{4}\pi \times 0.04^2 \times 40(38 \sin 45)$$

$$= 1.352 \text{ kN}$$

$$F_e = \sqrt{(F_x^{\ 2} + F_y^{\ 2})} = \sqrt{(3.36^2 + 1.352^2)} = 3.62 \text{ kN}$$

$$\alpha = \tan^{-1}(F_y/F_x) = \tan^{-1}(1.352/3.36) = 21.9°$$

From Newton's third law of motion, the force exerted by the fluid on the vane is equal and opposite to $F_e$ i.e. 3.62 kN $\searrow$

(b) Consider now the case where the vane moves to the right with velocity $\underline{U}$. A stationary observer will see a jet of water of velocity $\underline{u_1}$ entering the vane and a jet of water of velocity $\underline{u_2}$ leaving the vane. The velocity of the entering jet relative to the vane, $u_{R1}$, is easily obtained as $u_1 - U$, because the water and vane velocities occur in the same direction. It is assumed that the jet follows the contour of the vane, therefore the velocity, $u_{R2}$, and the angle, $\beta$, of the leaving jet are known. The velocity of the leaving jet, $\underline{u_2}$, is found by adding vectorially the vane velocity, $\underline{U}$, to the relative velocity, $\underline{u_{R2}}$.

For a moving control volume, and steady flow, the momentum equation 4.22 reduces to

$$\Sigma F = \int_S \underline{u}_S(\rho_S \underline{u}_{RS} \cdot d\underline{A}) \qquad \text{(i)}$$

Let the external force acting on the vane be $F_x$ as shown in figure 4.14b (negative in the direction chosen). The net pressure force on the control volume is zero. The velocity, $\underline{u}_{RS}$, associated with mass flow rate refers to velocities relative to the control surface i.e. $u_{R_1} = u_1 - U$. Applying equation (i) in the x direction

$$-F_x = \dot{m}(u_{x_2} - u_{x_1}) = \rho A(u_1 - U)(u_2 \cos \alpha - u_1) \qquad \text{(ii)}$$

From the velocity diagram at exit, $u_2 \cos \alpha = U - u_{R2} \cos \beta$ where $u_{R2} = k_f u_{R1}$. Equation (ii) gives

$$F_x = \rho A(u_1 - U)(k_f u_{R_1} \cos \beta - U + u_1)$$

$$= 10^3 \times \tfrac{1}{4}\pi \times 0.04^2(40 - 15)\left(\frac{38}{40} \times 25 \cos 45 - 15 + 40\right)$$

$$= 1.313 \text{ kN}$$

The same result is obtained using relative velocities in equation 4.23 or by imposing a velocity $\underline{U}$ to the left on the whole system to give a steady flow situation (see example 4.13).

*Example 4.10*

Water flows over the spillway shown in figure 4.15 at the rate of 15 $m^3/s$ per metre length. If the upstream depth is 5 m, determine the magnitude and direction of the force exerted on the spillway.

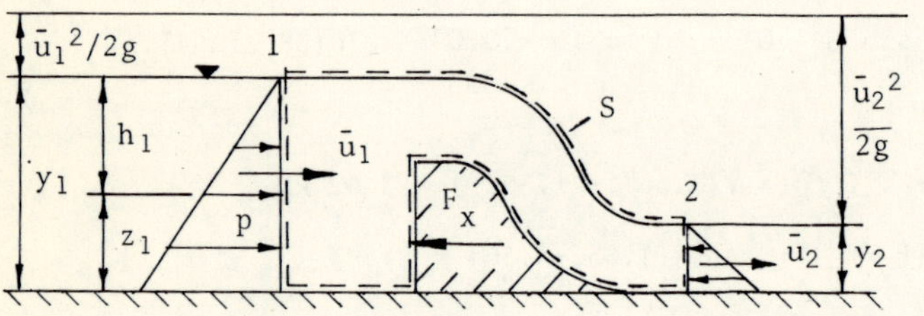


Figure 4.15

It is assumed that frictional effects are negligible and that at upstream and downstream stations the flow is one-dimensional with a hydrostatic pressure distribution. The Bernoulli equation gives

$$h_1 + \frac{\bar{u}_1^2}{2g} + z_1 = h_2 + \frac{\bar{u}_2^2}{2g} + z_2$$

or $$\frac{\bar{u}_1^2}{2g} + y_1 = \frac{\bar{u}_2^2}{2g} + y_2 \qquad \text{(i)}$$

The continuity equation gives

$$\dot{V} = \bar{u}_1 A_1 = \bar{u}_2 A_2 \qquad \text{(ii)}$$

From equations (i) and (ii)

$$\frac{\dot{V}^2}{2gy_1^2} + y_1 = \frac{\dot{V}^2}{2gy_2^2} + y_2$$

$$\frac{15^2}{2 \times 9.81 \times 5^2} + 5 = \frac{15^2}{2 \times 9.81 \times y_2^2} + y_2$$

$$y_2 = 1.76 \text{ m}$$

$$\bar{u}_2 = \frac{\dot{V}}{A_2} = \frac{15}{1.76} = 8.523 \text{ m/s}, \qquad \bar{u}_1 = \frac{15}{5} = 3 \text{ m/s}$$

Let the force exerted on the fluid by unit length of spillway be $F_x$ in the direction shown. The force due to hydrostatic effects is given by equation 2.16 i.e. $F = \frac{1}{2}\rho g y^2$. Applying the momentum equation 4.19 between stations 1 and 2, we obtain

$$\Sigma F = \tfrac{1}{2}\rho g y_1^2 - F_x - \tfrac{1}{2}\rho g y_2^2 = \dot{m}(\bar{u}_2 - \bar{u}_1)$$

$$\tfrac{1}{2} \times 10^3 \times 9.81 \times 5^2 - F_x - \tfrac{1}{2} \times 10^3 \times 9.81 \times 1.76^2$$

$$= 10^3 \times 15(8.523 - 3)$$

$$F_x = 24.46 \text{ N}$$

The force exerted by the fluid on the spillway is equal and opposite to $F_x$ i.e. 24.46 N to the right.

*Example 4.11*

Determine the magnitude and direction of the force required to hold the stationary nozzle shown in figure 4.16 in position when oil flows through it at a rate of 50 kg/s. The absolute pressure at station 1 is 500 kPa and the interior volume of the nozzle is 0.005 m$^3$. Neglect the head loss in the nozzle due to fluid friction and the effect of difference in elevation between entry and exit. $D_1$ = 100 mm, $D_2$ = 50 mm, $\theta$ = 30°, $\rho$ = 900 kg/m$^3$, $p_a$ = 100 kPa.

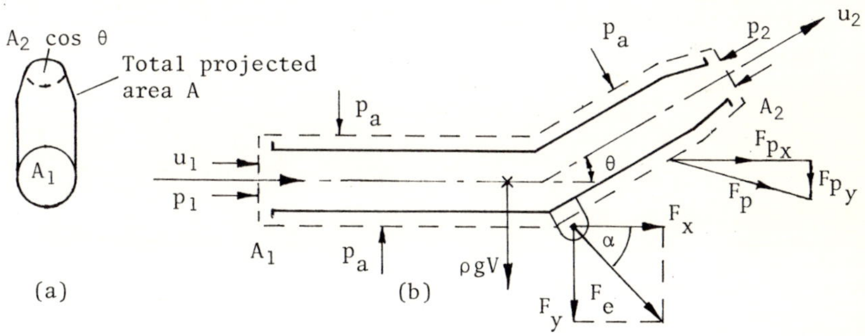


Figure 4.16

Consider a control volume enclosing the nozzle i.e. the control surface consists of the outside surface of the duct and entry and exit planes, $A_1$ and $A_2$, respectively. The momentum equation 4.19 gives

$$\Sigma \underline{F} = \dot{m}(\underline{u}_2 - \underline{u}_1) \qquad \text{(i)}$$

The pressures, $p_1$ and $p_2$, acting on areas $A_1$ and $A_2$, are *absolute* pressures. There is an external force, $\underline{F}_e$, acting on the nozzle and also a force, $F_p$, due to atmospheric pressure, $p_a$.

Applying equation (i) in the x direction, and using mean velocities ($\bar{u} = \dot{V}/A$), we obtain

$$p_1A_1 - p_2A_2 \cos\theta + F_{p_x} + F_x = \dot{m}(\bar{u}_{x_2} - \bar{u}_{x_1}) \qquad \text{(ii)}$$

Now from figure 4.16a, the force, $F_{p_x}$, is given by

$$F_{p_x} = p_a(A - A_1) - p_a(A - A_2 \cos\theta) = p_a(A_2 \cos\theta - A_1)$$

Substituting for $F_{p_x}$ in equation (ii), and rearranging, we obtain

$$F_x = \dot{m}(\bar{u}_2 \cos\theta - \bar{u}_1) - (p_1 - p_a)A_1 + (p_2 - p_a)A_2 \cos\theta \quad \text{(iii)}$$

Applying equation (i) in the y direction and including a vertical body force, $Fb_y = \rho g V$, due to the weight of the fluid inside the control volume, we obtain

$$F_y = \dot{m}(\bar{u}_2 \sin\theta - 0) - 0 + (p_2 - p_a)A_2 \sin\theta - \rho g V \quad \text{(iv)}$$

Note that equations (iii) and (iv) could be obtained directly from the vector equation 4.21.

Applying the continuity equation 3.5a

$$\bar{u}_1 = \frac{\dot{m}}{\rho_1 A_1} = \frac{50}{900 \times \frac{1}{4}\pi \times 0.1^2} = 7.1 \text{ m/s}$$

$$\bar{u}_2 = \frac{\dot{m}}{\rho_2 A_2} = \frac{50}{900 \times \frac{1}{4}\pi \times 0.05^2} = 28.4 \text{ m/s}$$

Applying the Bernoulli equation 4.7d to assumed frictionless flow through the passage, we obtain

$$\frac{p_1}{\rho} + \frac{\bar{u}_1^2}{2} = \frac{p_2}{\rho} + \frac{\bar{u}_2^2}{2}$$

$$p_2 = p_1 - \rho(\bar{u}_2^2 - \bar{u}_1^2)/2 = 5 \times 10^5 - 900(28.4^2 - 7.1^2)/2$$

$$= 163 \text{ kPa}$$

Substituting values in equations (iii) and (iv), we obtain

$$F_x = 50(28.4 \cos 30 - 7.1) - (5 - 1)10^5 \times \tfrac{1}{4}\pi \times 0.1^2$$

$$+ (1.63 - 1)10^5 \times \tfrac{1}{4}\pi \times 0.05^2 \cos 30$$

$$= -2.163 \text{ kN (i.e. } F_x \text{ acts } \leftarrow)$$

$$F_y = -50 \times 28.4 \sin 30 - (1.63 - 1)10^5 \times \tfrac{1}{4}\pi \times 0.05^2 \sin 30$$

$$- 900 \times 9.81 \times 0.005$$

$$= -816 \text{ N (i.e. } F_y \text{ acts } \uparrow)$$

$$F_e = \sqrt{(2163^2 + 816^2)} = 2.31 \text{ kN } \nwarrow$$

$$\alpha = \tan^{-1}(F_y/F_x) = \tan^{-1}(816/2163) = 20.6°$$

Note that $F_x$ and $F_y$ are negative. This means that the original directions chosen are opposite to the true directions. $F_e$ is the external force required to hold the nozzle stationary. By Newton's third law, the force exerted by the fluid on the nozzle is equal and opposite i.e. 2.31 kN $\searrow$.

*Example 4.12*

During a static test on a turbo-jet engine, air enters the 750 mm diameter intake with a uniform velocity of 75 m/s, temperature 15 °C and pressure 1 bar. The engine is fitted with an exhaust nozzle of radius R = 250 mm and during the test it is found that the distribution of exhaust velocity is closely approximated by

$$u/u_m = (1 - r^2/R^2)$$

where u is the velocity at any radius r and $u_m$ is the maximum gas velocity which, in this test, is found to be 500 m/s. The nozzle discharges at its design back pressure. If the increase in mass flow rate due to fuel addition is 1.5%, determine (i) the density of the exhaust gases, and (ii) the thrust of the engine. Assume atmospheric pressure 1.015 bar. $R_a$ = 287 J/kg K.

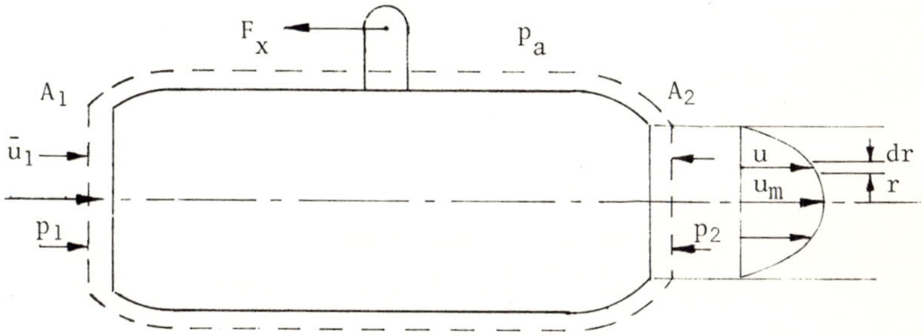


Figure 4.17

Consider a control volume enclosing the engine i.e. the control surface consists of the outside surface of the engine and the entry and exit planes, $A_1$ and $A_2$, respectively. The momentum equation 4.18, for steady flow, gives

$$\Sigma \underline{F} = \int_S \underline{u}(\rho \underline{u} \cdot d\underline{A}) \qquad \text{(i)}$$

Let the external force acting on the engine be $F_x$. The pressure forces acting on the entry and exit planes, $A_1$ and $A_2$, and atmospheric pressure, $p_a$, acting on the external surface of the engine, may be accounted for by replacing absolute pressures, $p_1$ and $p_2$, by gauge pressures as shown in example 4.11. Applying equation (i) to flow in the x direction

$$(p_1 - p_a)A_1 - (p_2 - p_a)A_2 - F_x = \int_{A_2} u(\rho u\, dA) - \int_{A_1} u(\rho u\, dA) \qquad \text{(ii)}$$

Air enters area $A_1$ with uniform velocity, $u_1$, and density, $\rho_1$, therefore

$$\int_{A_1} u(\rho u\, dA) = \rho_1 u_1{}^2 \int_{A_1} dA = \rho_1 u_1{}^2 A_1 \qquad \text{(iii)}$$

The exhaust nozzle discharges at its design back pressure i.e. without shock. Therefore $p_2 = p_a$. The velocity at any radius, r, is $u = u_m(1 - r^2/R^2)$. Considering flow through an element of area $dA = 2\pi r\,dr$, we have

$$\int_{A_2} u(\rho u\,dA) = \int_0^R \rho u_m^2(1 - r^2/R^2)^2 \times 2\pi r\,dr$$

$$= (\rho_2 \pi R^2 u_m^2)/3 \qquad \text{(iv)}$$

Substituting equations (iii) and (iv) in (ii) and rearranging, we obtain

$$F_x = (p_1 - p_a)A_1 - (\rho_2 \pi R^2 u_m^2)/3 + \rho_1 u_1^2 A_1 \qquad \text{(v)}$$

Applying the continuity equation between entry and exit planes $A_1$ and $A_2$, and including a 1.5% increase in mass flow rate due to fuel addition, we have

$$\int_{A_1} \rho u\,dA = 1.015\int_R \rho u\,dA$$

$$\int_0^R \rho u_m(1 - r^2/R^2)2\pi r\,dr = 1.015\rho_1 A_1 u_1$$

$$\frac{2\rho_2 \pi R^2 u_m}{4} = \frac{1.015 p_1 \pi D^2 u_1}{4R_1 T_1}$$

$$\rho_2 = \frac{1.015 \times 10^5 \times 0.75^2 \times 75}{2 \times 0.25^2 \times 500 \times 287 \times 288}$$

$$= 0.829 \text{ kg/m}^3$$

Substituting values in equation (v), we obtain

$$F_x = (1 - 1.015)10^5 \times \tfrac{1}{4}\pi \times 0.75^2 - \frac{0.829 \times \pi \times 0.25^2 \times 500^2}{3}$$

$$+ \frac{10^5 \times 75^2 \times \pi \times 0.75^2}{287 \times 288 \times 4}$$

$$= -11.23 \text{ kN (i.e. } F_x \text{ acts} \rightarrow)$$

The negative sign indicates that the external force exerted on the engine, to hold it stationary, is 11.23 kN to the right. By Newton's third law the force exerted by the fluid on the engine is equal and opposite i.e. the thrust of the engine is 11.23 kN to the left. Note from equation (v) that the thrust consists of (i) momentum thrust due to the change in momentum flux of the fluid between entry and exit, and (ii) pressure thrust due to pressure forces acting on the entry and exit planes. (See also example 4.14.)

*Example 4.13*

The engine of an aircraft moving through still air at 150 m/s delivers 2500 kW to an ideal propeller 3 m diameter. Determine (i) the slipstream velocity, (ii) the velocity through the propeller disc, (iii) the thrust, and (iv) the propulsive efficiency. $\rho_{air} = 1.2\ kg/m^3$.

Let the propeller move at velocity $u_1$ to the left through still air of pressure $p_a$ (figure 4.18a). At some distance ahead of and behind the propeller, the pressures $p_1$ and $p_4$ are equal to $p_a$. The pressure on the slipstream boundary is also $p_a$. Across the propeller itself the mean pressure increases from $p_2$ to $p_3$. We may obtain a steady state analysis by imposing a velocity, $u_1$, to the right on the whole system as shown in figure 4.18b. This gives a stationary control volume V.

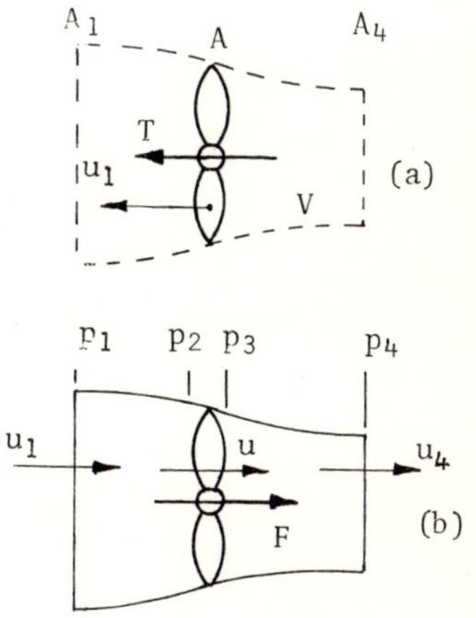

Figure 4.18

The force, F, exerted by the propeller on the fluid is obtained by applying the momentum equation 4.19, between stations 1 and 4.

$$F = \dot{m}(u_4 - u_1) = \rho u A(u_4 - u_1) \qquad \text{(i)}$$

where u is the mean velocity through the propeller ($u = u_2 = u_3$).

Applying the momentum equation between stations 2 and 3

$$F + (p_2 - p_3)A = 0 \qquad \text{(ii)}$$

Applying the Bernoulli equation between stations 1 and 2 and between stations 3 and 4 we obtain

$$p_1 + \tfrac{1}{2}\rho u_1^2 = p_2 + \tfrac{1}{2}\rho u^2 \qquad \text{(iii)}$$

and $$p_3 + \tfrac{1}{2}\rho u^2 = p_4 + \tfrac{1}{2}\rho u_4^2 \qquad \text{(iv)}$$

Combining equations (i), (ii), (iii) and (iv) and using $p_1 = p_4 = p_a$ we obtain

$$u = (u_1 + u_4)/2 \qquad \text{(v)}$$

The useful power output from the propeller is the thrust power, $P_o$, defined as

$$P_o = Fu_1 = \dot{m}(u_4 - u_1)u_1 \qquad \text{(vi)}$$

The input power, $P_i$, is that required to increase the velocity of the air in the slipstream from $u_1$ to $u_4$. From equations 4.39 and (v)

$$P_i = \dot{m}e_m = \tfrac{1}{2}\dot{m}(u_4{}^2 - u_1{}^2) = \dot{m}(u_4 - u_1)u \qquad \text{(vii)}$$

The propulsive efficiency of the propeller is given by

$$\eta = \frac{P_o}{P_i} = \frac{u_1}{u} \qquad \text{(viii)}$$

Substituting values in equation (vii)

$$2500 \times 10^3 = \tfrac{1}{2} \times 1.2 \times \frac{\pi}{4} \times 3^2 \left(\frac{u_4 + 150}{2}\right) (u_4{}^2 - 150^2)$$

$$u_4 = 162.1 \text{ m/s}$$

$$u = \frac{u_1 + u_4}{2} = \frac{150 + 162.1}{2} = 156 \text{ m/s}$$

The thrust, T, is given by

$$T = -F = -\rho uA(u_4 - u_1) = -1.2 \times 156 \times \frac{\pi}{4} \times 3^2(162.1 - 150)$$

$$= -16 \text{ kN}$$

$$\eta = \frac{Fu_1}{P_i} = \frac{16 \times 10^3 \times 150}{2500 \times 10^3} = 96\%$$

or $\quad \eta = \dfrac{u_1}{u} = \dfrac{150}{156} = 96\%$

*Example 4.14*

A rocket, of total mass 800 kg, is travelling vertically at an altitude where the pressure is 0.6 bar and the gravitational acceleration is 9.6 m/s$^2$. If the nozzle expands 10 kg/s of combustion products with an exit velocity and pressure of 1500 m/s and 0.85 bar, respectively, determine the thrust and acceleration of the rocket. The nozzle exit area is 0.25 m$^2$. The air resistance is estimated to be 1.5 kN.

The control surface is taken as the outside of the rocket itself and the plane of the nozzle exit. The rocket moves vertically upwards with velocity $\underline{U}$ relative to earth. The exhaust gases leave the nozzle at velocity $\underline{u}_j$ relative to the nozzle. In general, the absolute pressure at exit from the nozzle, $p_j$, is greater than atmospheric pressure, $p_a$ (i.e. the gas is underexpanded). For a moving control volume, equation 4.23 gives

$$\Sigma \underline{F} = M_V \frac{d\underline{U}}{dt} + \frac{d}{dt} \int_V \rho_V \underline{u}_{RV} \, dV + \int_S \underline{u}_{RS} (\rho_S \underline{u}_{RS} \cdot d\underline{A}) \qquad \text{(i)}$$

The sum of external forces $\Sigma F$ consists of a gravitational body force, $-F_b$, and a pressure force, $F_p = (p_j - p_a)A_j$, and a drag force, $-F_d$, on the rocket due to air resistance. Therefore

$$\Sigma F = -M_V g + (p_j - p_a)A_j - F_d \qquad \text{(ii)}$$

Term 1 on the right hand side of the equation represents the acceleration of the rocket. Note that $M_V$ is the instantaneous mass of the rocket and fuel.

Term 2 is the change of momentum in the control volume. $\int \rho_V \, dV$ is the instantaneous mass of the rocket and $\underline{u}_{RV}$ is the velocity relative to the control volume. For solid or liquid fuel, $\underline{u}_{RV}$ is zero and for steady flow of the gases its time rate of change is zero. For unsteady flow the mass of gas in negligible. Therefore

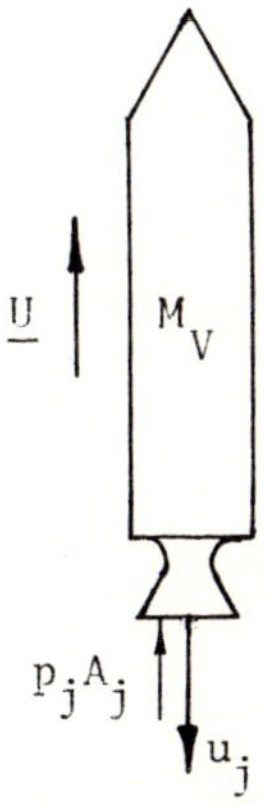


Figure 4.19

$$\frac{d}{dt}\int_V \rho_V \underline{u}_{RV} \, dV = 0 \qquad \text{(iii)}$$

Term 3 is the momentum flux. For a rocket the inlet area is zero. Assuming uniform flow at exit from the nozzle we have

$$\int_S \underline{u}_{RS}(\rho_S \underline{u}_{RS} \cdot d\underline{A}) = 0 - \rho_j u_j^2 A_j = -\dot{m}_p u_j \qquad \text{(iv)}$$

where $\dot{m}_p$ is the mass flow rate of propellant.

From equations (i), (ii), (iii) and (iv) we obtain

$$[\dot{m}_p u_j + (p_j - p_a)A_j] - (M_V g + F_d) = M_V \frac{dU}{dt} \qquad \text{(v)}$$

The first term in brackets is the *thrust*, T, of the rocket. The second term is the resistance due to weight and drag and the difference is equal to the product of mass and acceleration of the rocket.

Substituting values in equation (v)

$$[10 \times 1500 + 10^5(0.85 - 0.6)0.25] - [800 \times 9.81 + 1.5 \times 10^3]$$

$$= 800 \frac{dU}{dt}$$

$$21.25 \times 10^3 - 9.35 \times 10^3 = 800 \frac{dU}{dt}$$

$$dU/dt = 14.875 \text{ m/s}^2$$

$$\text{Thrust } T = [\dot{m}_p u_j + (p_j - p_a)] = 21.25 \text{ kN}$$

*Example 4.15*

The principle of a jet pump is shown in figure 4.20. Fluid is injected with high velocity along the axis of a pipe and the jet entrains a secondary fluid stream flowing through the pipe. The jet and secondary flows are thoroughly mixed at section 2.

In a particular jet pump, a jet of water of area 100 $mm^2$ and velocity 25 m/s is directed along the axis of a pipe of area 500 $mm^2$. The jet entrains a secondary stream of velocity 10 m/s. Neglecting wall skin friction and assuming one-dimensional flow, determine (i) the average velocity of flow at section 2, (ii) the pressure change between sections 1 and 2, and (iii) the rate of mechanical energy dissipation in watts. It may be assumed that the pressure of the jet and the secondary stream are the same at section 1.

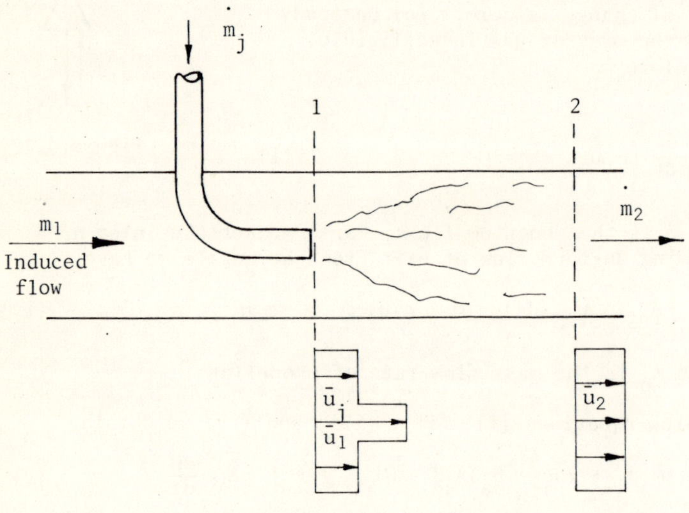


Figure 4.20

Consider a control volume enclosing the mixing region between stations 1 and 2. Skin friction is neglected, therefore there is no external force acting on the control volume. The momentum equation 4.18, for steady flow, gives

$$\Sigma F = \int_S \underline{u}(\rho\underline{u}\cdot d\underline{A}) \qquad \text{(i)}$$

$$p_1A_2 - p_2A_2 = \int_{A_2} u(\rho u\ dA) - \int_{A_1} u(\rho u\ dA) - \int_{A_j} u(\rho u\ dA)$$

Assuming one-dimensional flow, mean velocities ($\bar{u} = \dot{V}/A$) may be used

$$(p_1 - p_2)A_2 = \rho\bar{u}_2{}^2A_2 - \rho\bar{u}_1{}^2A_1 - \rho\bar{u}_j{}^2A_j \qquad \text{(ii)}$$

Applying the continuity equation 3.5b to multi-stream, incompressible flow

$$\bar{u}_1 A_1 + \bar{u}_j A_j = \bar{u}_2 A_2$$

$$10 \times 400 \times 10^{-6} + 25 \times 100 \times 10^{-6} = \bar{u}_2 \times 500 \times 10^{-6}$$

$$\bar{u}_2 = 13 \text{ m/s}$$

Substituting values in equation (ii)

$$(p_1 - p_2) \times 500 \times 10^{-6}$$

$$= 10^3[13^2 \times 500 - 10^2 \times 400 - 25^2 \times 100] \times 10^{-6}$$

$$p_1 - p_2 = -36 \text{ kPa}$$

Applying the energy equation 4.36 between stations 1 and 2, we have for the jet flow

$$\frac{p_1}{\rho} + \tfrac{1}{2}\bar{u}_j^2 = \frac{p_2}{\rho} + \tfrac{1}{2}\bar{u}_2^2 + (e_2 - e_j) \qquad \text{(iii)}$$

and for the secondary stream

$$\frac{p_1}{\rho} + \tfrac{1}{2}\bar{u}_1^2 = \frac{p_2}{\rho} + \tfrac{1}{2}\bar{u}_2^2 + (e_2 - e_1) \qquad \text{(iv)}$$

The rate of energy dissipation is given by

$$\frac{dE}{dt} = \dot{m}_1(e_2 - e_1) + \dot{m}_j(e_2 - e_j) \qquad \text{(v)}$$

From equations (iii), (iv), and (v) we have

$$\frac{dE}{dt} = \dot{m}_1\left(\frac{p_1 - p_2}{\rho} + \tfrac{1}{2}(\bar{u}_1^2 - \bar{u}_2^2)\right) + \dot{m}_j\left(\frac{p_1 - p_2}{\rho} + \tfrac{1}{2}(\bar{u}_j^2 - \bar{u}_2^2)\right)$$

$$= \frac{10^3 \times 400 \times 10}{10^6}\left(\frac{-36 \times 10^3}{10^3} + \tfrac{1}{2}(10^2 - 13^2)\right)$$

$$+ \frac{10^3 \times 100 \times 25}{10^6}\left(\frac{-36 \times 10^3}{10^3} + \tfrac{1}{2}(25^2 - 13^2)\right)$$

$$\frac{dE}{dt} = 198 \text{ W}$$

Note that the efficiency of a jet pump may be defined

$$\eta_p = \frac{\text{mechanical energy out}}{\text{mechanical energy in}} = 1 - \frac{\rho \, dE/dt}{\dot{m}_1 p_{o1} + \dot{m}_j p_{oj}}$$

*Example 4.16*

A centrifugal pump discharges 0.2 $m^3/s$ water when running at 1200 rev/min and the required shaft power is 65 kW. Gauges mounted on a horizontal line are connected to points on the suction and delivery pipes close to the pump and read 3 m below and 19 m above atmospheric pressure, respectively. The pump impeller has an outer diameter of 300 mm and a discharge area of 0.1 $m^2$. The blades are bent backwards so that the direction of the outlet relative velocity makes an angle of 145° with the tangent drawn in the direction of impeller rotation. The suction and delivery pipes are the same diameter. Determine (i) the overall efficiency of the pump, and (ii) the hydraulic efficiency assuming that water enters the impeller without whirl.

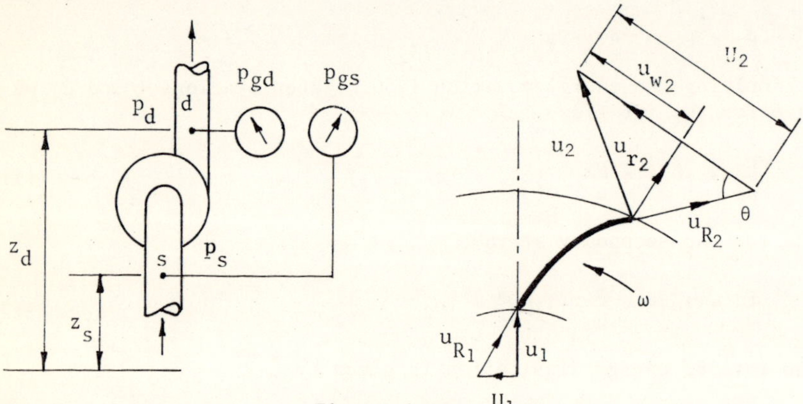


Figure 4.21

The useful energy transfer to the water is given by equation 4.39

$$e_m = \left(\frac{p_d}{\rho} + \tfrac{1}{2}\bar{u}_d^{\,2} + gz_d\right) - \left(\frac{p_s}{\rho} + \tfrac{1}{2}\bar{u}_s^{\,2} + gz_s\right) \qquad \text{(i)}$$

From the hydrostatic relationship, the readings on the pressure gauges are related to the actual pressures by

$$\frac{p_{gd}}{\rho} - \frac{p_{gs}}{\rho} = \frac{p_d}{\rho} - \left(\frac{p_s}{\rho} - g(z_d - z_s)\right)$$

$$= \left(\frac{p_d}{\rho} + gz_d\right) - \left(\frac{p_s}{\rho} + gz_s\right) \qquad \text{(ii)}$$

Therefore, if the pressure gauges are mounted on the same horizontal line the difference in the gauge readings $p_{gd} - p_{gs}$ is equal to the difference in piezometric pressure across the pump $p_d^* - p_s^*$. The suction and delivery pipes are the same diameter, therefore $\bar{u}_d = \bar{u}_s$. From equations (i) and (ii) we have

$$e_m = \frac{p_{gd}}{\rho} - \frac{p_{gs}}{\rho} = 9.81[19 - (-3)] = 215.8 \text{ J/kg}$$

The *overall efficiency*, $\eta_o$, is defined

$$\eta_o = \frac{\text{water power}}{\text{shaft power}} = \frac{\rho \dot{V} e_m}{P_s} = \frac{10^3 \times 0.2 \times 215.8}{65 \times 10^3} = 66.4\%$$

From equation 4.29a the energy transfer between impeller and fluid per unit mass of fluid is

$$e = \frac{\dot{E}}{\dot{m}} = u_{w_2} U_2 - u_{w_1} U_1 \qquad \text{(iii)}$$

The water enters the impeller without whirl ($u_{w1} = 0$) hence

$$e = u_{w_2} U_2 \qquad \text{(iv)}$$

Now $u_{r_2} = \bar{u}_{r_2} = \dfrac{\text{volumetric flow rate } \dot{V}}{\text{discharge area } 2\pi r_2 b} = \dfrac{0.2}{0.1} = 2$ m/s

and $U_2 = \dfrac{2\pi N r_2}{60} = \dfrac{2 \times \pi \times 1200 \times 0.15}{60} = 18.8$ m/s

From the exit velocity diagram (figure 4.21)

$$u_{w_2} = U_2 - u_{r_2} \cot\theta = 18.8 - 2 \cot 35 = 15.94 \text{ m/s}$$

From equation (iv)

$$e = u_{w_2} U_2 = 15.94 \times 18.8 = 300 \text{ J/kg}$$

Assuming no mechanical friction losses, the energy transfer between impeller and fluid, e, is equal to the shaft work $w_x$. From equation 4.41 the hydraulic efficiency, $\eta_H$, is given by

$$\eta_H = \frac{e_m}{w_x} = \frac{e_m}{e} = \frac{215.8}{300} = 72\%$$

*Example 4.17*

A Pelton wheel is situated 500 m below the water surface level in a reservoir. The connecting pipe is 600 mm diameter and 2000 m long and the friction factor may be assumed to be f = 0.005. The nozzle has a coefficient of velocity of 0.96 and the issuing jet diameter is 200 mm. Assuming that the buckets deflect the jet through 165°, that they run at 35 m/s and that the relative velocity at exit is 15% less than the relative velocity at inlet, determine (i) the water flow rate in $m^3$/s, and (ii) the power developed. Assume that the head loss due to friction in the pipe is given by $H_L = 4fL\bar{u}^2/2gD$.

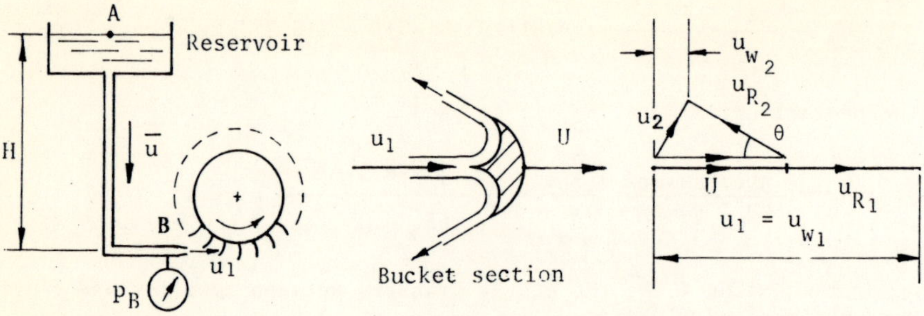


Figure 4.22

Let the mean velocity in the supply pipe (penstock) be $\bar{u}$. The Bernoulli equation with losses (equation 4.35) applied between point A on the reservoir water surface and point B just before the nozzle gives, in head units (see also equation 7.11)

$$\frac{p_a}{\rho g} + 0 + z_A = \frac{p_B}{\rho g} + \frac{\bar{u}^2}{2g} + z_B + \frac{4fL\bar{u}^2}{2gD} \qquad \text{(i)}$$

The Bernoulli equation 4.7c for ideal flow, applied across the nozzle gives

$$\frac{p_B}{\rho g} + \frac{\bar{u}^2}{2g} + z_B = \frac{p_a}{\rho g} + \frac{\bar{u}_1{}^2}{2g} + z_B \qquad \text{(ii)}$$

or

$$(\bar{u}_1)_{\text{ideal}} = \sqrt{(2g)}\left(\frac{p_B - p_a}{\rho g} + \frac{\bar{u}^2}{2g}\right)^{\frac{1}{2}} \qquad \text{(iii)}$$

The losses in the nozzle can be accounted for by the use of a nozzle coefficient, $C_V$. Thus the actual jet velocity is

$$\bar{u}_1 = C_V\sqrt{(2g)}\left(\frac{p_B - p_a}{\rho g} + \frac{\bar{u}^2}{2g}\right)^{\frac{1}{2}} \qquad \text{(iv)}$$

From equations (i) and (iv) and $z_A - z_B = H$ we have

$$\bar{u}_1 = C_V\sqrt{(2g)}\left(H - \frac{4fL\bar{u}^2}{2gD}\right)^{\frac{1}{2}} \qquad \text{(v)}$$

From continuity

$$\bar{u}_1 A_1 = \bar{u}A$$

$$\bar{u} = \bar{u}_1\left[\frac{A_1}{A}\right] = \bar{u}_1\left(\frac{200}{600}\right)^2 = \frac{\bar{u}_1}{9}$$

Substituting values in equation (v)

$$\bar{u}_1 = 0.96\sqrt{(2 \times 9.81)} \left(500 - \frac{4 \times 0.005 \times 2000 \times \bar{u}_1^2}{2 \times 9.81 \times 0.6 \times 9 \times 9}\right)^{\frac{1}{2}}$$

$$\bar{u}_1 = 71 \text{ m/s}$$

$$\dot{V} = \bar{u}_1 A = 71 \times \frac{\pi}{4} \times 0.2^2 = 2.23 \text{ m}^3/\text{s}$$

The Euler turbine equation 4.29b gives

$$\dot{E} = \dot{m}(u_{w_1} U_1 - u_{w_2} U_2)$$

For a Pelton wheel the tangential velocity U is constant. Therefore from the velocity diagram

$$\dot{E} = \dot{m}U(u_{w_1} - u_{w_2}) = \dot{m}U(u_{R_1} + u_{R_2} \cos \theta)$$

Now $u_{R_2} = ku_{R_1}$, $u_{R_1} = u_1 - U$ and $u_1 = \bar{u}_1$. Hence

$$\dot{E} = \rho \dot{V} U(\bar{u}_1 - U)(1 + k \cos \theta)$$

$$= 10^3 \times 2.23 \times 35(71 - 35)(1 + 0.85 \cos 15) = 5.2 \text{ MW}$$

This is the rate of energy transfer from fluid to rotor. If bearing losses are neglected, $\dot{E}$ is also the output shaft power.

Note that for a Pelton wheel the hydraulic efficiency is

$$\eta_H = \frac{U(u_{R_1} + u_{R_2} \cos \theta)}{\bar{u}_1^2/2} \quad \text{(based on jet kinetic energy)}$$

or $$\eta_H = \frac{U(u_{R_1} + u_{R_2} \cos \theta)}{g[H - 4fL\bar{u}^2/2gD]} \quad \text{(including nozzle efficiency)}$$

*Problems*

1 Determine the upward mass flow rate of water through a vertical venturi meter having inlet and throat diameters of 100 mm and 25 mm, respectively, when the difference in level registered by a mercury manometer is 100 mm. $C_d = 0.96$.

[2.348 kg/s]

2 Prove the following expression used for metering the mass flow rate of air to an internal combustion engine using a British Standard orifice

$$\dot{m} = C_d \rho_a A \sqrt{\left(2g \frac{h}{100} \frac{\rho_w}{\rho_a}\right)}$$

where h is the pressure head across the orifice in cm water, A is the area of the orifice and $\rho_a$ and $\rho_w$ are the densities of air and water, respectively. Explain why the orifice is usually used with an air tank.

The air consumption of a 4-stroke petrol-engine is measured by means of a circular orifice 40 mm diameter. The pressure drop across the orifice is 135 mm water and the coefficient of discharge is 0.6. Calculate the air consumption in kg/h. Ambient conditions are 758 mm Hg and 25 °C.

[151.9 kg/h]

3 A horizontal, open-circuit wind tunnel draws in surrounding air and then passes it through a contraction to the working section of cross-sectional area 0.25 $m^2$. A manometer connected to the working section reads 25 mm water vacuum at a particular operating condition. Determine the airspeed through the working section and the mass flow rate. $\rho_{air} = 1.2$ kg/$m^3$.

[20.22 m/s, 6.065 kg/s]

4 A circular orifice, of diameter 50 mm is made in the vertical side of a large tank mounted on rollers so that horizontal movement only is possible. Water enters the tank at the top, with negligible horizontal component, and maintains a head of 1.52 m above the centre of the orifice. The jet issuing from the orifice falls 305 mm in a horizontal distance of 1.22 m and the measured discharge is 7 kg/s. Determine the coefficients of discharge, velocity and contraction and the horizontal force required to maintain the tank in a horizontal position.

[0.653, 0.9, 0.728, 34.24 N]

5 Calculate the maximum crest height, z, and the minimum throat diameter D that will allow cavitation-free flow of water through the syphon shown in figure 4.23. The pipe diameter is 50 mm. Assume frictionless flow. $p_a$ = 100 kPa, $p_v$ = 2 kPa.

[6.49 m, 36.7 mm]

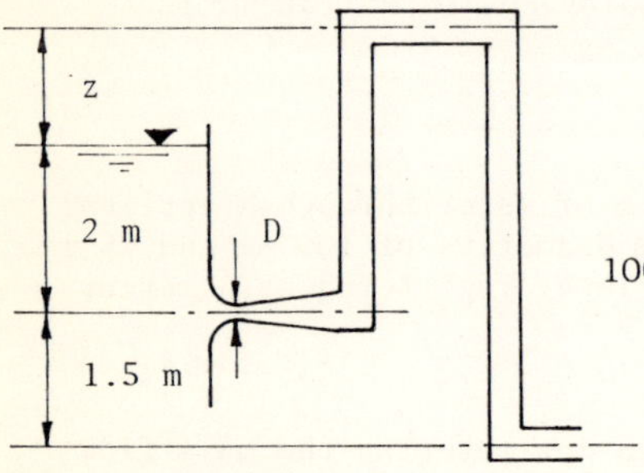


Figure 4.23

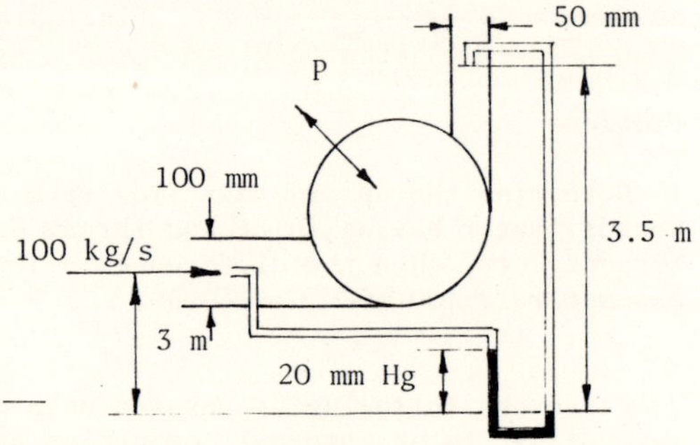


Figure 4.24

6 Water flows adiabatically through the rotodynamic machine shown in figure 4.24. Determine the direction and magnitude of power P. Neglect frictional effects. $\rho$ = 1000 kg/$m^3$.

[-247 W, i.e. pump]

7 A pitot-static tube is used to determine the maximum velocity of air-flow in a 200 mm diameter pipe. At a radius of 50 mm, the total head and static pressures recorded are 2 bar and 2.01 bar,

respectively. The velocity distribution in the pipe is known to be of the form $u = u_{max}(1 - r/R)^{1/7}$. The density of the air is 1.2 $kg/m^3$. Determine the maximum velocity and the discharge in the pipe.

[45.07 m/s, 1.388 kg/s]

8 An impeller 300 mm diameter, rotating concentrically about a vertical axis inside a closed cylindrical casing 1 m diameter, produces vortex motion in the water with which the casing is completely filled.

Inside the impeller the vortex motion is forced, while between the impeller and circular side of the casing the vortex is free. The pressure head at the side of the casing is 13 m water and the pressure head at the centre of the casing is 1.3 m water. Obtain the speed of the impeller in rev/min.

[698 rev/min]

9 A 150 mm diameter jet of water is deflected through 30° by a curved plate. Determine the magnitude and direction of the resultant force on the plate (i) when the plate is stationary, and (ii) when the plate has a velocity of 15 m/s parallel to and away from the jet. Assume a constant jet velocity of 20 m/s.

[3.66 kN, 75°, 229 N, 75°]

10 A pipe bend, whose axis lies in a horizontal plane, turns the direction of a flow of water through an angle of 30° while at the same time increasing the water speed by 50%. The bend discharges freely to atmosphere with a jet of 80 mm diameter and a flow rate of 0.04 $m^3/s$. Calculate the magnitude and direction of the resulting force exerted upon the bend due to the motion of the water.

[173.6 N, 66.5°]

11 A horizontal tapering pipe is used to convey sea water of density 1030 $kg/m^3$ from a pipe 100 mm diameter to another pipe 200 mm diameter in a desalination plant. Calculate the magnitude of the force exerted by the fluid on the taper if the flow rate is $5 \times 10^5$ kg/h and the pressure at entry to the taper is 5 bar. Energy losses in the taper may be ignored. Assume one-dimensional flow.

[14.48 kN]

12 Repeat problem 10, assuming uniform flow at entry to the tapering pipe and a velocity distribution of the form $u = u_{max}(1 - r^2/R^2)$ at exit.

[14.38 kN]

13 A jet of water issues horizontally from a pipe and impinges on a stationary vertical flat plate. The velocity distribution in the jet is of the form $u = u_{max}(1 - r/R)^{1/7}$. Determine the percentage error involved if the thrust on the plate is calculated assuming a uniform velocity in the jet.

[1.96%]

14 The discharge of water from a cylindrical duct 1 m diameter is

controlled by means of a conical valve which may be moved axially. The form of the flow at a particular setting is shown in figure 4.25. The pressure in the cylindrical duct just upstream of the conical valve is 105 kPa and atmospheric pressure is 1 bar. Neglecting frictional effects, determine the axial force exerted on the cone.

[3.47 kN]

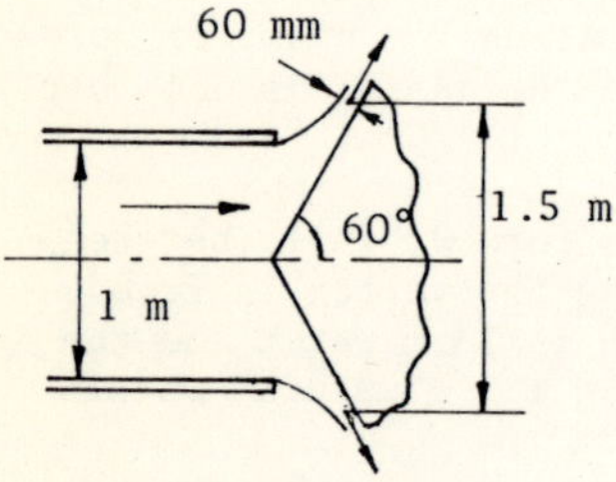


Figure 4.25

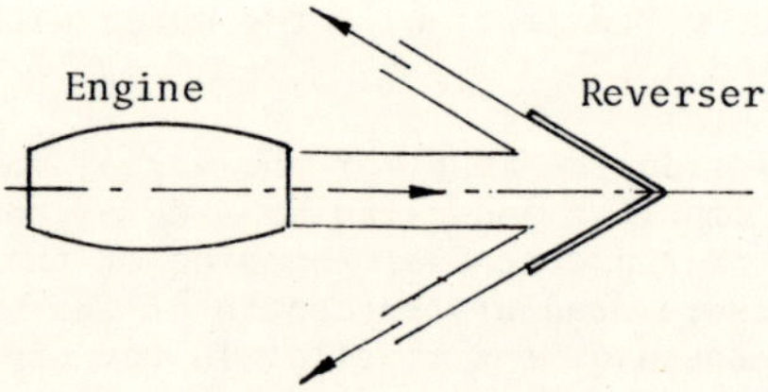


Figure 4.26

15 Show that the thrust, F, of a turbo-jet engine moving through the atmosphere with velocity, U, is given by

$$F = \dot{m}(\bar{u}_j - U) + (p_j - p_a)A_j$$

where $\dot{m}$ is the mass flow rate of fluid through the engine (assumed constant) and $\bar{u}_j$ is the jet velocity relative to the engine. It may be assumed that the intake pressure is atmospheric and the discharge pressure $p_j > p_a$.

16 Air enters the circular inlet of a stationary jet engine with a velocity of 100 m/s, temperature 290 K and pressure 1 bar. The exhaust gases leave in the form of a jet of circular cross-section with an average velocity of 700 m/s, temperature 850 K and pressure 1 bar. If the jet inlet diameter is 1.2 m and it is assumed that the increase in mass flow rate due to fuel addition is 2%, determine (i) the diameter of the exhaust jet, and (ii) the thrust produced by the engine. $R_a$ = 287 J/kg K, $R_g$ = 293 J/kg K.

[0.792 m, 83.44 kN]

17 The thrust reverser fitted to the jet engine of an aircraft consists of two plates which, in the closed position, deflect the jet of exhaust gases from the engine back through 150° as shown in figure 4.26. During engine tests, while the aircraft is standing still on the ground, the thrust reverser is moved into its operational position. Assuming the engine operating conditions are as in problem 16, and that frictional losses are negligible, determine (i) the thrust on the reverser itself and (ii) the net thrust on a two-engine aircraft.

[181.1 kN, 195.3 kN]

18 A sluice gate is used to control the flow of water in a horizontal rectangular channel, 5 m wide. The gate is lowered so that the stream flowing from under it has a depth of 1 m and a velocity

of 10 m/s. The depth upstream of the gate is 6 m. Assuming that the velocity distribution in the channel is uniform, and neglecting frictional losses, determine the force exerted by the water on the sluice gate.

[441.5 kN]

19 A desk fan of 300 mm diameter has air flowing through it at 60 $m^3$/min. If it is required to keep the fan stationary on the desk, calculate the least weight of the fan if the coefficient of friction between the base and the desk is 0.1. Assume that the air-stream area in front of the fan is 10% greater and the downstream area 10% smaller than the fan circle area. $\rho = 1.2$ kg/$m^3$.

[34.3 N]

20 A jet of incompressible fluid of velocity 2u and area A is directed along the axis of a pipe of area 2A. The jet entrains a secondary stream of velocity u. After mixing, the velocity of the fluid is uniform. Neglecting wall friction, show that the rate of energy degredation is $0.375\rho Au^3$.

21 A lawn sprinkler is fitted with two arms each fitted with a nozzle 20 mm diameter at 150 mm radius. The nozzles discharge tangentially and horizontally. Water enters along the axis of rotation at a rate of 0.003 $m^3$/s. Determine (i) the net torque on the rotor when it is held stationary and (ii) the power required in order to overcome air and bearing resistance when the sprinkler rotates at 30 rad/s.

[2.15 N m, 3.71 W]

22 The water surface in a lake supplying a Pelton wheel is 500 m above the wheel and the pipeline is 1000 m long and 500 mm diameter. The turbine is to develop 1.5 MW at 600 rev/min with an overall efficiency of 85%. The buckets deflect the jets through 165° and run at 0.46 jet speed and the relative velocity is reduced by 12% due to friction. Determine (i) the volumetric flow rate required, (ii) the wheel diameter, and (iii) the hydraulic efficiency. Assume $C_v = 0.98$, $f = 0.01$.

[1.964 $m^3$/s, 610 mm, 92%]

23 A centrifugal pump has an outer diameter of 500 mm and a discharge area 0.05 $m^2$. The blades curve backwards 135° from a tangent drawn in the direction of impeller rotation. The suction and delivery pipe diameters are equal at 200 mm. Pressure gauges are connected to inlet and exit points close to the pump and when the pump is delivering 0.3 $m^3$/s of water the readings on the gauges are 7 m of water below atmosphere and 36 m of water above atmosphere respectively. The gauges are situated on the same centre line 1.5 m above the water level in the supply sump. If the given water delivery occurs when the pump is running at 1000 rev/min and the required shaft power is 150 kW, determine (i) the overall efficiency, (ii) the hydraulic efficiency assuming no whirl velocity at entry to the impeller but a reduction of 10% in the theoretical whirl velocity at exit, and (iii) the friction head loss in the inlet pipe.

[84.5%, 89.5%, 0.82 m]

# 5 DIMENSIONAL ANALYSIS AND MODEL TESTING

In chapter 1 it was shown that any physical quantity may be expressed in terms of dimensions M, L and T. This forms the basis of dimensional analysis which is an extremely useful technique in situations where exact mathematical analysis is difficult or impossible. In this chapter, dimensional analysis will be introduced and its use in experimental work and the formulation of similarity laws will be considered.

## 5.1 DIMENSIONAL ANALYSIS

The physical relationship between the variables involved in any problem must be such that the governing equation is dimensionally homogeneous. Dimensional analysis is a method of obtaining a relationship between the variables without an exact mathematical analysis. This method is invaluable in complex fluid flow problems.

The relationship between the variables is not expressed in the form of a mathematical equation but in terms of dimensionless parameters, called Pi-groups, which may have physical significance. A dimensionless parameter is any quantity, physical constant or group of quantities formed in such a way that all the units identically cancel.

Dimensional analysis is based on the fact that if a physical relationship between the 'n' variables $\alpha_1, \alpha_2, \alpha_3 \ldots \alpha_n$ involved in a problem can be expressed by

$$\phi(\alpha_1, \alpha_2, \alpha_3 \ldots \alpha_n) = 0 \tag{5.1}$$

then the relationship can be expressed in terms of Pi-groups as follows.

$$\phi(\Pi_1, \Pi_2, \Pi_3 \ldots \Pi_{n-k}) = 0 \tag{5.2}$$

where k is the largest number of variables contained in the original list that will not combine into any dimensionless parameter. In most cases k is numerically equal to the number of independent dimensions required to define dimensionally all the variables involved.

The two main methods of dimensional analysis are the Rayleigh or Indicial method and the Buckingham Pi-theorem. These methods are described in the worked examples.

## 5.2 SIGNIFICANCE OF PI-GROUPS

In dimensional analysis, it is first necessary to select the variables involved in the problem. If relevant variables are omitted, an incomplete solution will be obtained. If irrelevant variables are included, unnecessary Pi-groups will be added to the solution. This tends to defeat the object of the analysis which is to obtain the minimum number of independent parameters required for correlation of experimental results.

A certain amount of experience is necessary before dimensional analysis can be applied to its best advantage. This is because the solution may be correct but expressed in terms of Pi-groups which have no recognisable physical significance. It may then be necessary to combine the Pi-groups to obtain new groups which have significance. For example, it is valid to take a solution of the form $\phi(\Pi_1, \Pi_2, \Pi_3, \Pi_4) = 0$ and rewrite it in terms of four other recognisable Pi-groups, obtained by combination of $\Pi_1$, $\Pi_2$, $\Pi_3$ and $\Pi_4$, say $\phi[\Pi_1, (\Pi_1 \times \Pi_2), (\Pi_2/\Pi_3), (\Pi_4/\Pi_1^2)] = 0$.

It is useful at this stage to identify some important Pi-groups which can be obtained by logical deduction, the general principle being that any ratio of like quantities must form a dimensionless parameter.

### 5.2.1 Shape Parameters

Any ratio of lengths of a body or system will express a geometrical relationship and imply that the shape is important rather than the size. For example, in pipe flow the ratios (length/diameter) and (surface roughness/diameter) are dimensionless shape parameters.

### 5.2.2 Method of Similitude

In this method, the forces important in a particular problem are identified and then expressed in terms of the parameters of the problem by physical or dimensional arguments. For example, the inertia force $F_i$ exerted on an element of fluid may be expressed as follows.

$$F_i \propto ma \propto \rho L^3(LT^{-2})$$

The velocity, u, has dimensions $LT^{-1}$ therefore $F_i$ can be expressed

$$F_i \propto \rho L^2 u^2$$

There are five other forces encountered in fluid mechanics and these are named below and then expressed in terms of relevant properties and a linear dimension L.

| | |
|---|---|
| Viscous force | $F_v \propto \tau L^2 \propto \mu(u/L)L^2 \propto \mu u L$ |
| Pressure force | $F_p \propto (\Delta p)L^2$ |
| Elastic force | $F_e \propto KL^2$ |
| Surface tension force | $F_s \propto \gamma L$ |
| Gravity force | $F_g \propto \rho g L^3$ |

From the above six forces it is possible to form fifteen independent dimensionless parameters from ratios of the forces, taken two at a time. Of these, the six most common are

Reynolds number $$\mathrm{Re} = \frac{\text{Inertia force}}{\text{Viscous force}} = \frac{\rho u L}{\mu}$$

Euler number $$\mathrm{Eu} = \frac{\text{Pressure force}}{\text{Inertia force}} = \frac{\Delta p}{\rho u^2/2}$$

Cauchy number $$\mathrm{Ca} = \frac{\text{Inertia force}}{\text{Elastic force}} = \frac{u^2}{K/\rho}$$

The square root of the Cauchy number is the Mach number, which is a ratio of velocities since the speed of sound $a = \sqrt{(K/\rho)}$.

Mach number $$M = \frac{u}{\sqrt{(K/\rho)}} = \frac{u}{a}$$

Weber number $$\mathrm{We} = \frac{\text{Inertia force}}{\text{Surface tension force}} = \frac{\rho u^2 L}{\gamma}$$

The square root of this number is also referred to as the Weber number

$$\mathrm{We} = \frac{u}{\sqrt{(\gamma/\rho L)}}$$

Froude number $$\mathrm{Fr} = \frac{\text{Inertia force}}{\text{Gravity force}} = \frac{u^2}{gL}$$

The square root of this number is also referred to as the Froude number

$$\mathrm{Fr} = \frac{u}{\sqrt{(gL)}}$$

Stokes number $$\mathrm{Sk} = \frac{\text{Pressure force}}{\text{Viscous force}} = \frac{(\Delta p)L}{\mu u}$$

## 5.3 SIMILARITY AND MODEL TESTING

It is necessary to consider the conditions under which different systems are physically similar to one another. The concept of similarity is important because it is the basis of model testing whereby tests are carried out on a model in order to predict the behaviour of a prototype which is a full-size ship, aircraft, river, estuary, pump, turbine or other device.

### 5.3.1 Mechanical Similarity

Similarity has precise meaning in geometry where figures can be similar although they differ in size and position. In any situation involving dynamic behaviour, however, similarity of shape is an insufficient condition for complete physical similarity and it is necessary to consider all types of similarity. Two systems are said to be physically similar if the ratio of the magnitudes of like quantities at corresponding points in the two systems is constant for all like quantities at all points. For this to be true the following forms of similarity must be satisfied when applied to a flow situation e.g. flow normal to a cylinder as shown in figure 5.1.

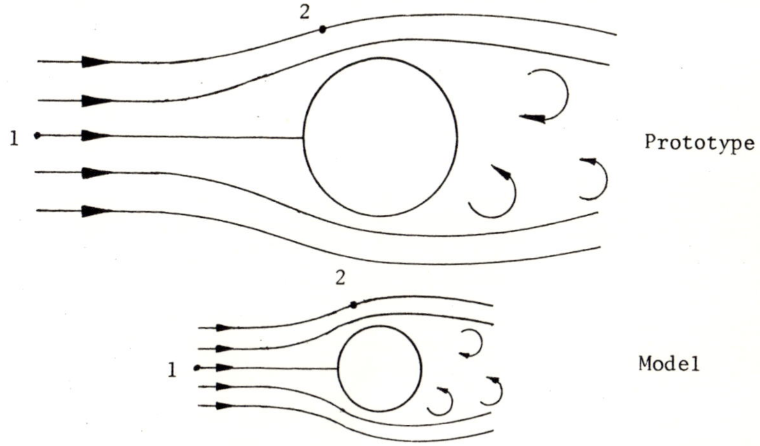


Figure 5.1

(a) Geometric Similarity: this is similarity of shape in which the ratio of any two dimensions in the model is the same as the ratio of the corresponding dimensions in the prototype. Thus

$$(L_1/L_2)_m = (L_1/L_2)_p$$

True geometric similarity implies similarity of surface roughness in which the ratio of the equivalent roughness, k, to a characteristic length, L, is the same in both model and prototype. Thus

$$(k/L)_m = (k/L)_p$$

(b) Kinematic Similarity: this is similarity of motion in which the flow pattern is the same for both model and prototype. This implies that the ratio of velocities (and accelerations) at any two points in the model is the same as the ratio of the velocities (and accelerations) at corresponding points in the prototype. Thus

$$(\underline{u}_1/\underline{u}_2)_m = (\underline{u}_1/\underline{u}_2)_p \text{ and } (\underline{a}_1/\underline{a}_2)_m = (\underline{a}_1/\underline{a}_2)_p$$

Since velocities and accelerations are vector quantities it is clearly necessary for the direction of motion at corresponding points to be the same. Kinematic similarity can only exist if the geometric similarity condition has been satisfied (except where scale effects exist).

(c) Dynamic Similarity: this is similarity with respect to the forces acting in a system. Dynamic similarity exists if the ratio of forces acting at any two points in the model is the same as the ratio of forces acting at corresponding points in the prototype. Thus

$$(\mathbf{F}_1/\mathbf{F}_2)_m = (\mathbf{F}_1/\mathbf{F}_2)_p$$

It is clearly necessary for the directions of action at corresponding points to be the same.

The flow pattern is determined by the forces which are acting. Therefore, if dynamic similarity exists in a situation where geometric similarity exists then the condition of kinematic similarity must be satisfied. Thus the condition of dynamic similarity is identical with that of physical similarity, if only mechanical aspects of flow are of interest.

The forces acting on any element of fluid may be due to viscosity, pressure, compressibility, surface tension or gravity. Therefore, the conditions for dynamic similarity may be expressed as

$$\frac{(\mathbf{F}_v)_p}{(\mathbf{F}_v)_m} = \frac{(\mathbf{F}_p)_p}{(\mathbf{F}_p)_m} = \frac{(\mathbf{F}_e)_p}{(\mathbf{F}_e)_m} = \frac{(\mathbf{F}_s)_p}{(\mathbf{F}_s)_m} = \frac{(\mathbf{F}_g)_p}{(\mathbf{F}_g)_m}$$

If the magnitudes and directions of the component forces acting on a fluid element are known then the resultant force may be determined from a vector diagram. Since the resultant force is the inertia force we may write

$$\mathbf{F}_v + \mathbf{F}_p + \mathbf{F}_e + \mathbf{F}_s + \mathbf{F}_g = \mathbf{F}_i$$

or

$$\frac{\mathbf{F}_v}{\mathbf{F}_i} + \frac{\mathbf{F}_p}{\mathbf{F}_i} + \frac{\mathbf{F}_e}{\mathbf{F}_i} + \frac{\mathbf{F}_s}{\mathbf{F}_i} + \frac{\mathbf{F}_g}{\mathbf{F}_i} = 1$$

It was shown earlier that the ratio of any two forces can give a non-dimensional quantity expressed as a named Pi-group. Hence we may write

$$\frac{K_1}{Re} + \frac{K_2}{Eu} + \frac{K_3}{Ca} + \frac{K_4}{We} + \frac{K_5}{Fr} = 1$$

where $K_1$, $K_2$, $K_3$, etc. are dimensionless coefficients.

It follows that for physical similarity the numerical value of each Pi-group must be the same for both model and prototype.

However, it is not always possible nor desirable to maintain the same value for each Pi-group for model and prototype, and there is a departure from complete similarity. This is known as scale effect.

First of all, it may not be possible to obtain the necessary degree of smoothness for the model surface i.e. $(k/L)_m \neq (k/L)_p$. Also, exact scaling down in size may result in a change in the nature of flow in the model. For example, in situations where the movement of solid particles is of interest e.g. sand deposits, scaling down of the particles would result in a fine powder whose behaviour is very different from the actual particles. A more pronounced type of departure from true flow results with river and estuary models where the direct scaling down of depth could result in surface tension or viscosity effects in the model which are not present in the prototype. Under these conditions it may be necessary to use distorted models to produce realistic flow behaviour.

Once the geometric scale factors have been resolved, it is necessary to consider dynamic similarity factors which determine the flow pattern. If a flow situation involves several different types of force it may not be possible to simultaneously satisfy equality of different Pi-groups. For example, in subsonic flow conditions it is usually considered that equality of Reynolds number is a sufficient requirement for dynamic similarity. However, for situations where a free surface is present, the Froude number is important and it may not be possible to maintain the same Reynolds number for the model. In supersonic flow the Mach number becomes more important than the Reynolds number. These points of conflict will be discussed in detail in the worked examples.

## 5.4 SIMILARITY FROM BASIC DIFFERENTIAL EQUATIONS

If the basic differential equations governing a problem may be written down then it is possible to obtain the Pi-groups which govern the flow conditions. This method has much to commend it because it enables a more rigorous statement of the problem to be made and leads to an appreciation of the physical nature of the problem. This is basically more useful than an arbitrary grouping of physical variables (without understanding) which occurs sometimes when using the Rayleigh or Buckingham methods. However, the differential equations for many flow problems are extremely complex and it is not easy to obtain the relevant Pi-groups.

*Example 5.1*

By application of the Rayleigh method of dimensional analysis, determine the height to which a liquid rises in a capillary tube. It may be assumed that the height, y, depends upon the density of the liquid, ρ, the radius of the tube, R, the surface tension of the liquid, γ, the angle of contact, θ, and gravitational acceleration g.

The relationship between the variables may be expressed by

$$y = \phi(\rho, R, \gamma, g, \theta)$$

With the Rayleigh method, it is assumed that this relationship signifies a series

$$y = K_1 \rho^{a_1} R^{b_1} \gamma^{c_1} g^{d_1} \theta^{e_1} + K_2 \rho^{a_2} R^{b_2} \gamma^{c_2} g^{d_2} \theta^{e_2} + \ldots$$

where each term has the same dimensions as y, and $K_1$, $K_2$ are numerical coefficients.

Selecting a typical term we have

$$y = K \rho^a R^b \gamma^c g^d \theta^e$$

Substituting the dimensions for each variable, we may form a dimensional equation

$$L \equiv (ML^{-3})^a L^b (MT^{-2})^c (LT^{-2})^d$$

Note that $\theta$ is itself a dimensionless parameter and does not appear in the equation.

Forming equations for the indices of M, L and T, we have three equations and four unknowns.

For M $\quad a + c = 0$

For L $\quad -3a + b + d = 1$

For T $\quad -2c - 2d = 0$

Expressing each of the unknowns in terms of index 'a' (chosen arbitrarily) we obtain $b = 1 + 2a$, $c = -a$, $d = a$. Therefore

$$y = K \rho^a R^{1+2a} \gamma^{-a} g^a \theta^e = KR \left[\frac{\rho g R^2}{\gamma}\right]^a \theta^e$$

This is as far as we can go with dimensional analysis. In order to obtain a more useful end result, experimental information is necessary such as the fact that $y \propto 1/R$. This gives $a = -1$. Hence

$$y = K \frac{\gamma}{\rho g R} \theta^e$$

It is of interest to compare this result with the mathematical solution obtained in example 1.3.

*Example 5.2*

The shear stress, $\tau_0$, at the surface of a pipe through which fluid is flowing under incompressible conditions is thought to depend upon the mean velocity, u, the pipe diameter, D, the surface roughness height, k, and the fluid properties density, $\rho$, and viscosity $\mu$. By application of the Rayleigh method of dimensional analysis, derive a non-dimensional expression for the shear stress in terms of the other variables.

The relationship between the variables may be expressed by

$$\tau_0 = \phi(u, D, k, \rho, \mu)$$

Applying the Rayleigh method

$$\tau_0 = K u^a D^b k^c \rho^d \mu^e$$

The dimensional equation is

$$ML^{-1}T^{-2} \equiv (LT^{-1})^a L^b L^c (ML^{-3})^d (ML^{-1}T^{-1})^e$$

Forming equations for the indices of M, L and T we have three equations and five unknowns.

For M $\quad d + e = 1$

For L $\quad a + b + c - 3d - e = -1$

For T $\quad -a - e = -2$

Expressing each of the unknowns in terms of b and e (chosen arbitrarily) we obtain $d = 1 - e$, $c = -e - b$, $a = 2 - e$. Therefore

$$\tau_o = K\, u^{2-e}\, D^b\, k^{-e-b}\, \rho^{1-e}\, \mu^e = K\, \rho u^2 \left(\frac{\mu}{\rho u k}\right)^e \left(\frac{D}{k}\right)^b$$

or $$\frac{\tau_o}{\rho u^2} = \phi\left[\frac{\rho u k}{\mu}\ ,\ \frac{k}{D}\right]$$

Thus, dimensional analysis has simplified the problem by reducing the six variables to three Pi-groups. However, the Pi-groups are not in the best form because it is customary, in problems on pipe flow, to use a Reynolds number based on diameter, D, rather than surface roughness, k. It is valid to change the form of the Pi-groups by the simple manipulation described in section 5.2.

$$\frac{\tau_o}{\rho u^2} = \phi\left[\left(\frac{\rho u k}{\mu}\right) \times \left(\frac{D}{k}\right),\ \frac{k}{D}\right] = \phi\left(\frac{\rho u D}{\mu}\ ,\ \frac{k}{D}\right)$$

The parameter $\tau_o/\rho u^2$ is usually written in the form $\tau_o/\frac{1}{2}\rho u^2$ and called a pipe friction factor, f. The Reynolds number $\rho u D/\mu$ is an important parameter and its value determines whether flow in the pipe is laminar or turbulent. The parameter k/D indicates the relative roughness of the pipe surface.

Thus $$f = \frac{\tau_o}{\frac{1}{2}\rho u^2} = \phi\left(Re,\ \frac{k}{D}\right)$$

It is now necessary to carry out experimental work on flow in pipes to determine the relationship between f and Re for different k/D values (see chapter 7).

*Example 5.3*

A bank of tubes is immersed in a stream of fluid. The frequency, n, at which vortices are shed from the cylinders depends upon the velocity, u, of the fluid, the cylinder diameter, D, the centre spacing, y, of the cylinders and fluid properties, dynamic viscosity, $\mu$, and density, $\rho$. By application of the Pi-theorem, show that

$$\frac{nD}{u} = \phi\left(\frac{\rho u D}{\mu},\ \frac{y}{D}\right)$$

In a particular heat exchanger, hydrogen at 400 K and 10 bar is to be used as the cooling fluid. The hydrogen flows at 15 m/s past a bank of tubes each 50 mm outside diameter.

In order to determine the frequency of vortex shedding, tests are carried out on a geometrically similar bank of tubes, 25 mm diameter, using air at 300 K and 1 bar.

For velocities up to 30 m/s the experimental data satisfy the relationship

$$n = 4.25u$$

Determine the frequency of vortex shedding in the hydrogen-cooled heat exchanger.

$R_a$ = 287 J/kg K, $R_h$ = 4124 J/kg K, $\mu_a = 1.846 \times 10^{-5}$ N s/m$^2$, $\mu_h = 10.87 \times 10^{-6}$ N s/m$^2$.

The relationship between the variables may be expressed by

$$n = \phi(u, \rho, \mu, D, y)$$

There are six variables in this problem, all of which may be expressed in terms of three primary dimensions M, L and T. Therefore, there will be 6 - 3 = 3 dimensionless parameters or Pi-groups in the solution.

Applying the Pi-theorem, it is first necessary to select three primary variables which between them include all the primary dimensions (M, L and T) involved. It is then necessary to show that these cannot be combined into any dimensionless parameter.

It is customary in fluid mechanics problems to select a characteristic length, as an indication of geometric similarity, a velocity or acceleration, as an indication of kinematic similarity and a variable including M as an indication of dynamic similarity. However this simple rule has exceptions and when thermal effects are present the selection of primary variables is more arbitrary. In the present problem we may select D, u and ρ as primary variables.

We first assume that the variables can be combined into a dimensionless parameter $\rho^a u^b D^c$ and attempt to evaluate the indices by dimensional reasoning. We may state

$$\frac{\rho^a u^b D^c}{1} \equiv \frac{(ML^{-3})^a \; (LT^{-1})^b \; L^c}{M^0 L^0 T^0}$$

Equating indices of M, L and T, we have

For M $\quad a = 0$

For L $\quad -3a + b + c = 0$

For T $\quad -b = 0$

Hence a = b = c = 0 and the solution is $\rho^0 u^0 D^0$. This means that there is no combination of the given variables that can give a dimensionless parameter, and the variables selected are independent.

Each of the remaining variables is now taken in turn and used to form a dimensionless Pi-group as follows

$$\Pi_1 = \frac{n}{\rho^a u^b D^c} \equiv \frac{T^{-1}}{(ML^{-3})^a (LT^{-1})^b\ L^c}$$

Equating indices

For M $\quad a = 0$

For L $\quad -3a + b + c = 0$

For T $\quad -b = -1$

Hence a = 0, b = 1, c = -1 and

$$\Pi_1 = \frac{nD}{u}$$

Similarly, taking the variable $\mu$

$$\Pi_2 = \frac{\mu}{\rho^a u^b D^c} \equiv \frac{ML^{-1}T^{-1}}{(ML^{-3})^a (LT^{-1})^b\ L^c}$$

Equating indices

For M $\quad a = 1$

For L $\quad -3a + b + c = -1$

For T $\quad -b = -1$

Hence a = 1, b = 1, c = 1 and

$$\Pi_2 = \frac{\mu}{\rho u D}$$

Similarly, taking the variable y

$$\Pi_3 = \frac{y}{\rho^a u^b D^c} \equiv \frac{L}{(ML^{-3})^a (LT^{-1})^b\ L^c}$$

Applying the above technique or by inspection

$$\Pi_3 = \frac{y}{D}$$

The solution may be stated in the form

$$\Pi_1 = \phi(\Pi_2, \Pi_3)$$

or $$\frac{nD}{u} = \phi\left(\frac{\rho u D}{\mu}, \frac{y}{D}\right)$$

From the gas law (equation 1.25c)

$$\rho_{air} = \left(\frac{p}{RT}\right)_{air} = \frac{10^5}{287 \times 300} = 1.16 \text{ kg/m}^3$$

$$\rho_{hyd} = \left(\frac{p}{RT}\right)_{hyd} = \frac{10 \times 10^5}{4124 \times 400} = 0.606 \text{ kg/m}^3$$

Geometric similarity conditions are satisfied i.e. $(y/D)_{model} = (y/D)_{prototype}$. Dynamic similarity conditions demand equality of Reynolds number i.e.

$$\left(\frac{\rho u D}{\mu}\right)_{model} = \left(\frac{\rho u D}{\mu}\right)_{prototype}$$

$$u_m = \frac{\rho_p D_p \mu_m u_p}{\rho_m D_m \mu_p} = \frac{0.606 \times 0.05 \times 1.846 \times 10^{-5} \times 15}{1.16 \times 0.025 \times 10.87 \times 10^{-6}} = 26.6 \text{ m/s}$$

Thus for dynamic similarity the tests in air must be carried out at 26.6 m/s. The actual tests were carried out at velocities up to 30 m/s therefore the given equation for frequency, n, is valid. Substituting values we obtain

$$n = 4.25u = 4.25 \times 26.6 = 113 \text{ Hz}$$

If similarity conditions are satisfied then the Strouhal number nD/u will be the same for model and prototype, ie.

$$\left(\frac{nD}{u}\right)_{model} = \left(\frac{nD}{u}\right)_{prototype}$$

$$n_p = \frac{D_m u_p n_m}{D_p u_m} = \frac{0.025 \times 15 \times 113}{0.050 \times 26.6} = 31.9 \text{ Hz}$$

*Example 5.4*

The volumetric flow rate, $\dot{V}$, of water over a spillway depends upon the height, H, of the water surface above the crest of the spillway, the length, L, of the spillway, the mean roughness height, k, of the concrete and fluid properties viscosity, $\mu$, density, $\rho$, and surface tension, $\gamma$. The acceleration due to gravity, g, is also an influencing variable. Show that the relationship between the variables may be expressed by

$$\frac{\dot{V}}{\sqrt{(gH)}H^2} = \phi\left(\frac{\rho\sqrt{(gH)}H}{\mu}, \frac{\rho(gH)H}{\gamma}, \frac{L}{H}, \frac{k}{H}\right)$$

A spillway having a length of 30 m is required to discharge 100 m$^3$/s of water. A model is to be tested and the rate of flow available in the laboratory is 0.25 m$^3$/s. Calculate the length of the model spillway assuming that the influence of surface roughness, viscosity and surface tension may be neglected.

The relationship between the variables may be expressed by

$$\dot{V} = \phi(H,\ L,\ k,\ \mu,\ \rho,\ \gamma,\ g)$$

The Pi-theorem method is preferable to the Rayleigh method because of the large number of variables. This method involves the selection of three primary variables and it is logical to choose H as a geometric term, g as a kinematic term and ρ as a dynamic term. Note that there is no velocity term in the variables, but in a free flow device the Bernoulli equation gives $u = \surd(2gH)$. It follows that the selection of H and g as primary variables is equivalent to choosing a velocity term. The quantity $\surd(gH)$ will then appear in the analysis. Applying the Pi-theorem we obtain

$$\frac{\dot{V}}{\surd(gH)H^2} = \phi\left(\frac{L}{H},\ \frac{k}{H},\ \frac{\mu}{\rho g^{\frac{1}{2}} H^{3/2}},\ \frac{\gamma}{H^2 g\rho}\right)$$

Note that the functional relationship is unchanged if the Pi-groups are inverted or rearranged. Thus

$$\frac{\dot{V}}{\surd(gH)H^2} = \phi\left(\frac{\rho\surd(gH)H}{\mu},\ \frac{\rho(gH)H}{\gamma},\ \frac{L}{H},\ \frac{k}{H}\right)$$

In the present problem, we are told that viscous and surface tension effects may be neglected. This is a reasonable assumption except for low heads. We are also told that surface roughness effects are absent. Thus, for kinematic similarity, we omit the Reynolds number $\rho\surd(gH)H/\mu$, the Weber number $\rho(gH)H/\gamma$, and k/H from the list of significant parameters, and state

$$\frac{\dot{V}}{\surd(gH)H^2} = \phi\left(\frac{L}{H}\right)$$

or $$\left(\frac{\dot{V}}{\surd(gH)H^2}\right)_{\text{model}} = \left(\frac{\dot{V}}{\surd(gH)H^2}\right)_{\text{prototype}}$$

and $$\left(\frac{L}{H}\right)_{\text{model}} = \left(\frac{L}{H}\right)_{\text{prototype}}$$

Combining the above groups for constant g we obtain

$$\left(\frac{\dot{V}^{5/2}}{L}\right)_{\text{model}} = \left(\frac{\dot{V}^{5/2}}{L}\right)_{\text{prototype}}$$

and $$L_m = \left(\frac{\dot{V}_m}{\dot{V}_p}\right)^{2/5} L_p = \left(\frac{0.25}{100}\right)^{2/5} \times 30 = 1.19\ \text{m}$$

Note that for true kinematic similarity the model should be operated under a head given by $H_m = H_p(L_m/L_p)$. This criterion is not too important if end effects on the spillway can be neglected.

*Example 5.5*

Assuming that the resistance to motion, F, of a body moving in a compressible fluid is a function of a characteristic dimension, L, of the body, the velocity u of the body and fluid properties density, ρ, viscosity, μ, and bulk modulus, K, show by dimensional analysis that

$$F = \rho u^2 L^2 \phi\left[\frac{\rho u L}{\mu},\ \frac{u}{\sqrt{(K/\rho)}}\right]$$

Discuss the significance of the non-dimensional groups obtained and state any difficulties that might arise in obtaining dynamic similarity when testing scale models of high speed aircraft in a wind tunnel.

A prototype aircraft is to fly at 500 m/s in air of pressure 0.8 bar, temperature 250 K and bulk modulus $1.12 \times 10^5$ N/m$^2$. Determine the pressure at which a compressed-air wind tunnel must be operated in order to obtain complete dynamic similarity for tests at prototype speed on a one-tenth scale model of the aircraft. If the drag on the model is found to be 200 N, determine the drag of the prototype aircraft. It may be assumed that the temperature of the air in the wind tunnel is 250 K and that viscosity is independent of pressure.

The relationship between the variables may be expressed by

$$F = \phi(L,\ u,\ \rho,\ \mu,\ K)$$

Selecting L, u and ρ as primary variables, as justified in example 5.3, and applying the Pi-theorem, we obtain

$$\frac{F}{\rho u^2 L^2} = \phi\left[\frac{\rho u L}{\mu},\ \frac{u}{\sqrt{(K/\rho)}}\right]$$

The group $F/\rho u^2 L^2$ is sometimes referred to as the drag coefficient, $C_D$, and written in the form $F/\frac{1}{2}\rho u^2 A$.

The group $\rho uL/\mu$ is the Reynolds number, Re, which is a criterion of dynamic similarity when viscous and inertia forces are significant.

The group $u/\sqrt{(K/\rho)}$ is the Mach number, M, which is a criterion of dynamic similarity when elastic and inertia forces are significant.

For complete dynamic similarity, Re and M should be the same for model and prototype. This is impossible if the fluid properties are the same for both. Therefore, Re is usually maintained constant for subsonic flow conditions and M maintained constant for supersonic flow.

It is possible to maintain both Re and M constant for model and prototype if a compressed air wind tunnel is used. The necessary tunnel pressure for model tests may be determined as follows

For complete dynamic similarity

$$\left(\frac{\rho u L}{\mu}\right)_{model} = \left(\frac{\rho u L}{\mu}\right)_{prototype} \qquad \text{(i)}$$

and
$$\left(\frac{u}{\sqrt{(K/\rho)}}\right)_{model} = \left(\frac{u}{\sqrt{(K/\rho)}}\right)_{prototype} \qquad \text{(ii)}$$

Mach number can be written in the form $u/\sqrt{(\gamma RT)}$ and in the present problem $u_m = u_p$ and $T_m = T_p$. It follows that equality of Mach number is obtained. The given value of K is superfluous.

For equality of Renolds number, equation (i) together with the gas law $p = \rho RT$ gives

$$\frac{p_m T_p}{p_p T_m} = \left(\frac{\mu_m}{\mu_p}\right)\left(\frac{L_p}{L_m}\right)\left(\frac{u_p}{u_m}\right)$$

In the present problem $u_m = u_p$ and $T_m = T_p$. It follows that since viscosity is temperature dependent, $\mu_m = \mu_p$. Therefore

$$\frac{p_m}{p_p} = \frac{L_p}{L_m}$$

and $p_m = p_p(L_p/L_m) = 0.8 \times 10 = 8$ bar

For dynamic similarity between model and prototype

$$\left(\frac{F}{\rho u^2 L^2}\right)_{model} = \left(\frac{F}{\rho u^2 L^2}\right)_{prototype}$$

$$F_p = \frac{\rho_p}{\rho_m}\left(\frac{L_p}{L_m}\right)^2\left(\frac{u_p}{u_m}\right)^2 F_m$$

$$= \frac{p_p}{p_m}\left(\frac{L_p}{L_m}\right)^2 F_m = \left(\frac{0.8}{8}\right) 10^2 \times 200 = 2 \text{ kN}$$

*Example 5.6*

By application of the Pi-theorem show that the resistance to motion of a surface vessel is given by

$$R = \rho u^2 L^2 \, \phi\left(\frac{\rho uL}{\mu}, \frac{u^2}{Lg}\right)$$

where it is assumed that the resistance, R, is a function of the vessel velocity, u, and length, L, and fluid properties density, ρ, and viscosity, μ.

A model of a ship made to 1/20 scale, with wetted area 4 $m^2$, is to be towed under test conditions in a tank of fresh water. The actual ship speed is to be 5.6 m/s. From tests carried out at the corresponding speed it is found that the measured resistance on the model is 46.6 N. From tests carried out on a thin plate of the same length as the model the skin resistance is found to be 1.5 $u^{1.95}$ $N/m^2$ of surface area. Assuming the skin resistance of the actual ship to be 1.4 $u^{1.9}$ $N/m^2$ determine the resistance of the ship in sea water at the design speed of 5.6 m/s. ρ sea water = 1025 $kg/m^3$.

The relationship between the variables may be expressed by

$$R = \phi(u, L, \rho, \mu, g)$$

Gravitational acceleration, g, is introduced as a variable because it is known that gravity forces are important in surface wave formation. It is assumed that capillary waves are insignificant therefore surface tension is not included. Selecting L, u and ρ as primary variables and applying the Pi-theorem, we obtain

$$\frac{R}{\rho u^2 L^2} = \phi\left(\frac{\rho uL}{\mu}, \frac{u^2}{Lg}\right)$$

The groups $R/\rho u^2 L^2$ and $\rho uL/\mu$ are recognisable as the drag coefficient and Reynolds number, Re, respectively. The group $u^2/Lg$ (or $u/\sqrt{(Lg)}$, as it is usually written) is the Froude number, Fr. For dynamic similarity

$$\left(\frac{u^2}{Lg}\right)_{\text{model}} = \left(\frac{u^2}{Lg}\right)_{\text{prototype}} \qquad \text{(i)}$$

and $$\left(\frac{\rho uL}{\mu}\right)_{\text{model}} = \left(\frac{\rho uL}{\mu}\right)_{\text{prototype}} \qquad \text{(ii)}$$

Now, if the *corresponding speed* of the model, $u_m$, is determined from condidion (i) it follows that

$$u_m = u_p \sqrt{(L_m/L_p)} \qquad \text{(iii)}$$

However, towing tests on surface vessels are usually carried out in water and the density and viscosity for model and prototype will be the same. Condition (ii) will then give

$$u_m = u_p(L_p/L_m) \qquad \text{(iv)}$$

It is clearly impossible to satisfy both conditions (iii) and (iv) simultaneously therefore Froude devised a method of dealing with the problem as follows.

(1) The corresponding speed of the model, $u_m$, is determined from condition (i) which ensures that dynamic similarity for wave-making resistance is obtained. Thus

$$u_m = u_p\sqrt{(L_m/L_p)} = 5.6\sqrt{(1/20)} = 1.25 \text{ m/s}$$

(2) The resistance to motion, R, is composed of (a) resistance due to skin friction, $R_f$, and (b) residual resistance due to eddy-making and wave-making, $R_r$.

$$R = R_f + R_r \qquad \text{(v)}$$

Resistance due to skin friction is a function of Reynolds number only and may be determined from flat plate data, which yields a formula of the form $R_f = CAu^n$. From the given information

$$(R_f)_{\text{model}} = 1.5\ u_m^{1.95}\ A_m = 1.5 \times 1.25^{1.95} \times 4 = 9.27 \text{ N}$$

and $$(R_f)_{\text{prototype}} = 1.4\ u_p^{1.9}\ A_p = 1.4 \times 5.6^{1.9} \times 4 \times 20^2 = 59 \text{ kN}$$

From condition (v) the residual resistance for the model is given by

$$(R_r)_m = R_m - (R_f)_m = 46.6 - 9.27 = 37.33 \text{ N}$$

Now if the residual resistance is a function of Froude number but not of Reynolds number

$$R_r = \rho u^2 L^2\ \phi(\text{Fr})$$

and $$\left(\frac{R_r}{\rho u^2 L^2}\right)_{\text{model}} = \left(\frac{R_r}{\rho u^2 L^2}\right)_{\text{prototype}}$$

$$(R_r)_p = \left(\frac{\rho_p}{\rho_m}\right)\left(\frac{u_p}{u_m}\right)^2\left(\frac{L_p}{L_m}\right)^2 (R_r)_m = \frac{1025 \times 5.6^2 \times 20^2 \times 37.33}{1000 \times 1.25^2 \times 1} = 307 \text{ kN}$$

The total resistance of the actual ship is

$$R_p = (R_f)_p + (R_r)_p = 59 + 307 = 366 \text{ kN}$$

*Example 5.7*

The performance of incompressible-flow rotodynamic turbines and pumps may be expressed in terms of the following variables: Power, P, volumetric flow rate, $\dot{V}$, specific energy change, gH, impeller diameter, D, characteristic flow passage dimension, B, surface roughness, k, impeller speed, N, fluid density, $\rho$, and fluid viscosity, $\mu$. By application of the Pi-theorem determine the dimensionless parameters involved and state the significance of each in terms of the requirements for similarity.

A centrifugal pump is required to run at 660 rev/min and deliver 13.5 $m^3/s$ of oil of density 950 $kg/m^3$ and viscosity 0.125 N $s/m^2$. In order to test the design, experiments are to be carried out on a quarter-scale model, at 1200 rev/min, using air of density 1.2 $kg/m^3$ and viscosity $1.8 \times 10^{-5}$ N $s/m^2$. Determine (i) the volumetric flow rate through the model for similarity conditions, (ii) the flow Reynolds numbers for model and prototype, and (iii) the power required to drive the actual oil pump at 660 rev/min if the power absorbed by the model at 1200 rev/min under similarity conditions is 100 W.

The relationship between the variables may be expressed by

$$\phi\ (D,\ B,\ k,\ N,\ \dot{V},\ \rho,\ \mu,\ gH,\ P) = 0$$

Selecting D, N and $\rho$ as primary variables and applying the Pi-theorem, we obtain

$$\phi\left(\frac{B}{D},\ \frac{k}{D},\ \frac{\dot{V}}{ND^3},\ \frac{\rho ND}{\mu},\ \frac{gH}{N^2D^2},\ \frac{P}{\rho N^3D^5}\right) = 0$$

Consider the conditions for similarity of flow in two machines.
(1) The first requirement is that of geometric similarity and for rotodynamic machines this means that, in addition to similarity of runner shape, all entry and exit passages which are integral parts of the machine must be the same shape for both machines. The relative roughness of solid surfaces must also be the same. Thus the shape parameters B/D and k/D must be maintained constant.
(2) The second requirement is that of kinematic similarity and this means that the velocity vector diagrams, at entry to and exit from the runner, must be the same shape for both machines. A typical mechanical part velocity is ND since the runner tip velocity is $\pi DN$. A typical fluid velocity is $\dot{V}/D^2$ since all passage areas are proportional to one another and to $D^2$. The ratio of these two velocities is $\dot{V}/ND^3$, which must be maintained constant for both machines for similar vector diagrams.
(3) The third requirement is that of maintaining similarity of flow in the various passages. This means that a constant value of flow Reynolds number must apply to both machines. Using $\dot{V}/D^2$ as a typical fluid velocity Re = $\dot{V}\rho/D\mu$. In the dimensional analysis, Reynolds number appeared in the form $\rho ND^2/\mu$ but this group may be combined with $\dot{V}/ND^3$ to give the flow Reynolds number $\dot{V}\rho/D\mu$.

Ideally if the above three criteria are met there will be complete dynamic similarity between the two machines. It is usually impossible to simultaneously satisfy conditions (2) and (3) but in the highly turbulent flow encountered in rotodynamic machines, variations in efficiency and values of other parameters with Reynolds number (scale effect) is small. Thus condition (3) may be relaxed. If the above criteria are met then the other dimensionless parameters will have the same values for the two machines, and

$$\left(\frac{gH}{N^2D^2}\right)_1 = \left(\frac{gH}{N^2D^2}\right)_2 = \text{Head Parameter}$$

$$\left(\frac{P}{\rho N^3D^5}\right)_1 = \left(\frac{P}{\rho N^3D^5}\right)_2 = \text{Power Parameter}$$

Also $$\left(\frac{\rho gH\dot{V}}{P}\right)_1 = \left(\frac{\rho gH\dot{V}}{P}\right)_2 = \text{Efficiency}$$

Note that the head parameter may be modified to give a pressure coefficient ($\Delta p/\rho N^2D^2$) by the substitution $\Delta p = \rho gH$.

A quarter-scale model pump is used in the tests therefore condition (1) for geometric similarity is satisfied. In order to satisfy condition (2) we have

$$\left(\frac{\dot{V}}{ND^3}\right)_{\text{model}} = \left(\frac{\dot{V}}{ND^3}\right)_{\text{prototype}}$$

or $$\dot{V}_m = \frac{N_m}{N_p}\left(\frac{D_m}{D_p}\right)^3 \dot{V}_p = \frac{1200}{660}\left(\frac{1}{4}\right)^3 13.5 = 0.383 \text{ m}^3/\text{s}$$

The scale, fluid properties and flow rate are now fixed for the model pump and it may be impossible to satisfy condition (3). The flow Reynolds numbers for model and prototype must be evaluated and compared.

$$(Re)_m = \left(\frac{\dot{V}\rho}{D\mu}\right)_m = \frac{0.383 \times 1.2}{D_m \times 1.8 \times 10^{-5}} = \frac{2.56 \times 10^4}{D_m}$$

$$(Re)_p = \left(\frac{\dot{V}\rho}{D\mu}\right)_p = \frac{13.5 \times 950}{D_p \times 0.125} = \frac{10.24 \times 10^4}{D_p}$$

The pump diameters are not given in the problem so the actual Reynolds number values cannot be determined. The ratio is

$$\frac{(Re)_m}{(Re)_p} = \frac{2.56 \times 10^4}{10.24 \times 10^4}\left(\frac{D_p}{D_m}\right) = 1$$

Thus, the flow Reynolds number is the same for model and prototype and complete similarity is obtained if the air flow in the model is incompressible. For complete similarity

$$\left(\frac{P}{\rho N^3 D^5}\right)_m = \left(\frac{P}{\rho N^3 D^5}\right)_p$$

$$P_p = \frac{\rho_p}{\rho_m}\left(\frac{N_p}{N_m}\right)^3\left(\frac{D_p}{D_m}\right)^5 P_m = \frac{950}{1.2}\left(\frac{660}{1200}\right)^3\left(\frac{4}{1}\right)^5 100 = 13.5 \text{ MW}$$

*Problems*

1 The velocity of propagation of waves in a liquid with a free surface is assumed to be a function of the wave length, $\lambda$, the depth of liquid, d, the density, $\rho$, and surface tension, $\gamma$. If the acceleration due to gravity is g show by application of (i) the Pi-theorem and (ii) the Rayleigh method, that

$$\frac{u^2}{\lambda g} = \phi\left(\frac{\gamma}{\lambda^2 g\rho}, \frac{d}{\lambda}\right)$$

2 When liquid flows through a vee-notch set in the side of a rectangular tank the volumetric flow rate $\dot{V}$ is assumed to depend upon the head of liquid, y, measured relative to the bottom of the notch, gravitational acceleration, g, the included angle of the notch, $\theta$, and on fluid properties surface tension $\gamma$, density $\rho$, and viscosity $\mu$. By application of the Pi-theorem, determine the relationship between the variables.

The flow of water through a model vee-notch of included angle 45° is found to be 0.005 $m^3/s$ under a head of 100 mm. Determine the flow of water through a geometrically similar vee-notch operating under a head of 1.5 m. Assume that the influence of surface roughness, viscosity and surface tension may be neglected.

[4.36 $m^3/s$]

3 It is desired to carry out model tests simulating the entry of a torpedo into the surface of the sea. The principal object of the experiments is to determine the maximum force $F_m$ applied to the torpedo during the deceleration, and it is known that the factors which influence this are the length L of the torpedo, the speed u of entry, the density $\rho$ of the water, the mean density $\rho_t$ of the torpedo, and the acceleration due to gravity, g. Derive three independent non-dimensional groups from the six variables.

In a particular experiment it is required to model the entry of a torpedo of length 4.3 m entering the water at 24.0 m/s. The model is 1.1 m long. Determine the speed at which it should enter the water. The value of $F_m$ measured on the model is 11.1 N. Calculate the corresponding value of $F_m$ for the full-scale torpedo.

$$\left[\frac{F_m}{\rho u^2 L^2} = \phi\left(\frac{\rho_t}{\rho}, \frac{Lg}{u^2}\right), \ 12.2 \text{ m/s}, \ 660 \text{ N}\right]$$

4 A fluid flows at volumetric rate $\dot{V}$ through a nozzle of diameter d located in a pipe of diameter D. The volumetric flow rate depends

on the diameters d and D, on the viscosity $\mu$ and density $\rho$ of the fluid, and on the pressure drop $\Delta p$ between the upstream length of pipe and nozzle throat. By application of the Pi-theorem, derive a relationship between the variables.

The flow of water through a nozzle of diameter 0.075 m in a pipe of diameter 0.15 m is 0.015 $m^3/s$ when the pressure drop is $10^4$ $N/m^2$. If air of density 1.20 $kg/m^3$ and viscosity $1.8 \times 10^{-5}$ Pa s flows under dynamically similar conditions through a nozzle and pipe of half the linear scale, determine the pressure drop and the volume rate of flow. Assume the viscosity and density of water to be $1.2 \times 10^{-3}$ Pa s and $10^3$ $kg/m^2$, respectively.

[7.5 kPa, 0.0938 $m^3/s$]

5 A circular cylinder rotates at N rev/min in a uniform normal stream of fluid of velocity u m/s. Assuming that the power, P, required to maintain rotation depends upon N and u and upon the cylinder diameter, D, and fluid properties viscosity, $\mu$, and density, $\rho$, obtain a dimensionless relationship between the variables.

Tests are to be carried out on a 50 mm diameter cylinder, rotating in a water stream, in order to predict the power required to rotate a 200 mm diameter cylinder in an air stream of 40 m/s. If the cylinders have the same length/diameter ratio and the tests are carried out under dynamically similar conditions, determine (i) the ratio of the rotational speeds for the cylinders, (ii) the water stream velocity, and (iii) the ratio of the powers required to rotate the two cylinders.

$\rho_a = 1.2$ $kg/m^3$, $\mu_a = 17.7 \times 10^{-6}$ Pa s, $\rho_w = 1000$ $kg/m^3$,
$\mu_w = 1.12 \times 10^{-3}$ Pa s

[0.822, 12.2 m/s, 1.465]

6 A motor vessel, 10 m long, is driven by a reaction jet at 90 m/s in such a manner that the viscous resistance is negligible compared with wave making resistance. A similar model of the vessel, 150 mm long is found to require a force of 0.85 N to drive it over the water at the corresponding speed. What is the corresponding speed and what thrust is exerted by the jet?

[11 m/s, 252 kN]

7 A multi-stage centrifugal pump, required to run at 1500 rev/min and pump water, consists of four identical impellers in series, each having a diameter of 300 mm. Tests are carried out on a single stage pump fitted with a geometrically similar impeller of diameter 150 mm. It is found that this pump delivers 0.01 $m^3/s$ of water against a head of 20 m at maximum efficiency conditions which occur at a speed of 500 rev/min. Assuming dynamically similar conditions for the two pumps, determine (i) the head the multi-stage pump will run against at maximum efficiency, and (ii) the discharge of water under these conditions.

[2880 m, 0.24 $m^3/s$]

8 A one-fifth scale Pelton wheel model is tested under a head of 40 m and the maximum overall efficiency is found to be 85% for a

discharge of 0.1 $m^3/s$. The full-scale Pelton wheel is required to work under a head of 400 m and to run at 500 rev/min. Determine (i) the speed of the model for dynamically similar conditions, (ii) the power developed by the full-scale turbine.

[790 rev/min, 26.5 MW]

9 The time, t, of formation of liquid drops is to be studied experimentally by an apparatus in which liquid at pressure, p, is forced through a capillary tube of diameter, D, and length, L. The relevant fluid properties are dynamic viscosity, μ, density, ρ, and surface tension, γ. Assuming the formation time, t, to be the dependent variable, show by dimensional analysis that

$$\frac{t\sqrt{(p/\rho)}}{D} = \phi\left(\frac{\sqrt{(\rho p)}D}{\mu}, \frac{\gamma}{pD}, \frac{L}{D}\right)$$

In a particular test on water at 250 Pa, the formation time is 1.5 s. Using the same apparatus, determine the required pressure for tests on benzene and the corresponding formation time. Discuss any difficulties in maintaining complete dynamic similarity. Assume the following property values. $\mu_w = 10.1 \times 10^{-4}$ Pa s, $\rho_w = 10^3$ kg/m$^3$ $\gamma_w = 0.073$ N/m. $\mu_b = 6.56 \times 10^{-4}$ Pa s, $\rho_b = 880$ kg/m$^3$, $\gamma_b = 0.029$ N/m.

[99.3 Pa, 2.23 s]

10 (a) An airscrew absorbs a power P and develops a thrust F. It is suggested that P and F depend upon the airscrew diameter D, the blade chord C, the number of blades B, the angular velocity N, the dynamic viscosity μ and density ρ of the fluid and the velocity U at which the airscrew moves through the fluid. By application of the Pi-theorem show that

$$\frac{F}{\rho U^2 D^2} = \phi\left(\frac{C}{D}, B, \frac{\rho UD}{\mu}, \frac{ND}{U}\right)$$

(b) A half-scale model airscrew is tested in a compressed-air wind tunnel which operates at twice sea-level density. The test conditions are those which are dynamically similar to operation at an altitude of 10 000 m where the density is 0.34 that at sea-level and the dynamic viscosity is 0.82 that in the test.

Determine the ratios of the power absorbed and the thrust developed by the model airscrew to those of the actual airscrew at an altitude of 10 000 m.

[0.105, 0.253]

# 6 FLOW OVER BODIES

Previous chapters have been devoted to the study of fluid properties and to the statics, kinematics and dynamics of fluids. The general equations for conservation of mass, momentum and energy have been derived and applied to real flow situations in which viscous effects were present but not fully understood.

It is now necessary to discuss the nature of real flow taking into account the viscosity of the fluid which is largely responsible for all observed phenomena.

## 6.1 CHARACTERISTICS OF REAL FLUID FLOW

It is customary to categorise fluid flow into two main classes for convenience in treating practical problems.
(1) Flow around bodies i.e. external flow around some object fully immersed in an otherwise unbounded fluid.
(2) Flow in ducts i.e. internal flow bounded by solid surfaces.

In fact, fluid behaviour is basically the same in both situations and is based on two concepts introduced in chapter 1. The first concept is that the flow may be *laminar* or *turbulent* in character. The second is that a flow field may be divided into (a) a region in which viscous effects are negligible and an *ideal* flow pattern assumed, and (b) a region in which viscous stress effects are significant because of the presence of a *boundary layer*.

### 6.1.1 Laminar Flow

In laminar flow, all fluid particles move in parallel paths and there is no transverse velocity component. The flow is stable against disturbances due to wall roughness or obstacles because viscous forces predominate and damp out any tendency for random motion. Individual fluid molecules may have irregular motion but such motion only influences viscosity. Newton's law of viscosity $\tau = \mu\, du/dy$ may be used to calculate shear stresses.

### 6.1.2 Turbulent Flow

In turbulent flow, fluid particles do not remain in layers but move in an irregular manner. The particles move in aggregates of varying sizes known as *eddies*.

At any point in the flow, the velocity fluctuates about a temporal mean velocity, $\underline{u}$, which may be resolved into mean components u, v and w. The irregular effect of turbulence may then be included by superimposing fluctuating velocity components u', v' and w' on the mean velocities. Thus, if $u_i$ is the instantaneous velocity component in the x direction, we may write $u_i = u + u'$. Similarly $v_i = v + v'$ and $w_i = w + w'$. A typical recording of $u_i$, taken with a hot wire anemometer, is shown in figure 6.1a.

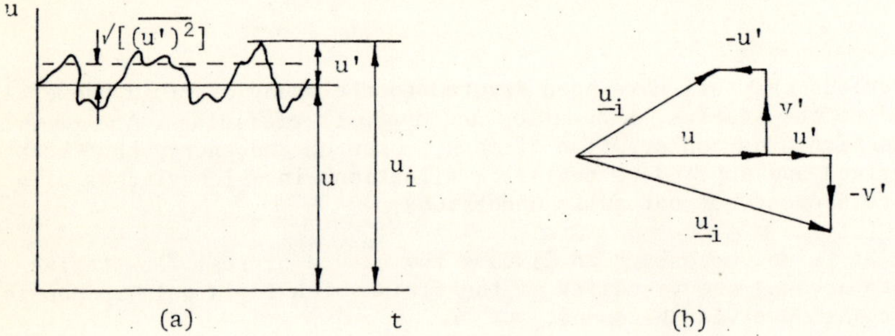


Figure 6.1

The instantaneous resultant velocity, $u_i$, for bulk flow in the x direction is composed of the vector sum of u and the fluctuating components u' and v' (say) shown in figure 6.1b.

The *intensity* of turbulence is expressed in terms of these fluctuating components. For flow in the x-direction, we define *intensity* as $\sqrt{[\overline{(u')^2}]}$ and *relative intensity* as $\sqrt{[\overline{(u')^2}]}/u$. Relative intensity varies from about 0.02%, in a low-turbulence wind tunnel, to about 3% in turbulent pipe flow, to about 50% in an air jet mixing with still air.

The *scale* of turbulence is measured in terms of the size of turbulent eddies. Large eddies are of the same order of size as a characteristic length in the flow and are clearly identifiable e.g. the eddies in the wake behind a cylinder (figure 6.3). On a small scale, individual eddies are difficult to identify and the term, eddy, refers to motion which is coherent over short distances.

Turbulence tends to decay in the absence of an external source of energy. This is shown in figure 6.3 where large eddies interact with one another and dissipate into a succession of smaller eddies which ultimately dissipate into energy at a molecular level i.e. internal energy.

Newton's law of viscosity cannot be used to calculate the shear stress in turbulent flow because the effective viscosity is now dependent on the turbulence itself. The shear stress at any point is given by $\tau = \mu(du/dy) + \rho\varepsilon(du/dy)$ where $\rho\varepsilon$ is the *eddy viscosity*. Other models of turbulence have been suggested to account for eddy stresses.

6.1.3 Boundary Layer

As real (viscous) fluid flows past a body (figure 6.2) fluid particles on the surface remain at rest and a high velocity gradient is set up near the boundary. This causes large viscous stresses and a region of retarded flow known as the *boundary layer*. The boundary layer grows in thickness along the body in the downstream direction and forms a wake. The wake mixes with the free-stream flow and is eventually dissipated some distance downstream. Inside the boundary layer and wake, flow is frictional and rotational, but outside these regions frictional effects are negligible and the flow pattern approximates to that for irrotational flow.

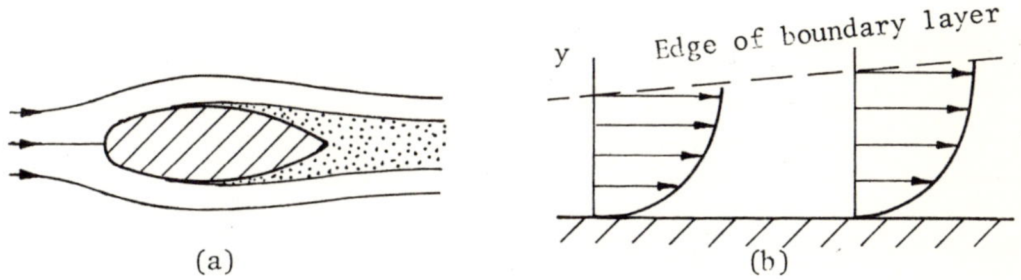


Figure 6.2

6.1.4 Separation

The wake is small for a streamlined body but a bluff body may produce a large wake. This wake region is not as orderly as that due to the retarded flow in the boundary layer but is irregular in character due to flow *separation* from the surface. Separation usually occurs in an adverse pressure gradient situation ($dp/dx > 0$) where the solid surface falls away from the preferred fluid flow direction. This is shown in figure 6.3 where the energy degredation in the boundary layer is accompanied by an increase in pressure. The fluid slows down and is eventually brought to rest at the separation point S after which there is some flow reversal. Orderly flow breaks down and is replaced by eddying. The streamline of discontinuity, A, sometimes becomes a *free streamline*, along which the pressure is constant.

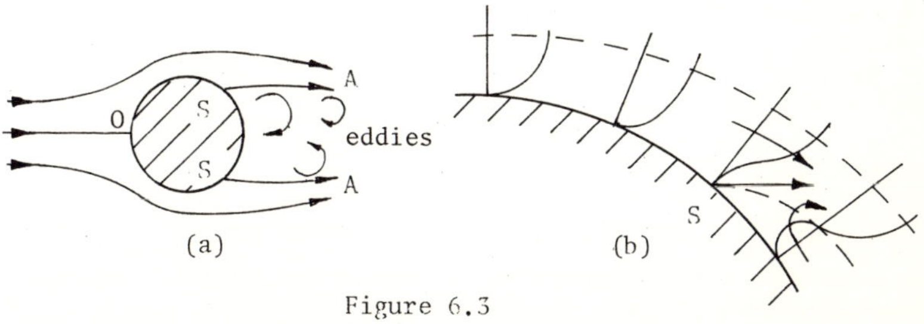


Figure 6.3

Boundary layer and separation effects also occur in duct flow.

6.1.5 Cavitation

In any flow it is important to remember that as the velocity increases the pressure decreases. This relationship is restricted for liquid flow because, if the pressure falls below the saturation pressure corresponding to the liquid temperature, the liquid will boil and pockets of vapour will form. This phenomenon is usually called *cavitation*. If the bubbles of vapour move to a high pressure region they collapse and produce high instantaneous forces. This can cause an efficiency drop and noise, vibration and mechanical damage. A special case of the free streamline, discussed in 6.1.4, can occur with liquid flow. If a cavitation zone exists behind the body then the constant pressure imposed is the vapour pressure.

## 6.2 INCOMPRESSIBLE FLOW AROUND BODIES

When relative motion of velocity, U, occurs between an immersed body and its surrounding fluid, a force, $\underline{F}$, is exerted on the body. This force may be resolved into two components (figure 6.4).

(a) A force component, $F_D$, parallel to the relative motion, known as the *drag*.

(b) A force component, $F_L$, perpendicular to the relative motion, known as the *lift*.

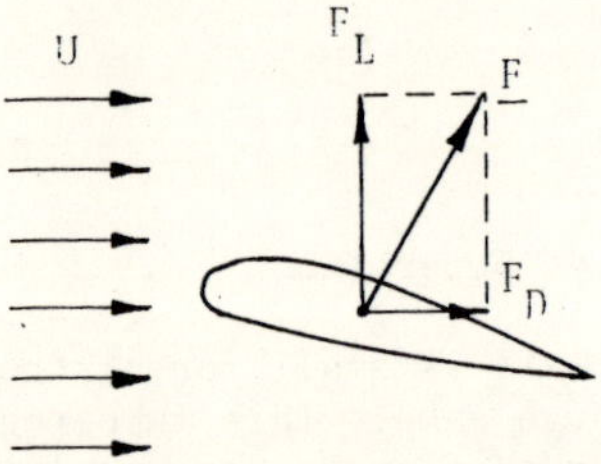

Figure 6.4

From the dimensional analysis of flow over a body (example 5.5) we obtain

$$F = \rho U^2 L^2 \phi(Re, M) \qquad (6.1)$$

If we restrict the analysis to incompressible flow, we may omit the Mach number effect and define a coefficient $C = \phi(Re)$. Replacing $L^2$ by a characteristic area, A, equation 6.1 may be applied to the lift and drag forces and written

$$F_D = \tfrac{1}{2}\rho U^2 C_D A \qquad (6.2)$$

and $$F_L = \tfrac{1}{2}\rho U^2 C_L A \qquad (6.3)$$

### 6.2.1 Drag

The drag on a body is due to the boundary layer effect and to flow separation.

Viscous or Friction drag, $(F_D)_f$, is due to tangential viscous forces acting on the body surface because of the boundary layer (figure 6.5).

$$(F_D)_f = \int_A \tau_o \, dA \sin\alpha \qquad (6.4)$$

Form or Pressure drag, $(F_D)_p$, is due to the normal pressure forces acting on the body surface. The pressure distribution is influenced by flow separation and wakes (figure 6.5).

$$(F_D)_p = \int_A p \, dA \cos \alpha \qquad (6.5)$$

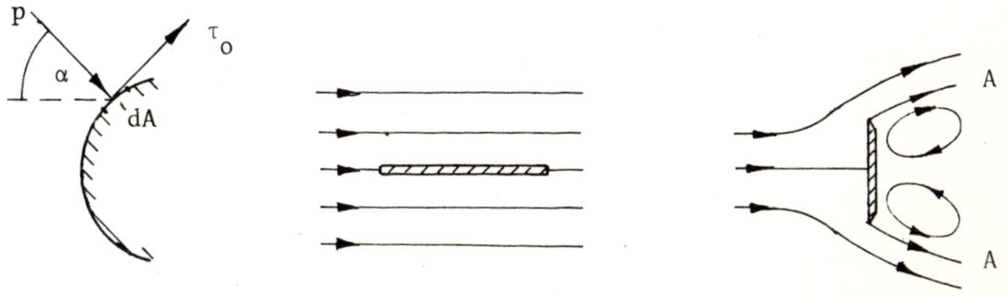


(a) (b) Friction drag only (c) Form drag only

Figure 6.5

Total or Profile drag, $F_D$, is the sum of the friction and form drag.

$$F_D = (F_D)_f + (F_D)_p \qquad (6.6)$$

6.2.2 Characteristics of Boundary Layers

The boundary layer thickness, $\delta$, is difficult to determine precisely because the velocity within it approaches the free stream velocity asymptotically. For convenience, $\delta$ is defined as that distance from the body surface where the velocity differs by 1% from the free stream velocity, U. i.e. $\delta$ is the value of y when u = 0.99U (figure 6.6a).

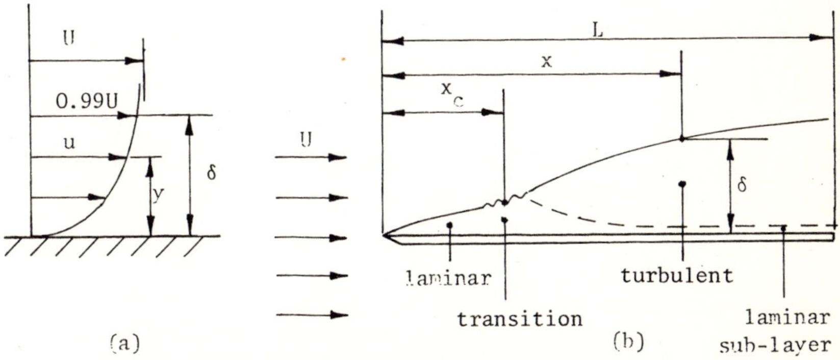


(a) (b)

Figure 6.6

The boundary layer on a flat surface in a parallel fluid stream (figure 6.6b) grows slowly from zero thickness at the leading edge and is initially laminar in character (even if the free stream is turbulent). There is then a flow region which is unstable in character and which marks a transition in the nature of the boundary layer. Downstream of the transition the boundary layer is turbulent in character but a thin laminar sub-layer exists next to the surface. The Reynolds number of flow, $Re_x$, is a variable with the characteristic length taken as the distance, x, from the leading edge. The critical $Re_x$ value for transition is usually taken as $5 \times 10^5$.

The laminar boundary layer may be analysed approximately by a momentum-integral method proposed by von Kármán. This involves the assumption of a particular velocity distribution and the calculation of the drag coefficient, $C_f$, and the boundary layer thickness, $\delta$. An exact mathematical analysis has also been performed by Blasius giving the following relationships.

$$C_f = \frac{1.328}{\sqrt{(Re_L)}} \tag{6.7}$$

$$\frac{\delta}{x} = \frac{4.91}{\sqrt{(Re_x)}} \qquad \text{(at } u = 0.99U\text{)} \tag{6.8}$$

The turbulent boundary layer cannot be analysed exactly and semi-empirical solutions must be used.

For $5 \times 10^5 < Re < 2 \times 10^7$ the velocity profile in the turbulent region is of the form $u/U = (y/\delta)^{1/7}$ and the following relationships hold.

$$\frac{\delta}{x} = \frac{0.377}{Re_x^{0.2}} \tag{6.9}$$

$$C_f = \frac{0.074}{Re_L^{0.2}} \tag{6.10}$$

For $Re > 2 \times 10^7$ the velocity distribution deviates from the 1/7 power law and the *Kármán-Schoenherr* equation then applies

$$\frac{1}{C_f} = 4.13 \log (Re_L \times C_f) \tag{6.11}$$

This formula is complex and an empirical interpolation formula by *Prandtl-Schlichting* may be used up to $Re = 10^9$.

$$C_f = \frac{0.455}{[\log Re_L]^{2.58}} \tag{6.12}$$

### 6.2.3 Separation and Form Drag

The existence of the boundary layer in an adverse pressure gradient leads to separation from bluff bodies and the formation of a wake. It is difficult to estimate the contribution of form drag alone, unless certain simplifying assumptions are made (see examples). In practice, it is more convenient to use $C_D$ - Re graphs, obtained by experiment, for bodies of specified shape. In order to reduce drag it is necessary to prevent boundary layer separation by streamlining or by boundary layer suction or blowing.

### 6.2.4 Lift

In fluid dynamics, we are particularly interested in shapes which yield high lift and low drag values e.g. aerofoils, hydrofoils and circular arcs. All these sections have in common the characteristic that the lift force increases as the angle of inclination (or angle of attack) to the fluid stream increases. There are, of course, limitations on the maximum angle in practice. The lift on an aerofoil may be explained in terms of the *Kutta-Joukowsky* theorem which demonstrates that a circulation, $\Gamma$, and a fluid velocity, U, are both necessary to produce a lift force ($F_L = \rho U\Gamma$). As the aerofoil moves from rest, the circulation is zero and the irrotational flow pattern tends to be set up (figure 6.7a). This pattern, which involves tortuous flow around the sharp trailing edge, cannot be maintained in a real fluid due to separation. Therefore the flow pattern of figure 6.7b forms. In the process, a circulation, $-\Gamma$, develops around the aerofoil and a *starting vortex*, of equal strength and opposite sign, is formed and then shed downstream (figure 6.7c). The circulation around the aerofoil and the lift are maintained.

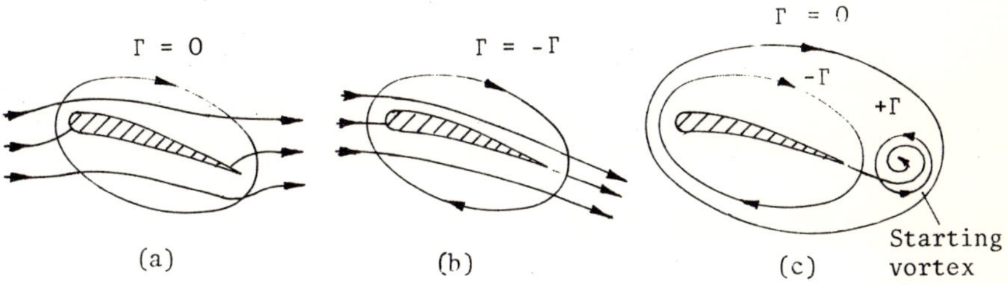


Figure 6.7

The same phenomenon causes the transverse vibration and 'singing' of telephone wires in a cross-wind. The alternate vortex shedding from the rear of the wire causes circulations of changing sign to be set up around the wire itself and this imposes an alternating transverse force. A rotating cylinder in cross-flow also experiences a lift force.

Lift may be explained in terms of the pressure distribution around the body. The crowding of streamlines on the upper surface of the aerofoil of figure 6.8 indicates an increase in velocity and a decrease in pressure. On the lower surface, the pressure increases. Thus a net upward force is exerted on the aerofoil. The pressure on the aerofoil may be expressed in terms of a pressure coefficient, $C_p$, defined

$$C_p = \frac{p - p_\infty}{\rho U^2/2}$$

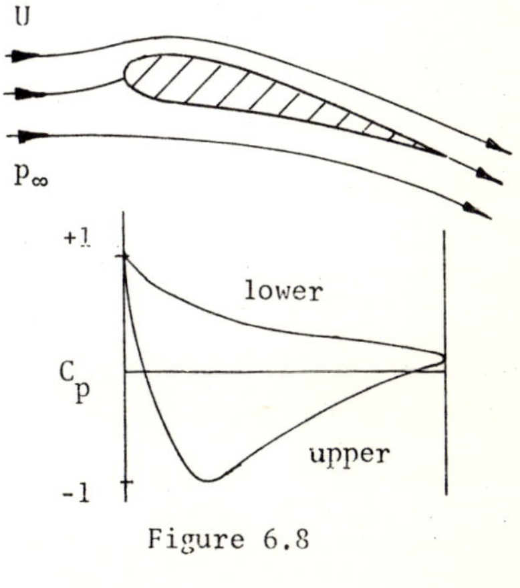


Figure 6.8

*Example 6.1*

A smooth rectangular plate, 600 mm wide by 25 m long, moves at 12 m/s in the direction of its length through oil. Determine (i) the length of the laminar boundary layer, (ii) the total drag force on the plate, and (iii) the thickness of the boundary layer at the trailing edge. Assume $\mu$ = 0.0128 Pa s , $\rho$ = 850 kg/m$^3$.

(i) Refer to figure 6.6b. Transition from a laminar boundary layer to a turbulent one occurs when $Re_x = 5 \times 10^5$. Therefore

$$x_c = \frac{\mu}{\rho U}(5 \times 10^5) = \frac{0.0128 \times 5 \times 10^5}{850 \times 12} = 0.625 \text{ m}$$

(ii) It is inaccurate to assume that the turbulent boundary layer grows from zero thickness at the transition point. In order to estimate the total drag force $F_D$ on a flat surface, the following expression should be used.

$$F_D = \begin{bmatrix} F_D \text{ for turbulent} \\ \text{B.L. on whole} \\ \text{plate} \end{bmatrix} - \begin{bmatrix} F_D \text{ for turbulent} \\ \text{B.L. up to } x_c \\ \phantom{x} \end{bmatrix} + \begin{bmatrix} F_D \text{ for laminar} \\ \text{B.L. up to } x_c \\ \phantom{x} \end{bmatrix}$$

In each case, equation 6.2 gives the drag force, $F_D = \frac{1}{2}\rho U^2 A C_f$. The value of $C_f$ is determined by the $Re_x$ value which is $5 \times 10^5$ for both laminar and turbulent layers up to $x_c$. At the end of the plate, the Reynolds number is

$$Re_L = \frac{\rho U L}{\mu} = \frac{850 \times 12 \times 25}{0.0128} = 1.99 \times 10^7$$

Therefore, equations 6.7 and 6.10 may be used to calculate the drag coefficients for the laminar and turbulent boundary layers, respectively.

$$F_D = [\tfrac{1}{2}\rho U^2 A_L \times 0.074\, Re_L^{-0.2}] - [\tfrac{1}{2}\rho U^2 A_x \times 0.074\, Re_x^{-0.2}] + [\tfrac{1}{2}\rho U^2 A_x \times 1.328\, Re_x^{-0.5}]$$

$$= \tfrac{1}{2} \times 850 \times 12^2 \left[\frac{0.6 \times 25 \times 0.074}{(1.99 \times 10^7)^{0.2}} - \frac{0.6 \times 0.625 \times 1.328}{(5 \times 10^5)^{0.5}} + \frac{0.6 \times 0.625 \times 1.328}{(5 \times 10^5)^{0.5}}\right] \times 2$$

$$= 4713 - 246 + 86 = 4.55 \text{ kN}$$

Note that if a wholly turbulent boundary layer is assumed, $F_D$ = 4.713 kN

(iii) Assuming a turbulent boundary layer on the whole plate, equation 6.9 gives

$$\delta = \frac{0.377L}{Re_L^{0.2}} = \frac{0.377 \times 25}{(1.99 \times 10^7)^{0.2}} = 327 \text{ mm}$$

*Example 6.2*

Using the momentum-integral method of boundary layer analysis, develop expressions for the drag coefficient, $C_f$, and the boundary layer thickness, $\delta$, for the laminar boundary layer on a flat plate assuming that the velocity profile is given by $u/U = [2(y/\delta) - (y/\delta)^2]$.

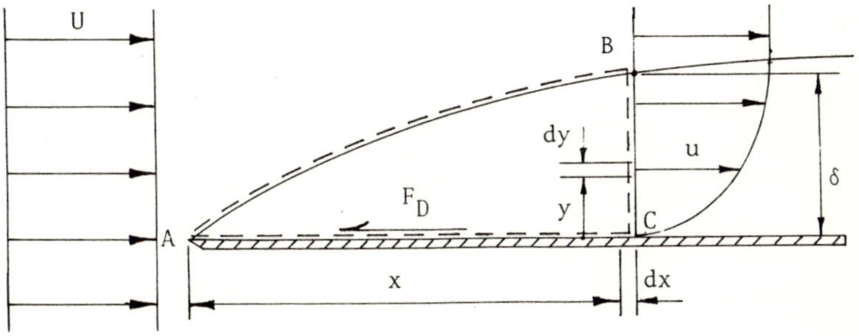


Figure 6.9

Consider a control volume ABC bounded by the edge of the boundary layer and the surface of the plate. For steady flow through the control volume, the momentum equation 4.18 reduces to

$$\Sigma \underline{F} = \int_S \underline{u}(\rho \underline{u} \cdot d\underline{A}) \qquad \text{(i)}$$

Assuming zero pressure gradient both inside and outside the boundary layer ($dp/dx = 0 = dp/dy$), the only force acting on the fluid within the control volume is the drag force $F_D$. Fluid enters the boundary layer, through surface AB, with main stream velocity, U, and flows through face BC with a velocity, u, which varies with distance, y, from the plate. Equation (i) gives

$$-F_D = \int_0^\delta u(\rho u \; dy) - \dot{m}U \qquad \text{(ii)}$$

Consider an elemental strip dy wide and of unit depth. The mass flow rate through the element is $\rho u \; dy$ and that through face BC (width $\delta$) is given by

$$\dot{m} = \int_0^\delta \rho u \; dy \qquad \text{(iii)}$$

From continuity, the mass flow rate is constant through the control volume. Substituting equation (iii) in (ii) we obtain

$$F_D = \int_0^\delta U(\rho u \; dy) - \int_0^\delta u(\rho u \; dy) = \int_0^\delta \rho u(U - u) \; dy \qquad \text{(iv)}$$

Substituting the given velocity distribution

$$F_D = \int_0^\delta \rho U[2(y/\delta) - (y/\delta)^2][U - U(y/\delta) + U(y/\delta)^2]dy$$

$$= \frac{2}{15} \rho U^2 \delta \qquad \text{(v)}$$

Differentiating

$$dF_D = \frac{2}{15}\rho U^2\, d\delta \qquad \text{(vi)}$$

The shear stress at any point in the laminar boundary layer is

$$\tau = \mu\frac{du}{dy} = \mu\frac{d}{dy}\left[U\frac{2y}{\delta} - U\left(\frac{y}{\delta}\right)^2\right] = \mu U\left(\frac{2}{\delta} - \frac{2y}{\delta^2}\right) \qquad \text{(vii)}$$

At the surface, $\tau = \tau_0$ when $y = 0$. Therefore

$$\tau_0 = \frac{2\mu U}{\delta} \qquad \text{(viii)}$$

The drag force on an element of surface dx wide and unit depth is

$$dF_D = \tau_0\, dx \qquad \text{(ix)}$$

Combining equations (vi), (viii) and (ix), rearranging and integrating, we obtain

$$\int_0^\delta \delta\, d\delta = \frac{15\mu}{\rho U}\int_0^x dx$$

and

$$\frac{\delta}{x} = \frac{5.48}{\sqrt{(Re_x)}} \qquad \text{(x)}$$

The general expression for drag force (equation 6.2) gives

$$F_D = \tfrac{1}{2}\rho U^2 A C_f$$

and

$$dF_D = \tfrac{1}{2}\rho U^2\, dx\, c_f \qquad \text{(xi)}$$

where $c_f$ is the *local* drag coefficient.

Combining equations (viii), (ix), (x) and (xi), we obtain

$$c_f = \frac{0.73}{\sqrt{(Re_x)}} \qquad \text{(xii)}$$

The total drag force on a plate of length L and area (L × 1) is

$$F_D = \tfrac{1}{2}\rho U^2 L C_f$$

where $C_f$ is the *mean* drag coefficient

$$C_f = \frac{F_D}{\tfrac{1}{2}\rho U^2 L} = \frac{\int_0^L \tau_0\, dx}{\tfrac{1}{2}\rho U^2 L}$$

Substituting equations (viii) and (x) we obtain

$$C_f = \int_0^L \frac{2\mu U\sqrt{(\rho U x/\mu)}\, dx}{5.48x} = \frac{1.46}{Re_L}$$

*Example 6.3*

(a) Define the term displacement thickness, $\delta^*$, used in boundary layer analysis.

(b) A honeycomb straightener is formed from flat strips of metal 150 mm long and 20 mm wide to give hexagonal passages of side length 20 mm. Water of kinematic viscosity $10^{-6}$ $m^2/s$ approaches the straightener with a velocity of 2 m/s. Neglecting the metal thickness and secondary flows, determine the pressure drop in the passage.

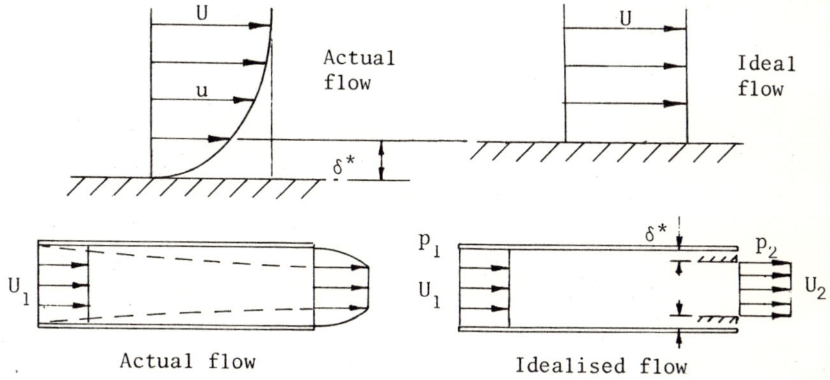


Figure 6.10

(a) The boundary layer thickness may be defined arbitrarily, as in 6.2.2, or in terms of its effect on the flow. The *displacement thickness*, δ*, is the distance the solid boundary would have to be displaced to give the same flow rate if the fluid were ideal i.e. the displacement thickness is the thickness of a stagnant layer with the same integrated velocity deficit as the actual boundary layer (figure 6.10).

$$U\delta^* = \int_0^\infty (U - u)dy$$

$$\delta^* = \int_0^\infty \left(1 - \frac{u}{U}\right)dy \qquad \text{(i)}$$

The displacement thickness at any distance x from the leading edge can be obtained by assuming a velocity distribution and integrating equation (i) as in example 6.2. For laminar flow ($Re < 5 \times 10^5$), an exact solution by Blasius gives

$$\delta^* = \frac{1.721x}{\surd(Re_x)} \qquad \text{(ii)}$$

(b) Assume the flow through the honeycomb to be the same as that over six individual flat plates of length L and width ℓ.

$$Re_L = \frac{UL}{\nu} = \frac{2 \times 0.15}{10^{-6}} = 3 \times 10^5$$

The boundary layer is laminar, therefore, equation (ii) applies.

$$\delta^* = \frac{1.721L}{\surd(Re_L)} = \frac{1.721 \times 0.15}{\surd(3 \times 10^5)} = 0.472 \text{ mm}$$

Now consider a flow situation in which fluid enters the honeycomb with uniform velocity, $U_1$, and leaves with ideal flow velocity, $U_2$, through an area reduced by the displacement thickness on each plate (figure 6.10).

$$\dot{V} = A_1U_1 = U_1 \times 2.6\ell^2$$

$$U_2 = \frac{\dot{V}}{A_2} = \frac{U_1 \times 2.6\ell^2}{(0.26\ell^2 - 6\delta^*\ell)} = \frac{2 \times 2.6 \times 0.02^2}{2.6 \times 0.02^2 - 6 \times 0.472 \times 10^{-3} \times 0.02}$$

$$= 2.12 \text{ m/s}$$

In the idealised flow situation assumed, we may apply the Bernoulli equation 4.7 between entry and exit of the honeycomb.

$$\frac{p_1}{\rho} + \tfrac{1}{2}U_1^2 = \frac{p_2}{\rho} + \tfrac{1}{2}U_2^2$$

$$p_1 - p_2 = \tfrac{1}{2}\rho(U_2^2 - U_1^2) = \frac{10^3}{2}(2.12^2 - 2^2) = 247 \text{ Pa}$$

*Example 6.4*

(a) Describe the flow pattern for flow normal to an infinite circular cylinder as the flow velocity is increased from zero to a high velocity.
(b) A screen in a 10 m wide stream, 2.5 m deep, consists of a number of vertical 25 mm diameter bars, spaced 0.1 m between centres. If the upstream water velocity is 2 m/s, estimate the drag force on the screen and the pressure loss coefficient. How do the screen conditions differ from those for infinite cylinders?
$\nu = 1.2 \times 10^{-6}$ m$^2$/s, $\rho = 1000$ kg/m$^3$.

(a) Refer to figure 6.11 which shows the flow patterns for different values of Reynolds number.

(i) For Re < 1, inertia forces are negligible and the streamlines close behind the cylinders giving a flow pattern which is similar to that for ideal flow.

(ii) For Re = 2-30, the boundary layer separates symmetrically at points S and two eddies are formed. For increasing Re, the eddies elongate.

(iii) For Re ≈ 40-70, a periodic oscillation of the wake occurs and for Re > 90 the eddies break away alternately from the cylinder and are washed downstream. (This causes vibration and 'singing' of telegraph wires in a cross-wind.)

(iv) For $250 < Re < 2 \times 10^5$, the frequency of vortex shedding increases until ultimately, at high Re values, the regular vortex street breaks down into random turbulence at the rear of the cylinder. At this stage, the boundary layer up to the point of separation is essentially laminar and this gives a wide wake and high form drag.

(v) For Re $\simeq 2 \times 10^5$, the boundary layer becomes turbulent and the separation point moves further round the cylinder surface giving a narrow wake and low form drag. Hence, $C_D$ drops suddenly in this region. Note that the turbulent boundary layer can be triggered off for lower Re values if the surface is roughened e.g. by sand particles, dimples or trip wires.

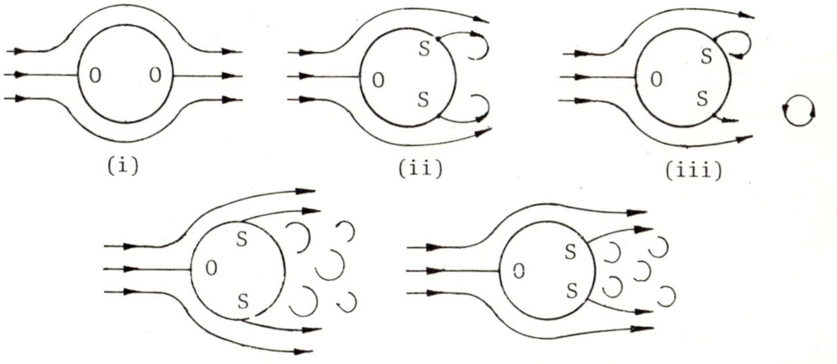


Figure 6.11

(b) The variation of drag coefficient, $C_D$, as a function of Re may be plotted on a graph similar to that for a sphere (figure 6.12). The graph refers to an infinite cylinder.

$$Re = \frac{UD}{\nu} = \frac{2 \times 0.025}{1.2 \times 10^{-6}} = 4.17 \times 10^4$$

From figure 6.12, $C_D = 1.35$. Note that the boundary layer is laminar. From equation 6.2 the total drag on the bars is

$$(F_D)_{total} = \tfrac{1}{2}\rho U^2 C_D AN = \tfrac{1}{2} \times 10^3 \times 2^2 \times 1.35 \times 2.5 \times 0.025 \times 100$$

$$= 16.88 \text{ kN}$$

The pressure loss coefficient, $(C_p)_L$, is defined

$$(C_p)_L = \frac{\Delta p}{\frac{1}{2}\rho U^2} = \frac{F_D/A}{\frac{1}{2}\rho U^2} = \frac{16.88 \times 10^3}{\frac{1}{2} \times 10^3 \times 2^2 \times 10 \times 2.5} = 0.338$$

The $C_D$ values given in figure 6.12 are those for an infinite cylinder. The actual flow differs because of end effects on the cylinder. At the junction with the stream bed there will be a non-uniform velocity distribution due to the boundary layer in the water near the bed. At the water free surface, wave effects may be present.

*Example 6.5*

(a) Sketch a graph of the variation in drag coefficient with Reynolds number for flow normal to a sphere. With reference to the nature of flow around the sphere, explain the reason for the sudden drop in $C_D$

when $Re \simeq 2 \times 10^5$.

(b) A spherical balloon of mass 0.82 kg and diameter 2 m ascends at a constant speed of 10 m/s through still air. Determine the drag coefficient. If the balloon is now tethered to the ground in a 20 m/s horizontal wind, determine the tension and angle of the retaining cable. $\mu_a = 1.8 \times 10^{-5}$ Pa s, $\rho_a = 1.25$ kg/m$^3$.

(a) Figure 6.12 gives the $C_D$-Re graph for a sphere. The change in flow pattern with increasing Re values is similar to that for flow normal to an infinite cylinder described in example 6.4. Initially, the curve follows Stokes' law $C_D = 24/Re$ for laminar flow. Then separation starts, followed by vortex shedding. For $Re > 3 \times 10^3$ a laminar boundary layer exists on the front surface of the sphere and $C_D \simeq$ const. For $Re \simeq 2 \times 10^5$, the laminar boundary layer changes to a turbulent one and the width of the wake reduces, causing a fall in $C_D$.

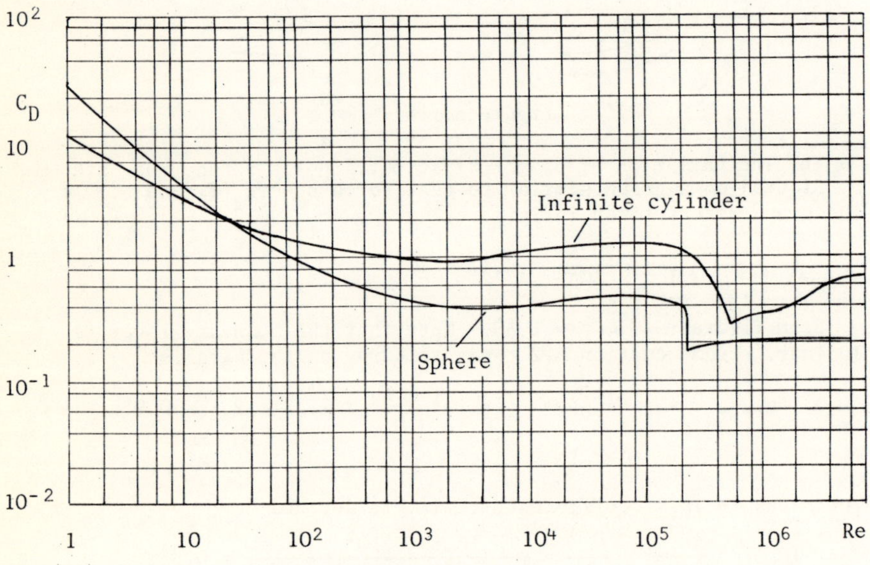


Figure 6.12

(b) The forces exerted on the balloon are the gravity force, $F_g$, the buoyancy force, $F_b$, and the drag force $F_D$.

$$F_g = mg = 0.82 \times 9.81 = 8.04 \text{ N}$$

$$F_b = \rho g \frac{4}{3} \pi R^3 = 1.25 \times 9.81 \times \frac{4}{3} \times \pi \times 1^3 = 51.4 \text{ N}$$

For ascent at constant velocity, $\Sigma F = 0$. Therefore

$$F_b - F_g - F_D = 0$$

$$F_D = 51.4 - 8.04 = 43.3 \text{ N}$$

From equation 6.2

$$F_D = \tfrac{1}{2}\rho U^2 C_D A$$

$$C_D = \frac{2F_D}{\rho U^2 A} = \frac{2 \times 43.3}{1.25 \times 10^2 \times \pi \times 1^2} = 0.22$$

The $C_D$ value can also be obtained from the $C_D$-Re graph (figure 6.12)

$$Re = \frac{\rho UD}{\mu} = \frac{1.25 \times 10 \times 2}{1.8 \times 10^{-5}} = 1.39 \times 10^6$$

At $Re = 1.39 \times 10^6$, $C_D = 0.22$. The boundary layer is turbulent.

When the balloon is tethered, $U_2 = 20$ m/s.

$$Re_2 = \frac{1.25 \times 20 \times 2}{1.8 \times 10^{-5}} = 2.78 \times 10^6. \quad C_{D_2} = 0.22$$

$$F_{D_2} = \tfrac{1}{2}\rho U_2{}^2 C_{D_2} A = \tfrac{1}{2} \times 1.25 \times 20^2 \times 0.22 \times \pi \times 1^2 = 172.8 \text{ N}$$

The cable tension is equal to the resultant force on the balloon (figure 6.13).

$$F_t = \sqrt{[(F_b - F_g)^2 + F_D{}^2]} = \sqrt{(43.3^2 + 172.8^2)} = 178 \text{ N}$$

$$\theta = \tan^{-1}[(F_b - F_g)/F_D] = \tan^{-1}(43.3/172.8) = 14.1°$$

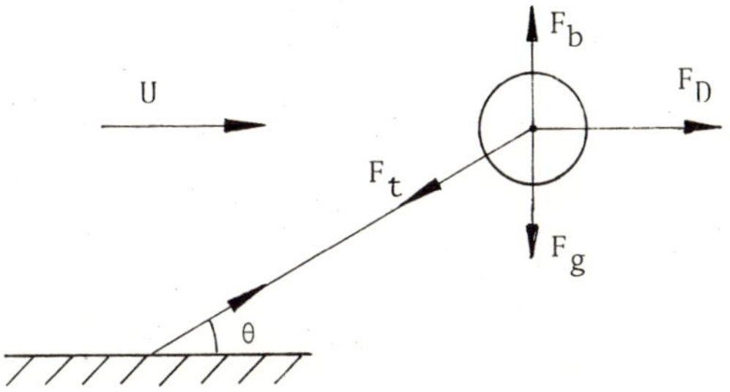


Figure 6.13

*Example 6.6*

During a test on a 150 mm chord wing section in a wind tunnel the velocity distribution across the wake of width $2Y_2$ is represented by

$$u = U[0.75 - 0.25 \cos(\pi y/Y_2)]$$

where u is the velocity at any point y from the centre line of the wake and U is the free stream velocity.

If the wake is 200 mm wide and the free stream velocity is 40 m/s determine the drag on the wing section per metre of span and the drag coefficient. Assume the pressure in the wake to be equal to the free stream pressure and the air density constant at 1.2 kg/m³.

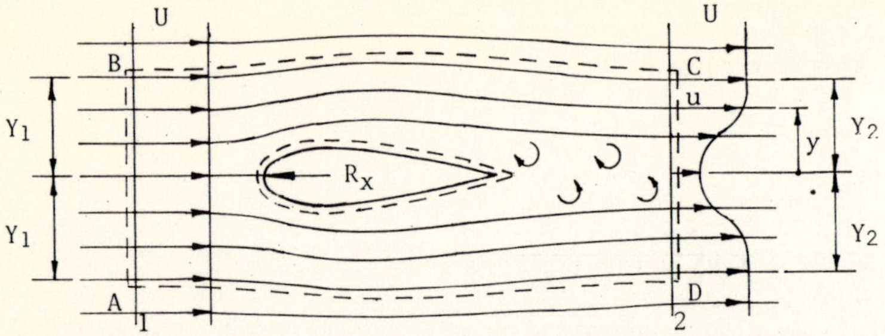


Figure 6.14

Consider a control volume ABCD, sufficiently far away from the wing section for transverse velocity gradients to be negligible. For steady flow, the momentum equation 4.18 reduces to

$$\Sigma \underline{F} = \int_S \underline{u}(\rho \underline{u} \cdot d\underline{A}) \tag{i}$$

The fluid pressure is uniform on the control surface and the only force acting on the fluid within the control volume is $R_x$. The flow is restricted to single entry with uniform velocity U and a single exit with variable velocity u. Considering unit length of wing, equation (i) gives

$$-R_x = 2\int_0^{Y_2} \rho u^2 \, dy - 2\rho U^2 Y_1 \tag{ii}$$

The flow rate through the control volume is

$$\dot{V} = 2UY_1 = \int_0^{Y_2} u \, dy \tag{iii}$$

Substituting equation (iii) and the given velocity distribution in (ii) we obtain

$$R_x = 2\rho\int_0^{Y_2} u(U - u)dy$$

$$= 2\rho\int_0^{Y_2}\Big[U[0.75 - 0.25\cos(\pi y/Y_2)] \times [U - 0.75U + 0.25U\cos(\pi y/Y_2)]dy\Big]$$

$$= 0.125\rho U^2 \times 2.5Y_2 = 0.125 \times 1.2 \times 40^2 \times 2.5 \times 0.1 = 60 \text{ N}$$

The force exerted by the fluid on the wing section is equal and opposite to $R_x$ i.e. 60 N to the right.

From equation 6.2

$$F_D = \tfrac{1}{2}\rho U^2 A C_D = -R_x \tag{v}$$

From equations (iv) and (v)

$$\tfrac{1}{2}\rho U^2 A C_D = 0.125\rho U^2 \times 2.5Y_2$$

$$C_D = \frac{2 \times 0.125 \times 2.5 \times 0.1}{0.15 \times 1} = 0.416$$

*Example 6.7*

A fluid of constant density, $\rho$, flows with velocity, U, over a sphere of radius, R. The magnitude of the velocity, $u_q$, at a point, q, just outside the boundary layer, is given by $u_q = 1.5U \sin\theta$ for $0 < \theta < \theta_1$ where the angle $\theta$ is defined by the radius to point q and that to the front stagnation point. For $\theta_1 < \theta < \pi$ the pressure on the surface of the sphere is constant and equal to that at $\theta = \theta_1$. Determine the drag coefficient when $\theta_1 = 90°$.

In this problem, the velocity, $u_q$, applies to a point just outside the boundary layer. Therefore, there is no degradation of energy between a point upstream and point q and the Bernoulli equation may be applied.

$$p + \tfrac{1}{2}\rho U^2 = p_q + \tfrac{1}{2}\rho u_q^2$$

The pressure at point, q, taking the upstream pressure, p, as reference and substituting $u_q = 1.5U \sin\theta$, is given by

$$p_q - p = \tfrac{1}{2}\rho U^2 - \tfrac{1}{2}\rho(1.5U \sin\theta)^2 = \tfrac{1}{2}\rho U^2(1 - 2.25 \sin^2\theta) \quad \text{(i)}$$

Now consider the pressure force, in the direction of flow, exerted on an element of area dA subtended by a cone of semi-angle $\theta$ (figure 6.15).

$$dA = (2\pi R \sin\theta)R\,d\theta$$

The pressure force on the element of area is given by

$$\begin{aligned} dF &= (p_q - p)\cos\theta\,dA \\ &= \tfrac{1}{2}\rho U^2(1 - 2.25\sin^2\theta)\cos\theta\,2\pi R^2 \sin\theta\,d\theta \\ &= \tfrac{1}{2}\rho U^2 2\pi R^2(\sin\theta\cos\theta - 2.25\sin^3\theta\cos\theta)\,d\theta \\ &= \tfrac{1}{2}\rho U^2 2\pi R^2(\sin\theta - 2.25\sin^3\theta)\,d(\sin\theta) \end{aligned}$$

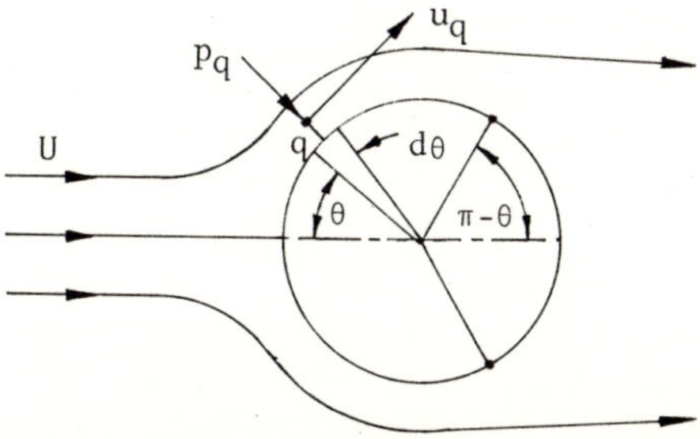


Figure 6.15

The total force on the upstream face, defined by $0 < \theta < \theta_1$, is

$$\begin{aligned} F_1 &= \tfrac{1}{2}\rho U^2 2\pi R^2 \int_0^{\sin\theta_1} (\sin\theta - 2.25\sin^3\theta)\,d(\sin\theta) \\ &= \tfrac{1}{2}\rho U^2 \pi R^2(\sin^2\theta_1 - 1.125\sin^4\theta_1) \end{aligned}$$

For $\theta > \theta_1$ the pressure on the surface of the sphere is constant at the value for $\theta = \theta_1$. From equation (i)

$$p_q - p = \tfrac{1}{2}\rho U^2(1 - 2.25 \sin^2 \theta_1)$$

The pressure force on an element of area for $\theta > \theta_1$ is

$$dF = \tfrac{1}{2}\rho U^2(1 - 2.25 \sin^2 \theta_1)\cos \theta \; 2\pi R^2 \sin \theta \, d\theta$$

$$= \tfrac{1}{2}\rho U^2 2\pi R^2(\sin \theta - 2.25 \sin^2 \theta_1 \sin \theta) \, d(\sin \theta)$$

The total force on the downstream face defined by $\theta_1 < \theta < \pi$ is

$$F_2 = \tfrac{1}{2}\rho U^2 2\pi R^2 \int_{\sin \theta_1}^{0} (\sin \theta - 2.25 \sin^2 \theta_1 \sin \theta) \, d(\sin \theta)$$

$$= \tfrac{1}{2}\rho U^2 \pi R^2(-\sin^2 \theta_1 + 2.25 \sin^4 \theta_1)$$

The net force on the sphere is given by

$$F_D = F_1 + F_2$$

$$= \tfrac{1}{2}\rho U^2 \pi R^2(\sin^2 \theta_1 - 1.125 \sin^4 \theta_1 - \sin^2 \theta_1 + 2.25 \sin^4 \theta_1)$$

$$= \tfrac{1}{2}\rho U^2 \pi R^2(1.125 \sin^4 \theta_1) \qquad \text{(ii)}$$

From equations 6.2 and (ii)

$$F_D = \tfrac{1}{2}\rho U^2 C_D A = \tfrac{1}{2}\rho U^2 \pi R^2(1.125 \sin^4 \theta_1)$$

Substituting $A = \pi R^2$ and the given value $\theta_1 = 90°$

$$C_D = 1.125 \sin^4(90) = 1.125$$

*Example 6.8*

(a) Sketch typical curves of lift coefficient and lift-drag for various values of angle of attack for an aerofoil section. Indicate on the curves the point of best efficiency and the stall point. Also give the reason for the fall-off of lift beyond the stall point.

(b) An aeroplane weighing 100 kN has a wing area of 45 $m^2$ and a drag coefficient (based on wing area) $C_D = 0.03 + 0.04 C_L^2$. Taking the density of air to be 1.2 kg/$m^3$ calculate for horizontal flight the speed and power when minimum power is required.

(a) Figure 6.16 gives typical performance curves. As the angle of attack increases, the circulation round the aerofoil and the lift force increase. The adverse pressure gradient near the tail on the upper surface also increases and separation occurs from the upper surface. As the angle of attack is increased further the separation point moves forward and the wake becomes wider with an increase in drag. The value of $\alpha$ at which the maximum value of lift coefficient $C_L$ occurs is known as the stall point and the flow separates from practically the whole of the upper surface.

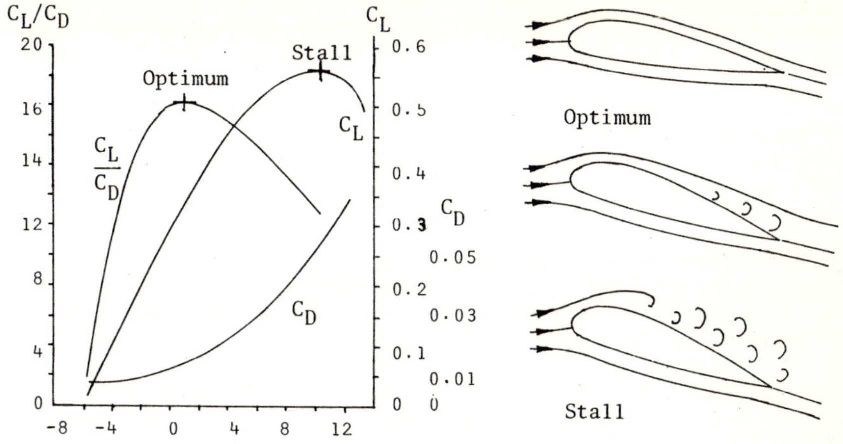


Figure 6.16

(b) From equation 6.3, the lift force, $F_L$, is given by

$$F_L = \tfrac{1}{2}\rho U^2 C_L A$$

$$C_L = \frac{F_L}{\tfrac{1}{2}\rho U^2 A} = \frac{2 \times 100 \times 10^3}{1.2U^2 \times 45} = \frac{3.7 \times 10^3}{U^2}$$

Now $C_D = 0.03 + 0.04C_L{}^2 = 0.03 + \dfrac{5.48 \times 10^5}{U^4}$ (i)

From equation 6.2, the drag force, $F_D$, is

$$F_D = \tfrac{1}{2}\rho U^2 C_D A = \tfrac{1}{2}\rho U^2 A(0.03 + 5.48 \times 10^5 \times U^{-4})$$

The power, P, is given by

$$P = F_D U = \tfrac{1}{2} \times 1.2 \times U^3 \times 45(0.03 + 5.48 \times 10^5 \times U^{-4})$$

$$= 0.81U^3 + 1.48 \times 10^7 U^{-1} \qquad \text{(ii)}$$

For minimum power $dP/dU = 0$

$$\frac{dP}{dU} = 3 \times 0.81U^2 - 1.48 \times 10^7 U^{-2} \qquad \text{(iii)}$$

$$\frac{d^2P}{dU^2} = 6 \times 0.81U + 2 \times 1.48 \times 10^7 U^{-3} \qquad \text{(iv)}$$

Equation (iv) is positive for all values of U, therefore $dP/dU = 0$ for minimum power. From equation (iii)

$$3 \times 0.81U^2 - 1.48 \times 10^7 U^{-2} = 0$$

$$U = 49.7 \text{ m/s}$$

From equation (ii)

$$P = 0.81 \times 49.7^3 + 1.48 \times 10^7 \times 49.7^{-1} = 397 \text{ kW}$$

*Example 6.9*

A racing car is fitted with an inverted aerofoil inclined at 10° to the horizontal. At this angle $C_D = 0.35$ and $C_L = 1.4$. The length of the aerofoil is 1.5 m and the chord is 1 m. The car length and body surface area are 4.5 m and 12 m$^2$ respectively, and the skin friction coefficient is given by $C_f = 0.0741/Re_L^{0.2}$ where $Re_L$ is based on car length. The car weight is 12.5 kN and the rolling resistance is 40 N per kN of normal force between the tyres and the road surface. Assuming that the form drag on the car is 450 N, determine the power required to maintain a constant velocity of 50 m/s. Assume ambient conditions of 10 kPa and 285 K. $R = 287$ J/kg K, $\mu = 17.5 \times 10^{-5}$ N s/m$^2$.

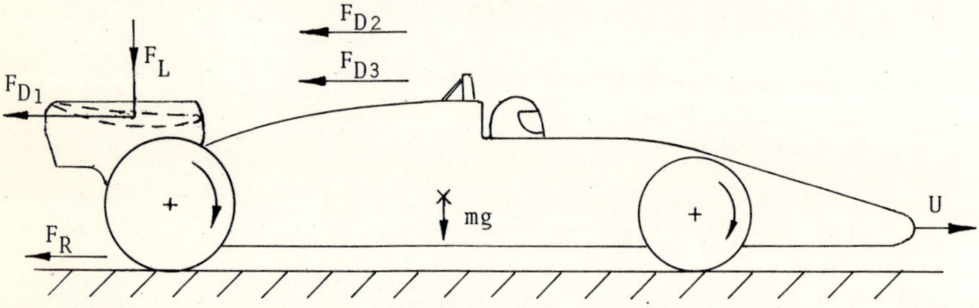


Figure 6.17

The density of the air is given by equation 1.25c

$$\rho = \frac{p}{RT} = \frac{10^5}{287 \times 285} = 1.22 \text{ kg/m}^3$$

The drag and lift forces on the aerofoil are calculated from equations 6.2 and 6.3

$$F_{D1} = \tfrac{1}{2}\rho U^2 A_A C_{D1} = 0.5 \times 1.22 \times 50^2 \times 1.5 \times 1 \times 0.35 = 802.5 \text{ N}$$

$$F_L = \tfrac{1}{2}\rho U^2 A_A C_L = 0.5 \times 1.22 \times 50^2 \times 1.5 \times 1 \times 1.40 = 3203 \text{ N}$$

The drag force on the car body due to skin friction is given by

$$F_{D2} = \tfrac{1}{2}\rho U^2 A_B C_{D2} = \tfrac{1}{2}\rho U^2 A_B \times 0.0741\ Re_L^{-0.2}$$

$$= \tfrac{1}{2}\rho U^2 A_B \times 0.0741\ (\rho UL/\mu)^{-0.2}$$

$$= \frac{1.22 \times 50^2 \times 12 \times 0.0741}{2[1.22 \times 50 \times 4.5/(17.5 \times 10^{-5})]^{0.2}} = 78 \text{ N}$$

The form drag is given as

$$F_{D_3} = 450 \text{ N}$$

Now the total normal force $F_N$ between the car wheels and the road surface is equal to the sum of the car weight and the downward 'lift' force of the aerofoil

$$F_N = F_L + mg = 3203 + 12500 = 15.71 \text{ kN}$$

The rolling resistance is given by

$$F_R = 40F_N = 40 \times 15.71 = 628.4 \text{ N}$$

The total resistance to motion is

$$F = F_{D_1} + F_{D_2} + F_{D_3} + F_R = 802.5 + 78 + 450 + 628.4 = 1959 \text{ N}$$

The power required to maintain a velocity of 50 m/s is

$$P = FU = 1959 \times 50 = 98 \text{ kW}$$

*Example 6.10*

(a) A surface vessel, fitted with an underwater hydrofoil, displaces 50 $m^3$ of sea water when stationary. When the vessel is in motion the hull partly rises out of the water due to lift on the hydrofoil. Determine for a vessel speed of 10 m/s (i) the volume displaced by the vessel, and (ii) the power (MW) required to propel the vessel if the hull drag is twice that of the hydrofoil. Hydrofoil details: length 3 m, chord 2 m, $C_L = 1.2$, $C_D = 0.2$, $\rho_{sw} = 1025$ kg/$m^3$.

(b) The peak velocity of flow over the bracket supporting the hydrofoil is 1.5 times the velocity remote from the arm. If the peak velocity occurs at a point 3 m below the water surface, determine the vessel speed at which cavitation can be expected to occur at a water temperature of 10 °C. Assume the sea water to have the properties of fresh water. $p_a$ = 100 kPa.

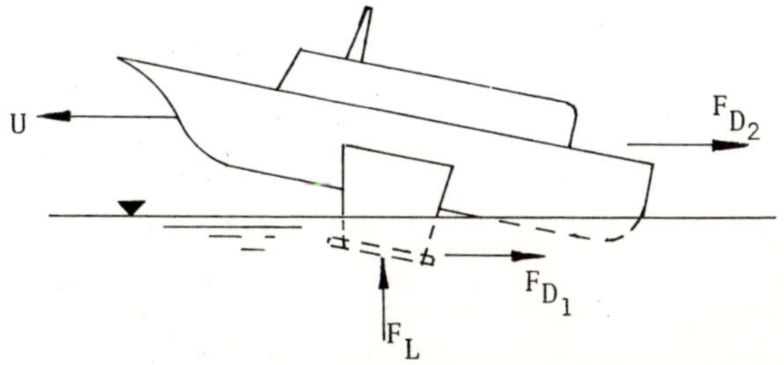


Figure 6.18

(a)(i) The lift force on the hydrofoil is given by equation 6.3

$$F_L = \tfrac{1}{2}\rho U^2 C_L A = \tfrac{1}{2} \times 1025 \times 10^2 \times 3 \times 2 \times 1.2 = 370 \text{ kN}$$

The volume of water, $V_1$, displaced by the stationary vessel is

50 $m^3$. This corresponds to a weight, mg, given by

$$mg = 50 \times 1025 \times 9.81 = 502.8 \text{ kN}$$

(ii) When the vessel is moving, the net downward force, F, is

$$F = mg - F_L = (502.8 - 370)10^3 = 132.8 \text{ kN}$$

The volume displaced by the moving boat, $V_2$, is given by

$$V_2 = \frac{mg - F_L}{\rho g} = \frac{132.8 \times 10^3}{1025 \times 9.81} = 13.2 \text{ m}^3$$

The drag force on the hydrofoil is given by equation 6.2

$$F_{D1} = \tfrac{1}{2}\rho U^2 C_D A = \tfrac{1}{2} \times 1025 \times 10^2 \times 3 \times 2 \times 0.2 = 61.5 \text{ kN}$$

The total drag on the vessel is

$$F_{D_t} = 3F_{D_1} = (F_{D_1} + F_{D_2})$$

The power required to propel the vessel is

$$P = F_{D_t} U = 3 \times 61.5 \times 10^3 \times 10 = 1.845 \text{ MW}$$

(b) Assume that the peak velocity occurs at a point, q, and that there is no degradation of energy between a point upstream and point q. Applying the Bernoulli equation between the two points we have

$$p + \tfrac{1}{2}\rho U^2 = p_q + \tfrac{1}{2}\rho u_q^2$$

Now at depth, h, the pressure $p = p_a + \rho gh$. The saturation pressure at 10 °C is $p_s$ = 1.227 kPa. Cavitation will occur when $p_q = p_s$. Also, we are told $u_q$ = 1.5U.

$$(p_a + \rho gh) + \tfrac{1}{2}\rho U^2 = p_s + \tfrac{1}{2}\rho (1.5U)^2$$

$$10^5 + 1025 \times 9.81 \times 3 + \tfrac{1}{2} \times 1025 \times U^2$$

$$= 1.227 \times 10^3 + \tfrac{1}{2} \times 1025 \times 2.25U^2$$

$$U = 14.3 \text{ m/s}$$

*Problems*

1 Find the ratio of the drag in water to the drag in air on a thin flat plate of length 1.2 m at zero incidence in a stream flowing at 5 m/s. The density and kinematic viscosity are to be taken as 1000 kg/$m^3$ and $1.1 \times 10^{-6}$ $m^2$/s for water, 1.2 kg/$m^3$ and $14.9 \times 10^{-6}$ $m^2$/s for air.

[1200:1]

2 Contrast the development cf a boundary layer on a flat plate with that inside a circular pipe.

A streamlined train is 110 m long, 2.75 m wide and has sides 2.75 m high. Assuming that the skin friction drag on sides and top is equivalent to that on a flat plate 8.25 m wide, calculate the power required to overcome the skin friction when the train moves at 160 km/h through air of density 1.2 $kg/m^3$ and viscosity $1.79 \times 10^{-5}$ N $s/m^2$. Use may be made of any of the following formulae. In the laminar boundary layer, the total force F between the boundary layer and the plate, per unit width of plate is $F = 2\rho U^2\delta/15$, where $\delta$ is the boundary layer thickness at distance x from the leading edge of the plate and U is the free stream velocity. The boundary layer thickness is given approximately by $\delta = 5.48x/Re_x^{0.5}$. In the turbulent boundary layer, $F = 7\rho U^2\delta/72$, and $\delta = 0.37x/Re_x^{0.2}$.

[68.4 kW]

3 Develop expressions for the coefficient of drag $C_D$ and the boundary layer thickness for the laminar boundary layer on a flat plate assuming that the velocity profile is (a) of the form $u/U = y/\delta$, and (b) of the form $u/U = \sin(\pi y/2\delta)$.

[$1.155Re^{-1/2}$, $3.464xRe_x^{-1/2}$, $1.31Re^{-1/2}$, $4.789xRe_x^{-1/2}$]

4 An open-return wind tunnel is fitted with a honeycomb straightener and a wire mesh screen at a station where the velocity is 5 m/s. The honeycomb straightener is formed from flat strips of metal 50 mm long and 10 mm wide to give hexagonal passages of side length 10 mm. The mesh is formed from vertical and horizontal interwoven wires 0.4 mm diameter with 800 wires per metre. Determine the pressure loss coefficient $\Delta p/\frac{1}{2}\rho v^2$ for (i) the honeycomb, and (ii) the wire mesh screen.

It may be assumed that for flow over a flat surface laminar flow exists when $Re_x < 5 \times 10^5$ for which $\delta^* = 1.721x/(Re)^{\frac{1}{2}}$. The area of a hexagon is $2.6L^2$ where L is the side length. The drag coefficient for flow perpendicular to a cylindrical wire may be assumed equal to that for flow over an infinite cylinder for which a $C_D$-Re graph is given (figure 6.12). $\rho_{air} = 1.2$ $kg/m^3$; $\mu_{air} = 17.7 \times 10^{-6}$ N $s/m^2$.

[0.49, 0.83]

5 (a) Describe briefly the nature of the flow about a sphere at (i) very low and (ii) very high values of Reynolds number.

(b) At very low Reynolds number the drag of a sphere of diameter D in steady motion is given by Stokes' law. $C_D = 24/Re$. A small spherical water droplet is observed to fall through the atmosphere at a constant speed of 50 mm/s. Calculate the diameter of the droplet.

(c) A 5 mm diameter spherical air bubble rises at constant velocity through water. Determine the bubble velocity. Assume $\rho_a = 1.2$ $kg/m^3$, $\rho_w = 1000$ $kg/m^3$, $\mu_a = 1.82 \times 10^{-5}$ Pa s , $\mu_w = 11.2 \times 10^{-5}$ Pa s .

[0.02 mm, 0.372 m/s]

6 A cylinder of diameter 50 mm is placed with its axis normal to an air stream of velocity 50 m/s. It is found that at a point where the width of the wake is equal to three cylinder diameters the velocity distribution is given by

$$u = U\left(0.8 - 0.2\cos\frac{2\pi y}{3D}\right)$$

where U is the free stream velocity and u is the velocity at any point distance y from a line through the cylinder centre. Assuming that the pressure in the wake is equal to the free stream pressure, determine the drag coefficient for the cylinder. If the cylinder is enclosed in a wind tunnel what effect would this have on the flow and the method of analysis?

[0.84]

7 A circular flat plate, radius R, is held in a uniform fluid stream, velocity U. The plane of the plate is perpendicular to the direction of the flow. The velocity just outside the boundary layer on the upstream face of the plate is given by $u = 1.34\ U(r/R)^2$ where r is the distance from the centre of the plate. Assuming that the pressure on the downstream face of the plate is uniform and equal to that at the edge of the upstream face, show that the drag coefficient for the plate, based on frontal area is 1.2.

8 A long, horizontal cylinder of semi-circular cross-section is mounted with its plane face vertical and its curved surface facing an airstream of 30 m/s which is normal to the plane face. The velocity $u_Q$ of the air immediately outside the boundary layer is given by $u_Q = 1.5U \sin\theta$ where U is the velocity of the undisturbed airstream and $\theta$ is the angle between the radius and the original flow direction. If the cylinder diameter is 100 mm, determine the resultant force acting on unit length of the curved surface. $\rho = 1.2$ kg/m$^3$. [13.5 N]

9 (a) The superposition of circulation of strength $\Gamma$ on uniform flow of U normal to a circular cylinder of radius, R, gives the velocity, $u_\theta$, at a point on the cylinder surface as $u_\theta = -2U \sin\theta - \Gamma/2\pi R$ (see example 3.8). Show that the lift force per unit length of cylinder is $\rho U\Gamma$.

(b) A Flettner ship is propelled by means of two rotating cylinders, each 2 m diameter and 5 m high (figure 6.19). The profile drag coefficient of the hull, $C_D$, is $3.5 \times 10^{-3}$ based on a wetted surface area, A, of 100 m$^2$. The drag of each rotor is equal to $1/\sqrt{3}$ of the 'lift' force produced by it. The velocity of the wind relative to earth is 12 m/s. Assuming that the wind velocity relative to the ship is due north, determine the angular speed of the rotors, in rev/min, required to propel the ship at a linear speed of 6 m/s, in the direction of the resultant force on the rotors.

[87.5 rev/min]

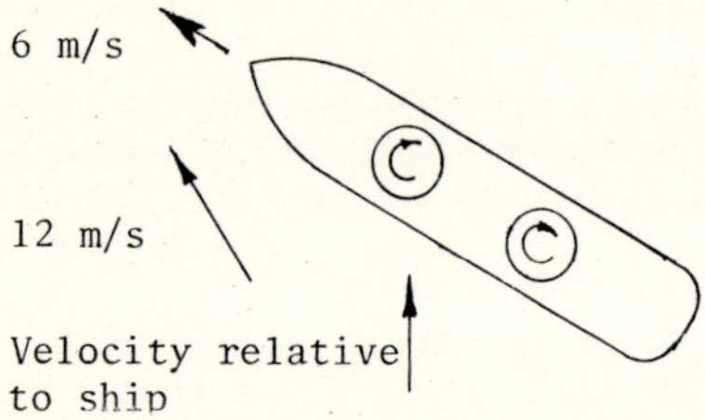


Figure 6.19

10 Wind tunnel tests on an aerofoil show that for small angles of incidence, $\alpha$, the lift coefficient, $C_L$, is equal to $5.5\alpha$ where $\alpha$ is measured in radians. A geometrically similar aerofoil having an

area 40 $m^2$ is required to sustain a weight of 20 kN at a height of 3000 m where the air pressure is 70.12 kPa and the temperature is -4.45 °C. Determine the relative air velocity required if the angle of incidence of the aerofoil is 3°.

[195 m/s]

11 A flat kite has a surface area of 0.5 $m^2$ and mass 0.45 kg. During a test flight it was noted that the air temperature was 15°C, the pressure 1 bar and the air speed 10 m/s. The tension in the anchor cord was 25 N when the cord was inclined at 45° to the horizontal and the kite inclined at 10° to the horizontal. Determine the coefficients of lift and drag for the kite. $R_{air}$ = 287 J/kg K.

[0.744, 0.594]

12 Modern ships are being provided with stabilisers which substantially consist of a pair of underwater wings protruding from the hull. Calculate the righting moment and drag of a pair of stabilisers fitted to a 200 MN displacement passenger ship travelling at a speed of 10 m/s. The stabiliser wings have 2 m chord and are protruding 4 m from the hull, where the width across the ship is 20 m. Assume $C_L$ = 1.0, $C_D$ = 0.1. Density of sea water is 1000 kg/$m^2$, $\nu$ = 1.2 $mm^2$/s.

[9.6 MN m, 80 kN]

13 A racing car, of length 4 m, surface area 12 $m^2$ and frontal area 1 $m^2$, is fitted with a single aerofoil at the rear 1.5 m long and two aerofoils near the nose each 500 mm long. The aerofoils have the same section and chord length 500 mm, but the angle of attack is different in each case. The lift and drag coefficients for the rear aerofoil are 0.4 and 0.03 respectively, whereas the coefficients for the nose aerofoils are 0.45 and 0.035 respectively. The skin-drag coefficient $C_f$ may be taken as $0.455(\log Re_L)^{-2.58}$ where $Re_L$ is based on car length, and the profile drag coefficient $C_D$ = 0.4 based on frontal area. Determine (i) the power required to propel the car at 40 m/s, and (ii) the increase in normal force between the wheels and the road due to the aerofoils. Assume $\rho_{air}$ = 1.18 kg/$m^3$, $\mu_{air}$ = $1.8 \times 10^{-5}$ N s/$m^2$.

[4.37 kW, 495 N]

14 Tests are carried out at the corresponding speed in fresh water on a 1/9 scale model boat fitted with an underwater hydrofoil. The length of the model is 1 m and the wetted surface area and projected area are 0.5 $m^2$ and 0.1 $m^2$, respectively. The model hydrofoil is 300 m long and 150 mm chord with lift and drag coefficients 0.3 and 0.03 respectively. The skin friction coefficient is given by $C_f = 0.455(\log Re_L)^{-2.58}$ and the form drag coefficient based on projected area is 0.05. If the total force exerted on the model boat is 300 N determine, for a prototype speed of 21 m/s in sea water, (i) the residual drags of the model and prototype, and (ii) the difference in displacement ($m^3$) of the prototype boat between stationary and 21 m/s conditions. It may be assumed that the drag and lift coefficients of the model hydrofoil apply to similar hydrofoils up to Re = $10^8$. For sea water $\rho$ = 1025 kg/$m^3$ and for fresh water $\rho$ = 1000 kg/$m^3$.

[105.5 N, 79 kN, 24.6 $m^3$]

# 7 FLOW IN DUCTS

In chapter 1, the Reynolds experiments on flow in a circular pipe were described and it was shown that flow is either laminar or turbulent. The nature of the two kinds of flow was treated in more detail in chapter 6 and boundary layer effects were discussed. It is now necessary to consider how these basic concepts can be applied to flow in ducts.

## 7.1 CHARACTERISTICS OF FLOW IN CIRCULAR PIPES

Reynolds deduced that flow is laminar when viscous forces predominate and turbulent when inertia forces predominate. The ratio of these forces gives the Reynolds number $Re = \rho \bar{u} D/\mu$ and this parameter can be used to predict the nature of flow (i.e. laminar or turbulent) for any fluid in any pipe. The value of Re at which the flow pattern changes from laminar to turbulent is called the critical Reynolds number and occurs at a nominal value of 2000.

The difference between laminar and turbulent flow is also apparent in (a) the settling length, (b) the velocity profile for fully developed flow and (c) the shear stress. These differences are shown in figure 7.1 for the same mean velocity of flow, $\bar{u}$.

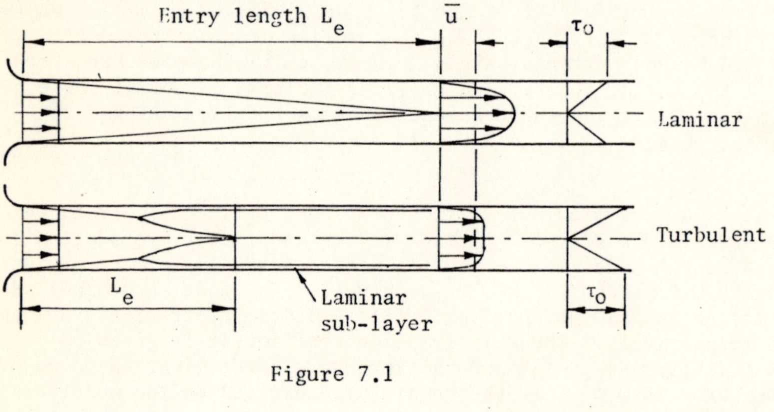


Figure 7.1

(a) The velocity profile is uniform at entry to a pipe. A boundary layer then forms and thickens in the flow direction. If $Re < 2000$ the boundary layer remains laminar in character until it occupies the whole cross-section of flow. If $Re > 2000$ a laminar boundary layer is formed initially, but transition to a turbulent boundary layer occurs, similar to that on a flat plate (figure 6.6). The length of pipe from inlet to the point at which the flow is fully developed in called the *entry* or *settling length*. For laminar flow the entry length is about 120 diameters and for turbulent flow about 50 diameters.

(b) For fully developed laminar flow, the velocity profile is parabolic and $u_{max} = 2\bar{u}$. For turbulent flow, the velocity profile is much flatter in the centre zone due to the mixing motion (momentum exchange) and $u_{max} \simeq 1.2\bar{u}$. Note that for turbulent flow the time mean profile is shown.

(c) The shear stresses and rate of dissipation of mechanical energy are much greater in turbulent flow than they are in laminar flow. For both types of flow, shear stress varies linearly from zero at the pipe centre to $\tau_0$ at the pipe wall.

## 7.2 FRICTION LOSSES IN DUCTS

Experiments show that, for fully developed flow in a constant-area duct, the dissipation or loss of mechanical energy is directly proportional to duct length. The steady-flow energy equations 4.35 and 4.36, together with equation 4.37, give

$$\frac{p_1}{\rho} + \overline{u_1}^2 + gz_1 = \frac{p_2}{\rho} + \overline{u_2}^2 + gz_2 + e_L \qquad (7.1)$$

The quantity, $e_L$, is the dissipation or loss of mechanical energy due to frictional effects. This dissipation may be expressed as a head loss, $H_L$, or a pressure drop, $\Delta p_L$, and evaluated from the *Darcy-Weisbach* equation

$$H_L = \frac{\Delta p_L}{\rho g} = \frac{e_L}{g} = \frac{4fL\bar{u}^2}{2gD_e} \qquad (7.2)$$

The friction factor, f, is defined in terms of the mean shear stress at the duct wall, $\tau_0$, or the dimensionless parameters of flow (see example 5.2).

$$f = \frac{\tau_0}{\rho\bar{u}^2/2} = \phi\left(Re, \frac{k}{D_e}\right) \qquad (7.3)$$

The equivalent diameter, $D_e$, is defined

$$D_e = \frac{4 \times \text{Cross-sectional area of duct}}{\text{Wetted perimeter of duct}} = \frac{4A}{P} \qquad (7.4)$$

(Note: Hydraulic radius $R_h = A/P$ is sometimes used instead of $D_e$)

For flow in a constant-area duct, the mean velocity is constant and $\overline{u_1} = \overline{u_2} = \bar{u}$. Equation 7.1 may be simplified and written in terms of specific energy, pressure or head units. In pressure units

$$p_1 + \rho g z_1 = p_2 + \rho g z_2 + \Delta p_L \qquad (7.5a)$$

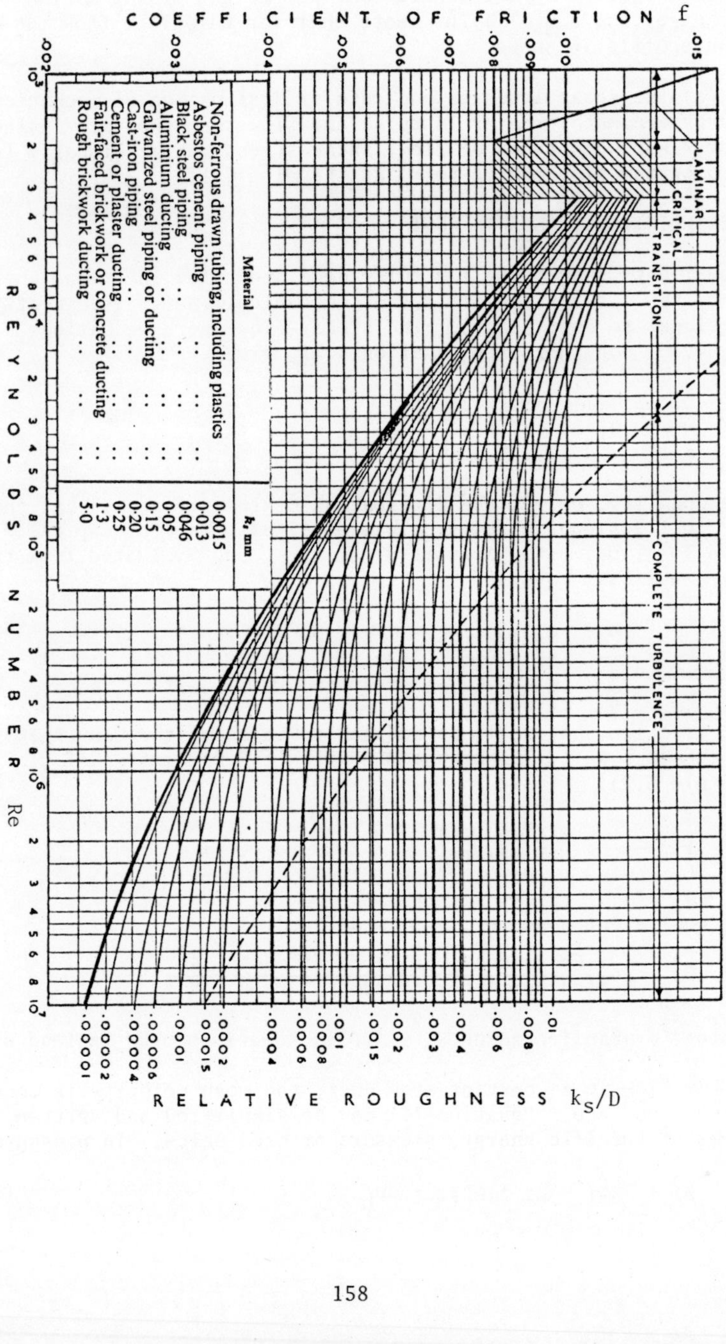
COEFFICIENT OF FRICTION f
.002
.003
.004
.005
.006
.007
.008
.009
.010
.015
LAMINAR
CRITICAL
TRANSITION
COMPLETE TURBULENCE
Material
$k_s$ mm
Non-ferrous drawn tubing, including plastics 0·0015
Asbestos cement piping 0·013
Black steel piping 0·046
Aluminium ducting 0·05
Galvanized steel piping or ducting 0·15
Cast-iron piping 0·20
Cement or plaster ducting 0·25
Fair-faced brickwork or concrete ducting 1·3
Rough brickwork ducting 5·0
REYNOLDS NUMBER Re
$10^3$
$10^4$
$10^5$
$10^6$
$10^7$
RELATIVE ROUGHNESS $k_s/D$
.01
.008
.006
.004
.003
.002
.0015
.001
.0008
.0006
.0004
.0002
.00015
.0001
.00006
.00004
.00002
.00001

Figure 7.2

(Reproduced by permission from the CIBS Guide)

or $p_1^* - p_2^* = \Delta p_L$ (7.5b)

Therefore, the dissipation or loss of mechanical energy in a constant-area duct expressed in pressure units is equal to the drop in piezometric pressure.

Equation 7.5a is frequently used in terms of head units

$$\frac{p_1}{\rho g} + z_1 = \frac{p_2}{\rho g} + z_2 + \frac{4fL\bar{u}^2}{2gD_e} \qquad (7.5c)$$

7.2.1 Friction Factor for Constant Diameter Pipes

For circular pipes of diameter D, equation 7.4 given $D_e = D$ and the Reynolds number of flow is $Re = \rho\bar{u}D/\mu$. Friction factor values are expressed in terms of Re.

For laminar flow ($Re < 2000$), the exact analysis of *Poiseuille* is possible, leading to the relationship

$$f = \frac{16}{Re} \qquad (7.6)$$

For turbulent flow ($2 \times 10^3 < Re < 10^5$), the empirical relationship of *Blasius* applies

$$f = \frac{0.0791}{Re^{0.25}} \qquad (7.7)$$

For turbulent flow ($Re > 10^5$), the *Kármán-Prandtl* empirical relationship applies

$$\frac{1}{\sqrt{f}} = 4 \log \left(\frac{Re\sqrt{(f)}}{1.26}\right) \qquad (7.8)$$

The value of f for laminar flow applies to pipes which are rough or smooth but the values of f for turbulent flow are only applicable to pipes which are mechanically smooth or which are hydraulically smooth because the roughness elements are effectively submerged by a laminar sub-layer. The values of f are more conveniently found from the f-Re curve of figure 7.2. The heavy line applies to smooth pipes.

For rough pipes, f is a function of both Re and a relative roughness factor k/D. The original experiments on rough pipes were carried out by *Nikuradse* (1932) on pipes artificially roughened by sand particles of uniform size k. However, this uniform roughness pattern is rarely encountered in practice and *Colebrook* and *White* (1938) carried out experiments on commercial pipes and obtained an empirical relationship using an effective sand roughness $k_s$.

$$\frac{1}{\sqrt{f}} = -4 \log \left(\frac{k_s}{3.7D} + \frac{1.26}{Re\sqrt{(f)}}\right) \qquad (7.9)$$

These results are superimposed on the f-Re diagram of figure 7.2 resulting in a composite graph sometimes referred to as the *Moody* diagram.

The f-Re diagram shows that for high Re values f becomes constant and is a function of $k_s/D$ alone. This region is known as the turbulent rough zone for which the following empirical relationship applies

$$\frac{1}{\sqrt{(f)}} = 4 \log \left(3.7 \frac{D}{k_s}\right) \tag{7.10}$$

7.2.2 Flow in Non-circular Ducts

For non-circular ducts, the equivalent diameter, $D_e$, is calculated from equation 7.4 and the Reynolds number of flow is $(Re)_e = \rho \bar{u} D_e/\mu$.

For turbulent flow, an approximate value of f may be calculated substituting $(Re)_e$ in the above equations for flow in a circular pipe. The accuracy improves as the shape of the duct tends towards the circular. Note that $A \neq (\pi/4)D_e^2$.

For laminar flow, it is inaccurate to evaluate f using equation 7.6 and $(Re)_e$. The loss associated with laminar flow must be calculated from first principles for the particular shape of duct. Information is given in reference 1.

## 7.3 MINOR LOSSES

Energy dissipation also occurs due to local disturbances in the flow such as entry to and exit from pipes, bends, changes of section, valves, etc. These 'fittings' losses are usually called *minor* losses because in a long pipe they are small in comparison with pipe friction losses. However, the significance of any loss is one of relative magnitude and minor losses should always be accounted for unless it can be shown that they can be neglected.

It is impossible to calculate the magnitude of the majority of minor losses therefore reference must be made to empirical data which expresses the loss in terms of a coefficient $k_m$ of the velocity head [i.e. $k_m(u^2/2g)$] or in terms of an equivalent length of pipe which would give a friction loss equal to the actual minor loss.

Some typical values of minor losses are quoted in the examples and a comprehensive list is given in reference 2.

## 7.4 PIPE SYSTEMS

Pipe friction losses and minor losses may be evaluated if the flow rate is known. In actual pipe systems, however, the flow rate is determined by the resistance of the pipes and fittings and by the form of the pipe combination (figure 7.3). The problem is evaluation of the total flow rate, or the flow rate in the individual pipes of a multiple pipe system.

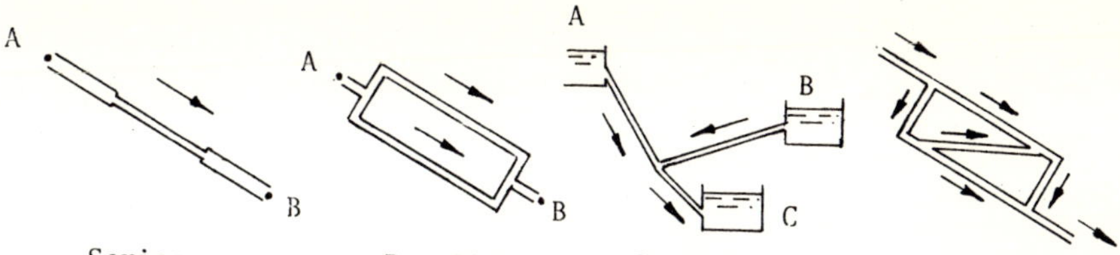


Figure 7.3

In series and parallel systems, solution is possible by application of the steady-flow energy and continuity equations. The energy equation 7.1 may be applied in head units form, between points A and B, giving

$$\frac{p_A}{\rho g} + \frac{\bar{u}_A^2}{2g} + z_A = \frac{p_B}{\rho g} + \frac{\bar{u}_B^2}{2g} + z_B + \Sigma\left(\frac{e_L}{g}\right) \qquad (7.11)$$

where $\Sigma\left(\frac{e_L}{g}\right) = \Sigma\left(\frac{4fL\bar{u}^2}{2gD}\right) + \Sigma\left(\frac{k_m\bar{u}^2}{2g}\right)$ is the sum of pipe friction and minor losses for the flow path considered.

This equation must be applied for each possible flow path a fluid particle can take between A and B. Note that this form of the energy equation is frequently referred to as the Bernoulli equation (with losses) in engineering literature.

For incompressible flow, the continuity equation may be applied in the form

$$\dot{V} = \Sigma(\bar{u}A) \qquad (7.12)$$

It is useful to draw the energy and piezometric pressure (hydraulic gradient) lines for flow in pipe systems.

Pipe systems involving branched pipes or networks are more complex and require solution by trial-and-error or iterative techniques.

## 7.5 EFFECT OF NON-UNIFORM VELOCITY DISTRIBUTION

Because of the variation in velocity across a duct, in real flow, it is necessary to reconsider the equations derived in chapters 3 and 4 and applied to one-dimensional flow using a mean velocity $\bar{u}$. The mass flow rate for flow through a duct is given by

$$\dot{m} = \int_A \rho u \, dA = \rho\bar{u}A \qquad (7.13)$$

The momentum flux, $\dot{M}$, is expressed in terms of a momentum correction factor $\beta$

$$\dot{M} = \int_A (\rho u \, dA)u = \beta\rho\bar{u}^2A \qquad (7.14)$$

The specific kinetic energy, $e_k$, is expressed in terms of a kinetic energy correction factor, $\alpha$

$$e_k = \frac{\int_A \frac{1}{2}(\rho u \, dA)u^2}{\int_A \rho u \, dA} = \tfrac{1}{2}\alpha\bar{u}^2 \tag{7.15}$$

*Example 7.1*

Oil flows downwards through a 50 m length of 100 mm diameter pipe, inclined at 30° to the horizontal. Working from first principles, determine (i) the maximum flow rate for which the flow remains laminar in character, and (ii) the corresponding pressure difference between the inlet and exit points. Assume $\mu = 0.285$ N s/m$^2$, $\rho = 950$ kg/m$^3$.

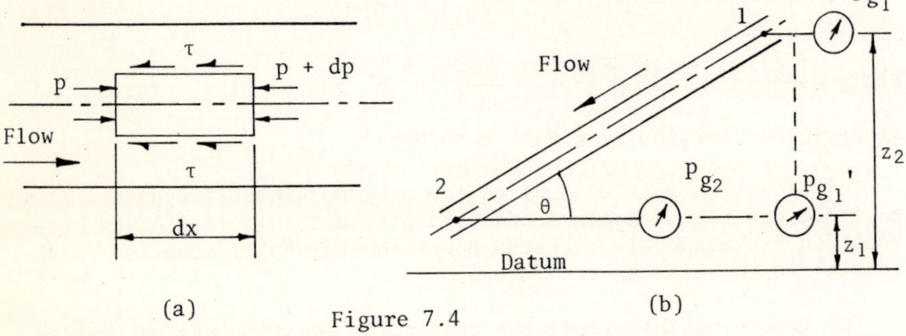


Figure 7.4

The first step is to derive an expression for the pressure drop in laminar flow. Consider the forces acting on a small cylindrical element of radius r, and length dx (figure 7.4a). For steady flow, the momentum equation 4.19 gives

$$\Sigma F = p\pi r^2 - (p + dp)\pi r^2 - \tau 2\pi r \, dx = 0$$

and $\tau = -\frac{1}{2}\left(\frac{dp}{dx}\right) r$

For fully developed flow, $(dp/dx) = (-\Delta p/L)$, where $\Delta p$ is the pressure drop (-ve) along a length of pipe, L, due to dissipation of mechanical energy. Therefore

$$\tau = \frac{1}{2}\left(\frac{\Delta p}{L}\right) r \tag{i}$$

For laminar flow $\tau = -\mu(du/dr)$. The negative sign is introduced because u decreases with increasing r. Substituting for $\tau$ in equation (i) we have

$$\frac{du}{dr} = -\frac{1}{2\mu}\left(\frac{\Delta p}{L}\right) r$$

Treating $(\Delta p/L)$ as a constant in the r direction and integrating with respect to r we obtain

$$u = -\frac{1}{4\mu}\left(\frac{\Delta p}{L}\right) r^2 + A$$

When r = D/2, u = 0, therefore $A = (\Delta p/L)D^2/16\mu$.

$\frac{dp}{dx}\frac{D^2}{16\mu}$

$$u = \frac{1}{16\mu}\left(\frac{\Delta p}{L}\right)(D^2 - 4r^2) \qquad \text{(ii)}$$

Thus the velocity profile for laminar flow in a pipe (axially symmetric Poiseuille flow) is a parabola in section and the true three-dimensional velocity profile is a paraboloid. When r = 0, $u = u_{max}$. Therefore

$$u_{max} = \frac{D^2}{16\mu}\left(\frac{\Delta p}{L}\right)$$

From the properties of a paraboloid, or from the use of equations 7.13 and (ii), the mean velocity $\bar{u}$ is

$$\bar{u} = \frac{u_{max}}{2} = \frac{\int_A \rho u \, dA}{\rho A} = \frac{D^2}{32\mu}\left(\frac{\Delta p}{L}\right) \qquad \text{(iii)}$$

Now consider the shear stress at the wall. When r = D/2, $\tau = \tau_0$. Therefore from equations (i) and (iii)

$$\tau_0 = \frac{8\mu\bar{u}}{D}$$

$\bar{u} = \frac{D^2}{32\mu}\cdot\frac{dp}{dx}$

The friction factor, f, defined by equation 7.3, is

$$f = \frac{\tau_0}{\rho\bar{u}^2/2} = \frac{8\mu\bar{u}/D}{\rho\bar{u}^2/2} = \frac{16}{Re}$$

(i) Consider flow through the pipe. For laminar flow, Re < 2000. Therefore the maximum flow rate is determined by $(\rho\bar{u}D/\mu) = 2000$ and

$$\bar{u} = \frac{2000\mu}{D\rho} = \frac{2000 \times 0.285}{0.1 \times 950} = 6 \text{ m/s}$$

$$\dot{V} = \bar{u}\tfrac{1}{4}\pi D^2 = \frac{6 \times \pi \times 0.1^2}{4} = 0.047 \text{ m}^3\text{/s}$$

(ii) The pressure drop, $\Delta p$, is given by equation (iii).

$$\Delta p = \frac{32\mu L\bar{u}}{D^2} = \frac{32 \times 0.285 \times 50 \times 6}{0.1^2} = 274 \text{ kPa}$$

In a horizontal pipe, $\Delta p$ = 274 kPa, is the *drop* in static pressure due to dissipation or loss of mechanical energy (equation 7.5a). In an inclined pipe, the loss is given by the *drop* in piezometric pressure, $p_1^* - p_2^* = p_{g1}' - p_{g2}$ = 274 kPa, recorded by gauges mounted on the same horizontal line (figure 7.5b). The *difference* in static pressures, $p_1 - p_2 = p_{g1} - p_{g2}$, is given by gauges mounted in line with points 1 and 2, respectively. This difference can be a drop or a rise in pressure. From equation 7.5a we obtain

$p_1 - p_2 = \Delta p_L - \rho g(z_1 - z_2)$

$= 274 \times 10^3 - 950 \times 9.81 \times 50 \sin 30 = 41$ kPa

*Example 7.2*

Show that for steady parallel laminar flow of fluid, a force balance on a small element of fluid yields the relationship

$\mu(d^2u/dy^2) = dp/dx$

A piston 100 mm diameter and 150 mm long has a mass of 10 kg. The piston is placed in a vertical cylinder 102 mm diameter containing oil of viscosity 0.25 Pa s. Calculate the time taken for the piston to fall 100 mm. It may be assumed that the piston and cylinder are concentric throughout the piston motion.

Refer to figure 7.5 which shows laminar flow of fluid between two parallel surfaces of infinite width normal to the plane of the diagram. The lower surface is stationary and the upper surface moves with velocity U.

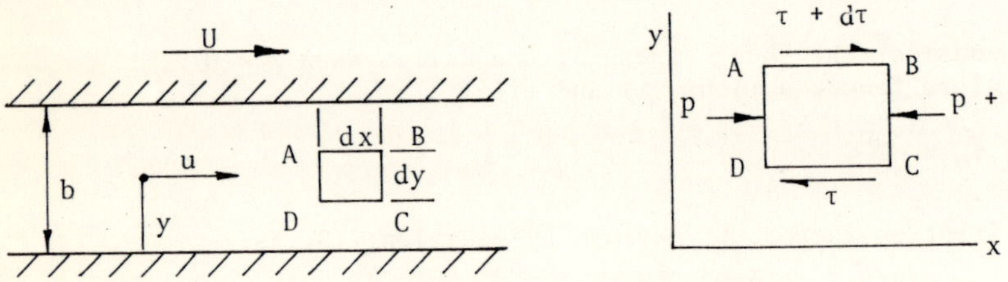


Figure 7.5

Consider a small rectangular element ABCD of unit width. The only forces acting on the element are surface forces due to viscous stresses and pressure. For steady flow, the momentum equation 4.19 gives

$$\Sigma F_{\text{viscous}} + \Sigma F_{\text{pressure}} = 0$$

$$(\tau + d\tau)dx - \tau\, dx - (p + dp)dy + p\, dy = 0$$

$$\frac{d\tau}{dy} = \frac{dp}{dx} \qquad \text{(i)}$$

Newton's law of viscosity $\tau = \mu(du/dy)$ may be applied to parallel laminar flow. Substituting in (i) we obtain

$$\frac{d}{dy}\left(\mu\frac{du}{dy}\right) = \frac{dp}{dx}$$

or $$\frac{d^2u}{dy^2} = \frac{1}{\mu}\frac{dp}{dx} \qquad \text{(ii)}$$

Now consider flow in the annular gap between a piston and a cylinder (figure 7.6). If the gap is small compared with the diameter, the flow can be treated as flow between parallel flat surfaces.

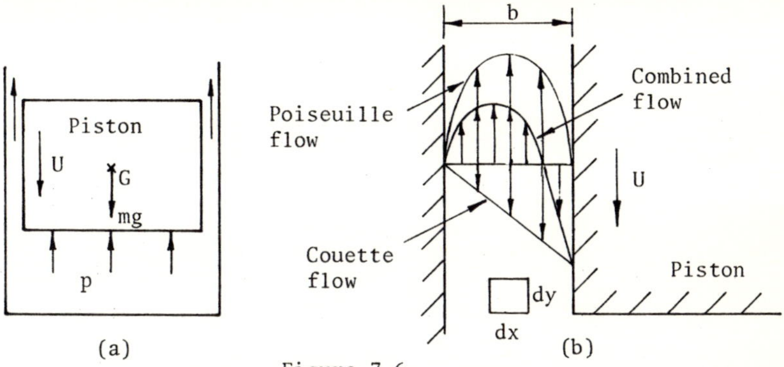


Figure 7.6

Equation (ii) for flow in the y direction gives

$$\frac{d^2v}{dx^2} = \frac{1}{\mu}\left(\frac{dp}{dy}\right)$$

This equation may be integrated with respect to x, treating (dp/dy) as a constant in the x direction. Hence

$$v = \frac{1}{2\mu}\left(\frac{dp}{dy}\right) x^2 + Ax + B$$

The boundary conditions are $v = 0$ when $x = 0$, and $v = -U$ when $x = b$. Therefore

$$v = \frac{1}{2\mu}\left(\frac{dp}{dy}\right)(x^2 - xb) - \frac{Ux}{b}$$

The velocity equation consists of two parts. The first term gives *Poiseuille* flow upwards through the annular gap due to the pressure gradient dp/dy. The second term gives *Couette* flow downwards due to the piston motion. The velocity profiles for the individual and combined flows are shown in figure 7.6b. The volumetric flow through the annular gap of width $\pi D$ is given by

$$\dot{V} = \int_0^b v\pi D\, dx = \pi D \int_0^b \frac{1}{2\mu}\left(\frac{dp}{dy}\right)(x^2 - xb)\, dx - \pi D \int_0^b \frac{Ux\, dx}{b}$$

$$\dot{V} = \pi D\left(-\frac{1}{12\mu}\left(\frac{dp}{dy}\right) b^3 - \frac{Ub}{2}\right) \qquad \text{(iv)}$$

The volumetric flow rate through the annular gap and the pressure gradient are determined by the displacement of the piston. Therefore

$$\dot{V} = \tfrac{1}{4}\pi D^2 U \qquad \text{(v)}$$

and $\frac{dp}{dy} = -\frac{\Delta p}{L} = -\left(\frac{mg}{\pi D^2/4}\right)\frac{1}{L}$ (-ve in flow direction) (vi)

From equations (iv), (v) and (vi)

$$U = \frac{1}{(1 + 2b/D)} \frac{4mgb^3}{3\mu\pi D^3 L}$$

$$= \frac{1}{(1 + 2/100)} \frac{4 \times 10 \times 9.81 \times 1^3 \times 10^{-9}}{3 \times 0.25 \times \pi \times 1^3 \times 10^{-3} \times 0.15}$$

$$= 0.00109 \text{ m/s}$$

The time, t, for the piston to fall a distance, y, is

$$t = \frac{y}{U} = \frac{0.1}{0.00109} = 92$$

Note that if $D >> b$ the contribution of Couette flow is negligible to a first approximation.

*Example 7.3*

Fluid of invariable density, $\rho$, and viscosity, $\mu$, flows through the annular passage between two horizontal concentric pipes under steady, laminar flow conditions. The inner and outer radii of the wetted surfaces of the annular passages are $R_1$ and $R_2$, respectively. Determine the radius at which the maximum velocity occurs.

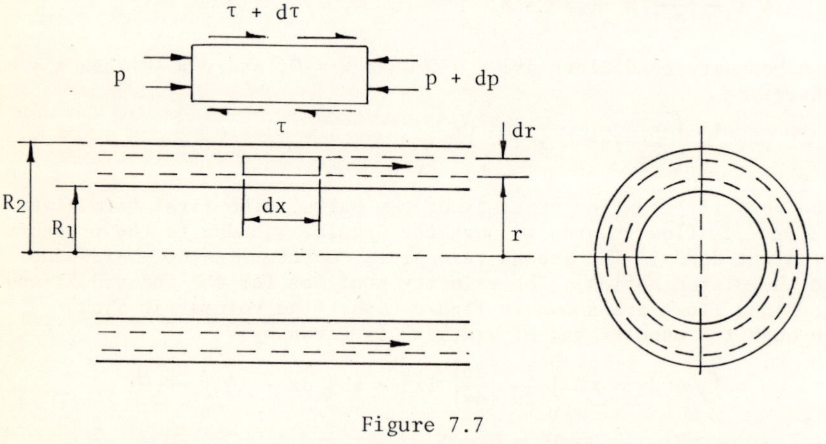


Figure 7.7

Figure 7.7 shows flow through an annulus in which the clearance between the two surfaces is too large for the assumption of simple Poiseuille flow between parallel surfaces to apply. Consider an annular element of radius r, thickness dr and length dx. For steady flow, the momentum equation 4.19 gives

$$\Sigma F_{viscous} + \Sigma F_{pressure} = 0$$

$$(\tau + d\tau)2\pi(r + dr)dx - \tau 2\pi r\, dx - (p + dp)2\pi r\, dr + p\, 2\pi r\, dr = 0$$

Neglecting the product of small quantities this reduces to

$$\tau\ dr\ dx + r\ d\tau\ dx - r\ dp\ dr = 0$$

$$\tau + \frac{d\tau}{dr} r = r\left(\frac{dp}{dx}\right)$$

or $$\frac{d(\tau r)}{dr} = r\left(\frac{dp}{dx}\right)$$

For laminar flow, $\tau = \mu(du/dr)$, therefore

$$\frac{d}{dr}\left(r\frac{du}{dr}\right) = \frac{1}{\mu}\left(\frac{dp}{dx}\right) r$$

Treating (dp/dx) as a constant and integrating

$$r\frac{du}{dr} = \frac{1}{2\mu}\left(\frac{dp}{dx}\right) r^2 + A$$

$$u = \frac{1}{4\mu}\left(\frac{dp}{dx}\right) r^2 + A \ln r + B$$

The boundary conditions are $u = 0$ at $r = R_1$ and $u = 0$ at $r = R_2$. Therefore

$$0 = \frac{1}{4\mu}\left(\frac{dp}{dx}\right) R_1^2 + A \ln R_1 + B \qquad \text{(i)}$$

and $$0 = \frac{1}{4\mu}\left(\frac{dp}{dx}\right) R_2^2 + A \ln R_2 + B \qquad \text{(ii)}$$

From equations (i) and (ii), we have

$$A = \frac{R_1^2 - R_2^2}{4\mu}\left(\frac{dp}{dx}\right)\frac{1}{\ln(R_2/R_1)}$$

and $$u = \frac{1}{4\mu}\left(\frac{dp}{dx}\right) r^2 + \frac{R_1^2 - R_2^2}{4\mu}\left(\frac{dp}{dx}\right)\frac{\ln r}{\ln(R_2/R_1)} + B \qquad \text{(iii)}$$

The constant B could be evaluated by substituting for A in equation (i) or (ii). However, in this problem we are asked to determine the maximum value of u, which occurs when (du/dr) = 0. Differentiating equation (iii) with respect to r, we obtain

$$\frac{du}{dr} = \frac{1}{2\mu}\left(\frac{dp}{dx}\right) r + \frac{R_1^2 - R_2^2}{4\mu}\left(\frac{dp}{dx}\right)\frac{1}{r \ln(R_2/R_1)} = 0$$

This gives the radius for maximum velocity as

$$r^2 = \frac{R_1^2 - R_2^2}{2 \ln(R_2/R_1)}$$

*Example 7.4*

Fluid of constant density, $\rho$, flows upwards in a vertical pipe of radius R. At entry to the pipe the velocity has a uniform value of $\bar{u}$, but at a distance, L, from entry the flow is fully developed and the velocity distribution is of the form $u = u_m(1 - r/R)^{1/7}$, where u is the velocity at any radius r and $u_m$ is the velocity at the pipe centre line. Derive an expression for the average wall shear stress, $\tau_o$, in terms of $\rho$, g, R, $\bar{u}$, L, and the pressure drop, $\Delta p$.

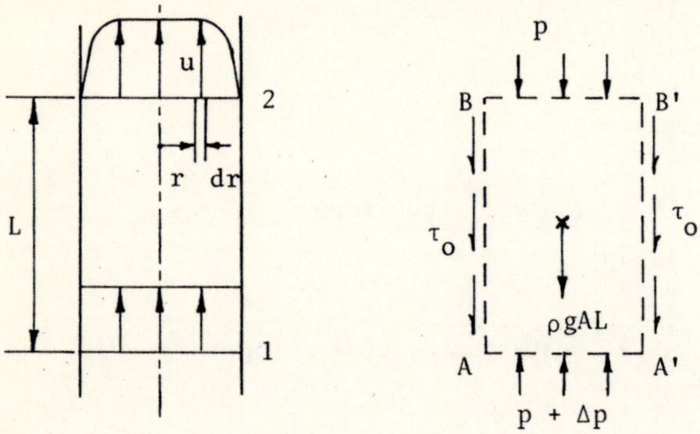


Figure 7.9

Consider the control volume ABB'A' (figure 7.8). For steady flow, the momentum equation 4.18 reduces to

$$\Sigma F = \int_S \underline{u}(\rho \underline{u} \cdot d\underline{A}) \quad \text{(i)}$$

At section BB' the velocity profile is non-uniform and to take account of this we must consider the flow through an annular element of area $2\pi r\ dr$. At any radius r, the velocity is given by $u = u_m(1 - r/R)^{1/7}$. Equation (i) gives

$$\Sigma F_{pressure} + \Sigma F_{viscous} + \Sigma F_{gravity} = \int_{A_2} u(\rho u\ dA) - \int_{A_1} \bar{u}(\rho \bar{u}\ dA)$$

$$(p + \Delta p)\pi R^2 - p\pi R^2 - \tau_o 2\pi RL - \rho g \pi R^2 L$$

$$= \int_0^R \rho u_m{}^2 (1 - r/R)^{2/7}\, 2\pi r\ dr - \rho\pi R^2 \bar{u}^2$$

Making the substitution $r = R - y$, from which $dr = -dy$ we obtain

$$\Delta p - \frac{2\tau_o L}{R} - \rho g L = \frac{\rho}{R^2}\left[\frac{-2u_m{}^2}{R^{2/7}} \int_0^R y^{2/7}(R - y)dy - R^2\bar{u}^2\right]$$

$$\Delta p - \frac{2\tau_o L}{R} - \rho g L = \frac{98}{144}\rho u_m{}^2 - \rho\bar{u}^2 \quad \text{(ii)}$$

It is now necessary to express $u_m$ in terms of $\bar{u}$. From equation 7.12 we have $\dot{m} = \int_{A_1} \rho u\ dA = \int_{A_2} \rho u\ dA$.

$$\dot{m} = \int_{A_1} \rho u\ dA = \rho\bar{u}\pi R^2 \quad \text{(iii)}$$

and $\dot{m} = \int_{A_2} \rho u\ dA = \int_0^R \rho u_m (1 - r/R)^{1/7}\, 2\pi r\ dr = \rho\, \frac{98}{120} u_m \pi R^2$ (iv)

From equations (ii), (iii) and (iv) we obtain

$$\Delta p - \frac{2\tau_o L}{R} - \rho g L = 1.02\rho\bar{u}^2 - \rho\bar{u}^2$$

$$\tau_o = \frac{R\Delta p}{2L} - \frac{0.01\rho\bar{u}^2 R}{L} - \frac{\rho g R}{2}$$

*Example 7.5*

Fluid of constant density flows through an abrupt enlargement between two concentric pipes. Stating any assumptions made, derive an expression for the total head loss and express this in terms of a loss coefficient $k_m$.

Fluid of constant density flows through two abrupt enlargements between three concentric pipes of cross-sectional areas $A_1$, $A_2$ and $A_3$ respectively. Show that for the overall pressure rise $(p_3 - p_1)$ to be a maximum the cross-sectional area of the intermediate pipe must be given by

$$A_2 = 2A_1A_3/(A_1 + A_3)$$

Neglect frictional losses and interference effects due to the intermediate pipe.

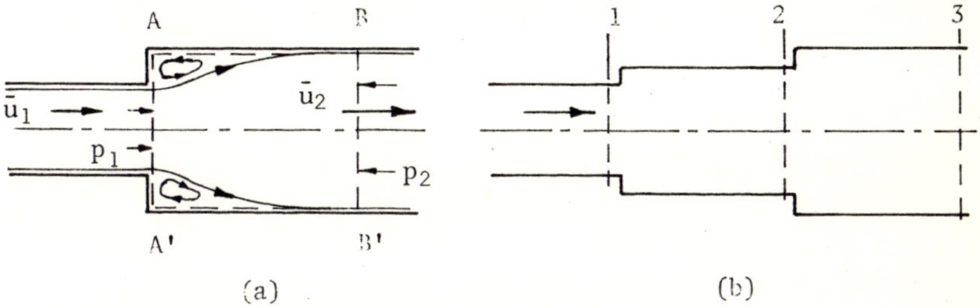


Figure 7.9

Consider the control volume ABB'A' (figure 7.9) and assume that the pressure just after the enlargement is equal to that just before i.e. $p_1 = p_A$. For steady flow, the only forces acting on the control volume are pressure forces and viscous forces. Neglecting the viscous forces over the short length of pipe and applying the momentum equation 4.19 we have

$$\Sigma F = p_1A_2 - p_2A_2 = \rho\bar{u}_2A_2(\bar{u}_2 - \bar{u}_1)$$

$$p_1 - p_2 = \rho\bar{u}_2(\bar{u}_2 - \bar{u}_1) \qquad \text{(i)}$$

Applying the energy equation 7.11

$$\frac{p_1}{\rho g} + \frac{\bar{u}_1^2}{2g} = \frac{p_2}{\rho g} + \frac{\bar{u}_2^2}{g} + k_m\frac{\bar{u}_1^2}{2g} \qquad \text{(ii)}$$

where $k_m(\bar{u}_1^2/2g)$ is the mechanical energy loss (head loss) due to the enlargement based, by convention, on the kinetic energy in the approach pipe. From equations (i) and (ii) we obtain

$$k_m\frac{\bar{u}_1^2}{2g} = \left(1 - \frac{A_1}{A_2}\right)^2\frac{\bar{u}_1^2}{2g} \qquad \text{(iii)}$$

$$k_m = (1 - A_1/A_2)^2 \qquad \text{(iv)}$$

Note that for a pipe discharging into a large tank $A_2 = \infty$ and $k_m = 1$.

Consider now two expansions in series (figure 7.9b) and apply equation (i) to each expansion

$$p_1 - p_2 = \rho(\bar{u}_2^2 - \bar{u}_2\bar{u}_1) \qquad \text{(v)}$$

and $$p_2 - p_3 = \rho(\bar{u}_3^2 - \bar{u}_3\bar{u}_2) \qquad \text{(vi)}$$

The continuity equation gives

$$\dot{V} = A_1\bar{u}_1 = A_2\bar{u}_2 = A_3\bar{u}_3 \qquad \text{(vii)}$$

From equations (v), (vi) and (vii)

$$p_3 - p_1 = \rho\dot{V}^2\left(\frac{1}{A_2A_1} - \frac{1}{A_2^2} + \frac{1}{A_3A_2} - \frac{1}{A_3^2}\right) \qquad \text{(viii)}$$

The pressure rise $(p_3 - p_1)$ is a maximum when $d(p_3 - p_1)/dA_2 = 0$. Therefore

$$\frac{d(p_3 - p_1)}{dA_2} = \rho\dot{V}^2\left(-\frac{1}{A_2^2A_1} + \frac{2}{A_2^3} - \frac{1}{A_3A_2^2}\right) = 0$$

$$A_2 = \frac{2A_1A_3}{A_1 + A_3} \qquad \text{(ix)}$$

$$\frac{d^2(p_3 - p_1)}{dA_2^2} = \frac{2}{A_2^3A_1} - \frac{6}{A_2^4} + \frac{2}{A_2^3A_3} = \frac{1}{A_2^3}\left(\frac{2}{A_1} - \frac{6}{A_2} + \frac{2}{A_3}\right)$$

Substitute for $A_2$ from equation (ix)

$$\frac{d^2(p_3 - p_1)}{dA_2^2} = \frac{1}{A_2^3}\left(\frac{A_1 + A_3}{A_1A_3}\right)(-1) \qquad \text{(i.e. -ve)}$$

The pressure rise is a maximum.

*Example 7.6*

Air flows at 1.0 $m^3/s$ through a 50° conical diffuser from a 150 mm diameter pipe to a 300 mm diameter pipe. The value of k in the pressure loss term $\frac{1}{2}\rho u_1^2 k[1 - (A_1/A_2)]^2$ is 1.134. The velocity distribution may be assumed uniform at entry to the diffuser and at points downstream greater than six pipe diameters but at exit from the diffuser the velocity profile is given by $u = 2\bar{u}(1 - r^2/R^2)$. Determine (i) the diffuser efficiency, and (ii) the friction head loss between diffuser exit and the station downstream where the static pressure is equal to the static pressure at diffuser exit. Assume incompressible flow with $\rho = 1.2$ kg/m .

For ideal (frictionless) flow the Bernoulli equation may be applied between the diffuser inlet and exit

$$p_1 + \tfrac{1}{2}\rho\bar{u}_1^2 = p_2' + \tfrac{1}{2}\rho\bar{u}_2^2$$

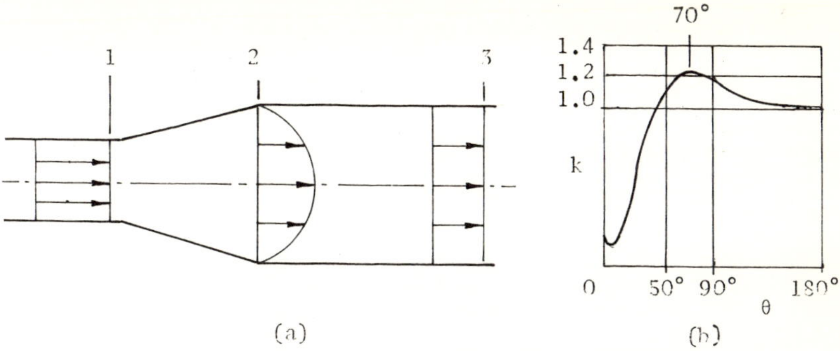


Figure 7.10

Rearranging and applying the continuity equation $\dot{V} = Au$

$$p_2' - p_1 = \tfrac{1}{2}\rho\bar{u}_1^2\left[1 - \left(\frac{A_1}{A_2}\right)^2\right] \qquad \text{(i)}$$

Applying the Bernoulli equation with losses between inlet and exit

$$p_1 + \tfrac{1}{2}\rho\bar{u}_1^2 = p_2 + \tfrac{1}{2}\rho\bar{u}_2^2 + \Delta p_L$$

Rearranging and substituting $\Delta p_L = \tfrac{1}{2}\rho\bar{u}_1^2 k\left(1 - \frac{A_1}{A_2}\right)^2$

$$p_2 - p_1 = \tfrac{1}{2}\rho\bar{u}_1^2\left[1 - \left(\frac{A_1}{A_2}\right)^2\right] - \tfrac{1}{2}\rho\bar{u}_1^2 k\left(1 - \frac{A_1}{A_2}\right)^2 \qquad \text{(ii)}$$

or $\quad p_2 - p_1 = (p_2' - p_1) - \Delta p_L \qquad$ (iii)

We may define a *diffuser efficiency*, $\eta_D$, and a *pressure recovery factor*, $c_{pr}$, as follows

$$\eta_D = \frac{p_2 - p_1}{p_2' - p_1} = \frac{p_2 - p_1}{(\rho\bar{u}_1^2/2)[1 - (A_1/A_2)^2]} \qquad \text{(iv)}$$

$$c_{pr} = \frac{p_2 - p_1}{\rho\bar{u}_1^2/2} \qquad \text{(v)}$$

Substituting values in equation (ii)

$$p_2 - p_1 = \frac{1.2 \times 1^2 \times 4^2}{2 \times \pi^2 \times 0.15^4}\left[\left(1 - \frac{0.15^4}{0.3^4}\right) - \left(1 - \frac{0.15^2}{0.3^2}\right)^2 1.134\right]$$

$$= 576 \text{ Pa}$$

$$\eta_D = \frac{2 \times 576 \times \pi^2 \times 0.15^4}{1.2 \times 1^2 \times 4^2(1 - 0.15^4/0.3^4)} = 32\%$$

Applying the Bernoulli equation with losses between stations 2 and 3 and including kinetic energy correction factors

$$p_2 + \tfrac{1}{2}\alpha_2\rho\bar{u}_2^2 = p_3 + \tfrac{1}{2}\alpha_3\rho\bar{u}_3^2 + \Delta p_f$$

or $$p_3 - p_2 = \tfrac{1}{2}\rho\bar{u}_2^2(\alpha_2 - \alpha_3) - \Delta p_f \qquad \text{(vi)}$$

For uniform flow at station 3, $\alpha_3 = 1$ and for a profile $u = 2\bar{u}\ (1 - r^2/R^2)$ at station 2 we have from equation 7.15.

$$e_k = \frac{\int_A \tfrac{1}{2}(\rho u\ dA)u^2}{\int_A \rho u\ dA} = \tfrac{1}{2}\alpha_3\bar{u}^2$$

$$\alpha_3 = \frac{8\bar{u}\int_0^R (1 - r^2/R^2)^3\ 2\pi r\ dr}{\bar{u}\pi R^2} = 2$$

Substituting values in equation (vi) for the case $p_3 = p_2$ we have

$$\Delta p_f = \frac{1.2 \times 1^2 \times 4^2}{2 \times \pi^2 \times 0.3^4}\ (2 - 1) = 120\ \text{Pa}$$

*Example* 7.7

(a) Describe the ideal and actual nature of flow through a 90° bend in a rectangular duct.
(b) A horizontal curved duct, 1.2 m square, has a mean radius of 4 m. Determine the mass flow rate of air at 1.1 bar and 290 K if the difference in pressure between the inner and outer walls is 15 mm $H_2O$. Assume frictionless, free vortex flow conditions with negligible change in density. $R_{air} = 287$ J/kg K.

(a) Refer to figure 7.11a. Motion along a curved streamline gives an acceleration towards the centre of curvature and a radial pressure gradient (section 4.3.1). Ideal flow round a bend conforms to a free or irrotational vortex, with a velocity distribution ur = const. Thus, between entry to the bend and the mid-point of the bend, the fluid close to the inner wall accelerates and the fluid close to the outer wall decelerates. The reverse happens between mid-point and exit.

In real flow, the velocity near the walls is reduced by viscous action (figure 7.11b). As a result of this, the increase in pressure between the inner and outer radii is smaller at the upper and lower walls (PU and RS) than it is along the centre line QT. This causes a secondary flow in the form of a double eddy which persists well downstream of the bend. In addition to the secondary flow, separation occurs in the regions of adverse (positive) pressure gradient AB and CD.

The losses in a bend are normally expressed in the form $H_L = k_m\bar{u}^2/2g$, where $k_m$ varies between 0.1 for a large bend and 1.1 for a right-angled bend. The losses due to separation and secondary flow in a right-angled bend can be reduced considerably by the use of a cascade of vanes (figure 7.11c).

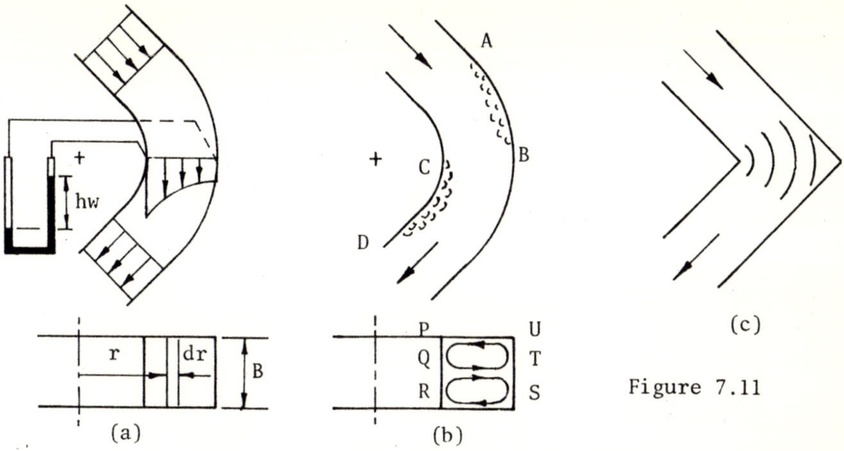


Figure 7.11

Assuming frictionless, free vortex flow conditions (ur = C) and applying the Bernoulli equation between the inner and outer radii, $R_1$ and $R_2$, we have

$$p_1 + \tfrac{1}{2}\rho u_1{}^2 = p_2 + \tfrac{1}{2}\rho u_2{}^2$$

and $$p_2 - p_1 = \tfrac{1}{2}\rho C^2\left(\frac{1}{R_1{}^2} - \frac{1}{R_2{}^2}\right) \qquad \text{(i)}$$

The pressure between the inner and outer walls is given by the manometer reading, $h_w$.

$$p_2 - p_1 = \rho_w g h_w = 10^3 \times 9.81 \times 15 \times 10^{-3} = 147.1 \text{ Pa} \qquad \text{(ii)}$$

The density of the air is given by equation 1.25c

$$\rho = \frac{p}{RT} = \frac{1.1 \times 10^5}{287 \times 290} = 1.322 \text{ kg/m}^3 \qquad \text{(iii)}$$

Rearranging equation (i) and substituting (ii) and (iii) we have

$$C = \left(\frac{2(p_2 - p_1)}{\rho(1/R_1{}^2 - 1/R_2{}^2)}\right)^{\frac{1}{2}} = \left(\frac{2 \times 147.1}{1.322(1/3.4^2 - 1/4.6^2)}\right)^{\frac{1}{2}} = 75.3$$

Consider the mass flow rate through a small element of area B dr. From equation 7.13 we obtain

$$\dot{m} = \int_{R_1}^{R_2} \rho u B\, dr = \rho B \int_{R_1}^{R_2} C r^{-1}\, dr = \rho B C \ln (R_2/R_1)$$

$$= 1.322 \times 1.2 \times 75.3 \ln (4.6/3.4) = 36.1 \text{ kg/s}$$

*Example 7.8*

Water flows from a tank A to a tank B through a 70 m length of 300 mm diameter pipe, followed by a 35 m length of 150 mm diameter pipe connected in series. The 300 mm pipe rises over a barrier at a point 35 m along the pipe and the centre of the pipe at this point lies 1 m below the water level in tank A. The difference in water levels between tanks A and B is 8 m. Determine (i) the volumetric flow rate, and (ii) the absolute pressure at a point 35 m along the 300 mm diameter pipe. Sketch the energy line and hydraulic gradient for the pipe system. Assume a roughness height, $k_s = 0.33$, and complete turbulence in both pipes. The minor losses, obtained from reference 2, are as follows: pipe 1 entry $0.43\ \bar{u}_1^2/2g$, sudden contraction $0.37\ \bar{u}_2^2/2g$, pipe 2 exit $\bar{u}_2^2/2g$. Atmospheric pressure is 1 bar.

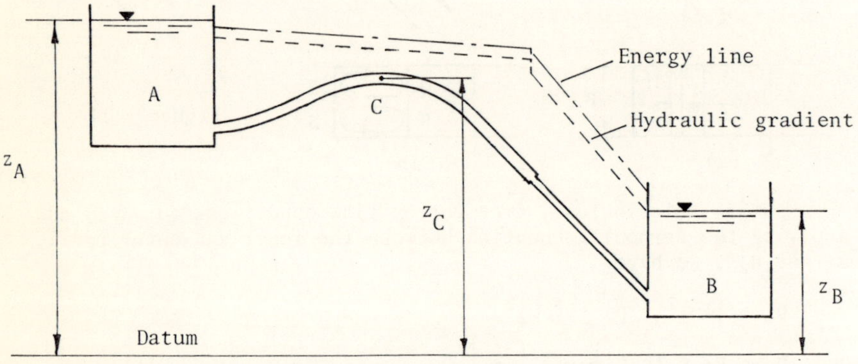


Figure 7.12

Assuming complete turbulence, the friction factor f is a function of $k_s/D$ only and can be found from figure 7.2.

For the 300 mm pipe, $k_s/D_1 = 0.33/300 = 0.0011$. $f_1 = 0.005$.
For the 150 mm pipe, $k_s/D_2 = 0.33/150 = 0.0022$. $f_2 = 0.006$.

(i) Applying the energy equation 7.11 between points on the water surface in tanks A and B we have

$$\frac{p_A}{\rho g} + \frac{\bar{u}_A^2}{2g} + z_A = \frac{p_B}{\rho g} + \frac{\bar{u}_B^2}{2g} + z_B + \Sigma\left(\frac{4fL\bar{u}^2}{2gD}\right) + \Sigma\left(\frac{k_m\bar{u}^2}{2g}\right)$$

and $$\frac{p_a}{\rho g} + 0 + z_A = \frac{p_a}{\rho g} + 0 + z_B + 0.43\frac{\bar{u}_1^2}{2g} + \frac{4f_1L_1\bar{u}_1^2}{2gD_1}$$

$$+ 0.37\frac{\bar{u}_2^2}{2g} + \frac{4f_2L_2\bar{u}_2^2}{2gD_2} + \frac{\bar{u}_2^2}{2g} \qquad \text{(i)}$$

The continuity equation 7.12 gives

$$\dot{V} = \bar{u}_1A_1 = \bar{u}_2A_2$$

$$\bar{u}_2 = A_1\bar{u}_1/A_2 = \pi 150^2\bar{u}_1/\pi 0.075^2 = 4\bar{u}_1 \qquad \text{(ii)}$$

Substituting for $\bar{u}_2$ in equation (i) and rearranging

$$\bar{u}_1^2(0.43 + 4f_1L_1/D_1 + 5.92 + 64f_2L_2/D_2 + 16) = 2g(z_A - z_B)$$

$$\bar{u}_1^2(0.43 + 4 \times 0.005 \times 70/0.3 + 5.92 + 64 \times 0.006 \times 35/0.15 + 16)$$

$$= 2 \times 9.81 \times 8$$

$$\bar{u}_1^2(0.43 + 4.68 + 5.92 + 89.6 + 16) = 157$$

This equation clearly shows the relative magnitudes of major (pipe friction) and minor (fittings) losses.

$$\bar{u}_1 = 1.14 \text{ m/s}$$

$$\dot{V} = \bar{u}_1 A_1 = 1.14 \times \pi \times 0.15^2 = 0.0805 \text{ m}^3/\text{s}$$

(ii) To determine the pressure at point C, 35 m along pipe 1, we must apply equation 7.11 between a point on the water surface in tank A, and point C.

$$\frac{p_a}{\rho g} + 0 + z_A = \frac{p_C}{\rho g} + \frac{\bar{u}_1^2}{2g} + z_C + 0.43\frac{\bar{u}_1^2}{2g} + \frac{4f_1L_1'\bar{u}_1^2}{2gD_1}$$

$$p_C = \rho g(z_A - z_C) - \tfrac{1}{2}\rho\bar{u}_1^2(0.43 + 4f_1L_1'/D_1 + 1) + p_a$$

$$= 10^3 \times 9.81 - \tfrac{1}{2} \times 10^3 \times 1.14^2(1.43 + 4 \times 0.005 \times 35/0.3) + 10^5$$

$$= 107.35 \text{ kPa}$$

Note that the absolute pressure, $p_C$, falls as the elevation of the crest increases. In extreme cases of low pressure (2.337 kPa at 20 °C) a vapour lock can form and prevent flow.

(iii) The total energy head at any point in a fluid pipeline is given by $H = p/\rho g + \bar{u}^2/2g + z$. Thus, an *energy line* may be drawn on the pipeline system as shown in figure 7.12. Pipe friction losses are shown by the slope of the energy line whereas minor losses are shown by step changes in the energy line.

The *hydraulic gradient*, or *piezometric line*, is the line drawn through points of $(p/\rho g + z)$ and at any point it lies below the energy line by the velocity head $\bar{u}^2/2g$.

The energy line must fall continuously due to the various losses but the hydraulic gradient may rise at enlargements of section. The usefulness of the hydraulic gradient is that it indicates the pressure variation in the pipeline and if the hydraulic gradient falls below the pipe centre line at any point the pressure is below atmospheric pressure. Thus, the hydraulic gradient can be used for stressing purposes or to predict the formation of vapour locks.

*Example 7.9*

(a) Discuss the Nikuradse experiments on rough pipes and explain why the rough pipe curves on the Moody diagram differ from those obtained by Nikuradse.
(b) It is required to supply ventilation air through a 750 mm square duct at the rate of 120 $m^3$/min. The duct is 200 m long and includes

four mitred bends, each of which causes a head loss of $0.4\bar{u}^2/2g$, a butterfly valve which in the fully open position causes a head loss of $\bar{u}^2/g$ and an inlet filter which causes a head loss of $2\bar{u}^2/g$. The outlet point is 5 m above the inlet and the pressures at inlet and outlet are equal. (i) Determine the power required to circulate the air. (ii) Due to faulty maintenance the inlet filter develops a leak which reduces its head loss to $\bar{u}^2/2g$ but which allows contaminating particles to be drawn into the ducting system. These encrust the inner surface of the duct and give an effective relative roughness of $k_s/D = 0.001$. Determine the fan power required to maintain the original air flow. For air assume $\rho = 1.2$ kg/m$^3$ and $\mu = 18 \times 10^{-6}$ N s/m$^2$.

(a) Nikuradse carried out his work on pipes artificially roughened with sand grains of uniform size k. He obtained results for six values of k/D ranging from 1/30 to 1/1014. The test data produced a series of smooth curves on a f-Re graph as shown in figure 7.13a. It should be noted that (i) the laminar portion is unaffected by roughness, (ii) above a certain value of Re, for each k/D value, f becomes independent of Re and (iii) as k/D increases, departure from the smooth pipe curve occurs for decreasing Re values.

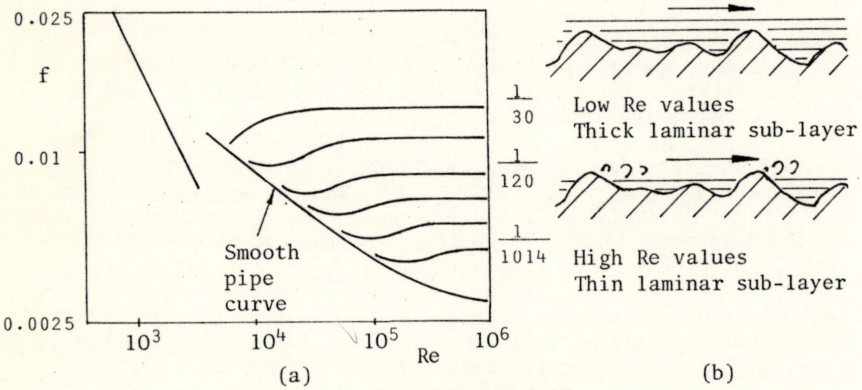


Figure 7.13

The physical explanation for the effect of surface roughness is connected with the fact that a laminar sub-layer always exists next to the wall in turbulent flow and the thickness of this sub-layer decreases with increasing Re (figure 7.13b). If the roughness elements are completely immersed in this laminar sub-layer the pipe is said to be hydraulically smooth and the smooth pipe law is applicable. As Re increases and the sub-layer thickness decreases, the roughness elements are progressively exposed to the main turbulent flow and resistance is then partly due to form drag and the curve departs from the smooth pipe curve. For high Re values the roughness elements become completely exposed to the main flow and resistance is due entirely to form drag.

Now in the Nikuradse experiments, all the roughness elements were the same size and so they were exposed simultaneously to the main stream flow. This produced a fairly sudden break away from the smooth pipe curves and inflexion points occurred.

The Moody diagram (figure 7.2) is based on results for commercial pipes where the roughness elements are irregular and exposure to the main stream is also irregular with increasing Re. Thus the rough pipe curves deviate from the smooth pipe curve gradually.

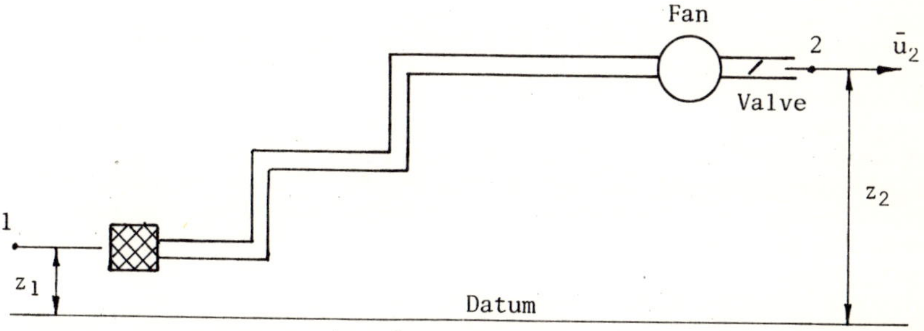


Figure 7.14

(b)(i) For incompressible, adiabatic flow with work transfer, the energy equation 4.32b gives, in head units

$$\frac{p_1}{\rho g} + \frac{\bar{u}_1{}^2}{2g} + z_1 = \frac{\dot{W}_x}{\dot{m}g} + \frac{p_2}{\rho g} + \frac{\bar{u}_2{}^2}{2g} + z_2 + \frac{1}{g}(e_2 - e_1) \qquad \text{(i)}$$

Now $(e_2 - e_1)/g$ is the sum of pipe friction and minor losses and $W_x/\dot{m}g = H_{fan}$ is the increase in head across the fan. Applying the convention that work done on a system is negative, equation (i) may be written

$$\frac{p_1}{\rho g} + \frac{\bar{u}_1{}^2}{2g} + z_1 + H_{fan} = \frac{p_2}{\rho g} + \frac{\bar{u}_2{}^2}{2g} + z_2 + \Sigma\left(\frac{4fL\bar{u}^2}{2gD}\right) + \Sigma\left(\frac{k_m\bar{u}^2}{2g}\right) \text{(ii)}$$

Applying equation (ii) between a point 1 sufficiently far from the inlet filter for the velocity to be negligible, and a point 2 at exit we obtain

$$\frac{p_a}{\rho g} + 0 + z_1 + H_{fan} = \frac{p_a}{\rho g} + \frac{\bar{u}_2{}^2}{2g} + z_2 + \frac{4f_1L\bar{u}^2}{2gD_e} + \frac{2\bar{u}^2}{g} + \frac{1.6\bar{u}^2}{2g} + \frac{\bar{u}^2}{g}$$

$$H_{fan} = (z_2 - z_1) + (8.6 + 4f_1L/D_e)\bar{u}^2/2g$$

and $$\dot{W}_x = \dot{m}gH_{fan} = \dot{m}g\left[(z_2 - z_1) + (8.6 + 4f_1L/D_e)\bar{u}^2/2g\right] \qquad \text{(iii)}$$

Note that the fan work consists of two parts. The first part $g(z_2 - z_1)$ is the work required to lift the fluid through an elevation $(z_2 - z_1)$ and the second part, $gH_L$, is the work required to overcome the frictional resistance of the pipe and fittings.

Now $D_e = \frac{4A}{p} = \frac{4 \times 0.75 \times 0.75}{4 \times 0.75} = 0.75$ m

$$Re = \frac{\rho \bar{u} D_e}{\mu} = \frac{\rho \dot{V} D_e}{A\mu} = \frac{1.2 \times 120 \times 0.75}{60 \times 0.75^2 \times 18 \times 10^{-6}} = 1.78 \times 10^5$$

For a smooth duct, the f-Re chart (figure 7.2) gives $f_1 = 0.004$. The fan power $P = \dot{W}_x$ is obtained from equation (iii)

$$P_1 = \rho \dot{V} g H_{fan}$$

$$= \frac{1.2 \times 120 \times 9.81}{60}\left\{5 + \frac{120^2(8.6 + 4 \times 0.004 \times 200/0.75)}{(0.75^2)^2 \times 60^2 \times 2 \times 9.81}\right\}$$

$$= 313 \text{ W}$$

(ii) The fittings' losses now change and in addition the friction factor for the duct increases due to encrusting. From the f-Re chart, for $k_s/D = 0.001$ and $Re = 1.78 \times 10^5$, we obtain $f_2 = 0.0053$.

$$P_2 = \rho \dot{V} g\left[(z_2 - z_1) + \frac{\bar{u}^2}{2g} + \frac{4fL\bar{u}^2}{2gD_e} + \frac{\bar{u}^2}{2g} + \frac{4 \times 0.4\bar{u}^2}{2g} + \frac{\bar{u}^2}{g}\right]$$

$$= \frac{1.2 \times 120 \times 9.81}{60}\left\{5 + \frac{120^2(5.6 + 4 \times 0.0053 \times 200/0.75)}{(0.75^2)^2 \times 60^2 \times 2 \times 9.81}\right\}$$

$$= 288 \text{ W}$$

*Example 7.10*

Two reservoirs having a difference in surface level of 100 m are connected by a series pipe system which consists of an initial 800 m length of 400 mm diameter pipe followed by a 200 m length of 200 mm diameter pipe. Assuming a constant friction factor of 0.01, determine (i) the initial volumetric flow rate, and (ii) the percentage change in the volumetric flow rate from the upper reservoir if the downstream 200 mm diameter pipe is now tapped by side pipes so that one-quarter of the water entering the pipe is withdrawn uniformly over this length. Neglect all losses other than those due to pipe friction.

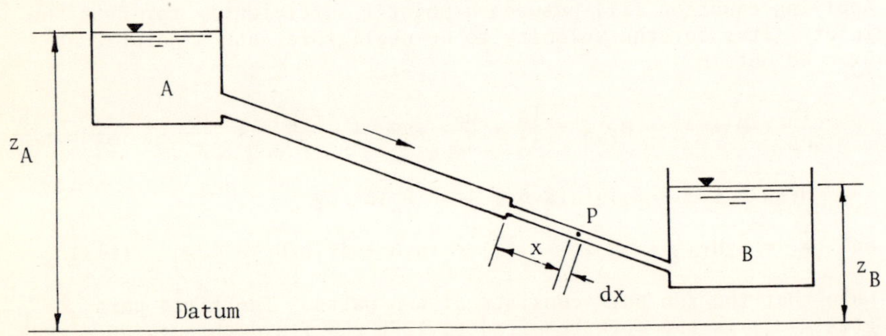


Figure 7.15

(i) Applying the energy equation 7.11 between points on the water surfaces in tanks A and B we obtain

$$\frac{p_a}{\rho g} + 0 + z_A = \frac{p_a}{\rho g} + 0 + z_B + \frac{4fL_1\bar{u}_1^2}{2gD_1} + \frac{4fL_2\bar{u}_2^2}{2gD_2} \qquad \text{(i)}$$

Substituting $\bar{u}_1 = \dot{V}/A_1$ and $\bar{u}_2 = \dot{V}/A_2$ and rearranging

$$\dot{V} = \left(\frac{2g(z_A - z_B)(\pi/4)^2}{4f(L_1/D_1^5 + L_2/D_2^5)}\right)^{\frac{1}{2}} = \left(\frac{2 \times 9.81 \times 100 \times \pi^2 \times 0.25^2}{4 \times 0.01(800/0.4^5 + 200/0.2^5)}\right)^{\frac{1}{2}}$$

$$= 0.207 \text{ m}^3/\text{s}$$

(ii) Let q $\text{m}^3/\text{s}$ of water per metre of pipe be drawn off uniformly from the second pipe. Let the flow in the first pipe be $\dot{V}_1$. At a point, P, distance x from entry to the second pipe, the water velocity is given by

$$\bar{u} = \frac{\dot{V}_1 - qx}{\pi D_2^2/4}$$

From the Darcy equation 7.2 the head losses in pipe 2 are

$$(H_L)_{dx} = \frac{4f\,dx}{2gD_2}\left(\frac{\dot{V} - qx}{\pi D_2^2/4}\right)^2$$

$$\text{and} \quad H_{L2} = \frac{4f}{2g(\pi/4)^2 D_2^5}\int_0^{L_2} (\dot{V}_1 - qx)^2\, dx$$

$$= \frac{4f}{2g(\pi/4)^2 D_2^5}\left(\dot{V}^2 L_2 - qL_2^2\dot{V}_1 + \frac{q^2L_2^3}{3}\right)$$

Substituting the given information $q = (\dot{V}_1/4)/L_2$ we have

$$H_{L2} = \frac{4fL_2\dot{V}^2 \times 37}{2g(\pi/4)^2 D_2^5 \times 48}$$

Applying the energy equation 7.11 again between points on the water surfaces in tanks A and B we obtain

$$\frac{p_a}{\rho g} + 0 + z_A = \frac{p_a}{\rho g} + 0 + z_B + \frac{4fL_1\dot{V}_1^2}{2g(\pi/4)^2 D_1^5} + \frac{4fL_2\dot{V}_1^2 \times 37}{2g(\pi/4)^2 D_2^5 \times 48}$$

$$\dot{V} = \left(\frac{(z_A - z_B)2g(\pi/4)^2}{4f(L_1/D_1^5 + 37L_2/48D_2^5)}\right)^{\frac{1}{2}}$$

$$= \left(\frac{100 \times 2 \times 9.81 \times (\pi/4)^2}{4 \times 0.01[800/0.4^5 + (37 \times 200)/(48 \times 0.2^5)]}\right)^{\frac{1}{2}} = 0.232 \text{ m}^3/\text{s}$$

$$\text{Percentage increase in flow} = \frac{(0.232 - 0.207)100}{0.207} = 12.1\%$$

*Example 7.11*

Two pipes, each 30 m long and 50 mm and 100 mm diameter, respectively, are connected in parallel between two reservoirs whose difference of water level is 8 m. Determine (i) the flow in $\text{m}^3/\text{s}$ for each pipe and draw the corresponding hydraulic gradients, and (ii) the diameter of a single pipe, 30 m long, which will give the same flow as the two pipes in (i). Assume an entry loss of $0.5\bar{u}^2/2g$ and a friction factor $f = 0.008$ in each case.

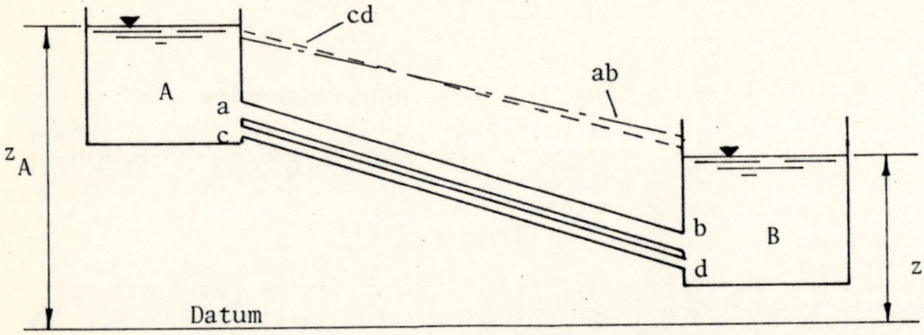


Figure 7.16

(i) In any problem on parallel flow it is necessary to apply the energy equation 7.11 between points on the water surfaces in tanks A and B for each possible flow path a fluid particle may take. In this problem there are two possible flow paths ab and cd. The exit loss for each pipe is $\bar{u}^2/2g$.

<u>Path ab</u> $$\frac{p_a}{\rho g} + 0 + z_A = \frac{p_a}{\rho g} + 0 + z_B + 0.5\frac{\bar{u}_1^2}{2g} + \frac{4fL\bar{u}_1^2}{2gD_1} + \frac{\bar{u}_1^2}{2g} \qquad \text{(i)}$$

<u>Path cd</u> $$\frac{p_a}{\rho g} + 0 + z_A = \frac{p_a}{\rho g} + 0 + z_B + 0.5\frac{\bar{u}_2^2}{2g} + \frac{4fL\bar{u}_2^2}{2gD_2} + \frac{\bar{u}_2^2}{2g} \qquad \text{(ii)}$$

From equations (i) and (ii)

$$\bar{u}_1 = \left(\frac{2g(z_A - z_B)}{0.5 + 4fL/D_1 + 1}\right)^{\frac{1}{2}} = \left(\frac{2 \times 9.81 \times 8}{0.5 + 4 \times 0.008 \times 30/0.1 + 1}\right)^{\frac{1}{2}}$$

$$= 3.76 \text{ m/s}$$

and $$\bar{u}_2 = \left(\frac{2 \times 9.81 \times 8}{0.5 + 4 \times 0.008 \times 30/0.05 + 1}\right)^{\frac{1}{2}} = 2.76 \text{ m/s}$$

From the continuity equation 7.12

$$\dot{V} = \Sigma(\bar{u}A) = \bar{u}_1A_1 + \bar{u}_2A_2 = 3.76 \times \tfrac{1}{4}\pi \times 0.1^2 + 2.76 \times \tfrac{1}{4}\pi \times 0.05^2$$

$$= 0.0349 \text{ m}^3/\text{s}$$

(ii) Applying the energy equation 7.11 for flow through a single pipe of diameter D

$$\frac{p_a}{\rho g} + 0 + z_A = \frac{p_a}{\rho g} + 0 + z_B + 0.5\frac{\bar{u}_3^2}{2g} + \frac{4fL\bar{u}_3^2}{2gD_3} + \frac{\bar{u}_3^2}{2g}$$

$$\frac{\bar{u}_3^2}{2g}\left(0.5 + \frac{4fL}{D_3} + 1\right) = z_A - z_B$$

Substituting values and $\bar{u}_3 = \dot{V}/\tfrac{1}{4}\pi D_3^2$

$$1.5 + \frac{4 \times 0.008 \times 30}{D_3} = \frac{8 \times 2 \times 9.81 \times 0.25^2 \times \pi^2 \times D_3^4}{0.0349^2}$$

$$8.05 \times 10^4 D_3^5 - 1.5\, D_3 - 0.96 = 0$$

By successive approximations $D_3 = 0.107$ m

Note that the head loss at entry is greater for the 100 mm diameter pipe than for the 50 mm diameter pipe ($\bar{u}_1 > \bar{u}_2$). Thus the hydraulic gradients show that the friction loss must be greater for the smaller pipe.

*Example 7.12*

In example 7.11 a burst occurs in the 100 mm diameter pipe and in order to effect a repair the centre 10 m of this pipe is removed. During repair the total flow is diverted over this length through the 50 mm diameter pipe. Determine the reduction in flow due to the diversion. Neglect minor losses and comment on the accuracy of the end result.

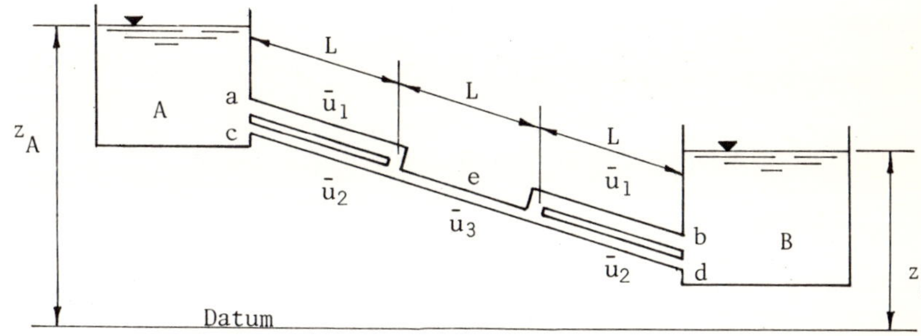


Figure 7.17

Refer to figure 7.17 which shows the configuration during the diversion. There are now four possible flow paths a fluid particle may take, viz. aeb, aed, ced and ceb, of which aed and ceb are essentially the same when minor losses are neglected. Applying the energy equation 7.11 between points on the water surfaces in tanks A and B for each possible flow path we obtain

<u>Path aeb</u>

$$\frac{4f(2L)\bar{u}_1^2}{2gD_1} + \frac{4fL\bar{u}_3^2}{2gD_2} = z_A - z_B$$

$$\frac{2\bar{u}_1^2}{D_1} + \frac{\bar{u}_3^2}{D_2} = \frac{g(z_A - z_B)}{2fL} \qquad \text{(i)}$$

<u>Path aed</u>

$$\frac{4fL\bar{u}_1^2}{2gD_1} + \frac{4fL\bar{u}_3^2}{2gD_2} + \frac{4fL\bar{u}_2^2}{2gD_2} = z_A - z_B$$

$$\frac{\bar{u}_1^2}{D_1} + \frac{\bar{u}_3^2}{D_2} + \frac{\bar{u}_2^2}{D_2} = \frac{g(z_A - z_B)}{2fL} \qquad \text{(ii)}$$

Path ced

$$\frac{4f(2L)\bar{u}_2^2}{2gD_2} + \frac{4fL\bar{u}_3^2}{2gD_2} = z_A - z_B$$

$$\frac{2\bar{u}_2^2}{D_2} + \frac{\bar{u}_3^2}{D_2} = \frac{g(z_A - z_B)}{2fL} \qquad \text{(iii)}$$

From equations (i) and (iii)

$$\bar{u}_1^2 = D_1\bar{u}_2^2/D_2 \qquad \text{(iv)}$$

The continuity equation 7.12 gives

$$\dot{V}_3 = \Sigma(\bar{u}A) = \bar{u}_1\tfrac{1}{4}\pi D_1^2 + \bar{u}_2\tfrac{1}{4}\pi D_2^2 = \bar{u}_3\tfrac{1}{4}\pi D_3^2$$

$$\bar{u}_1D_1^2 + \bar{u}_2D_2^2 = \bar{u}_3D_3^2 \qquad \text{(v)}$$

From equations (iv) and (v)

$$\bar{u}_2D_1^2\sqrt{(D_1/D_2)} + \bar{u}_2D_2^2 = \bar{u}_3D_2^2$$

$$\bar{u}_2 = \left(\frac{\bar{u}_3D_2^2}{D_1^2\sqrt{(D_1/D_2)} + D_2^2}\right) = \left(\frac{0.05^2}{0.1^2\sqrt{(0.1/0.05)} + 0.05^2}\right)\bar{u}_3$$

$$= 0.15\bar{u}_3 \qquad \text{(vi)}$$

From equations (iii) and (vi)

$$\frac{2 \times 0.15^2\bar{u}_3^2}{D_2} + \frac{\bar{u}_3^2}{D_2} = \frac{g(z_A - z_B)}{2fL}$$

$$\bar{u}_3 = \left(\frac{9.81 \times 8 \times 0.05}{2 \times 0.008 \times 10(0.045 + 1)}\right)^{\frac{1}{2}} = 4.84 \text{ m/s}$$

$$\dot{V}_3 = \bar{u}_3A_3 = 4.84 \times \tfrac{1}{4}\pi \times 0.05^2 = 0.0095 \text{ m}^3\text{/s}$$

Reduction in flow = 0.0349 - 0.0095 = 0.0254 $m^3/s$

In obtaining this solution, minor losses were neglected in order to simplify the analysis. However, it was shown in example 7.11 that minor losses are significant (up to 10% of pipe friction losses) and the flow during the diversion will be less than the value of 0.0093 $m^3/s$ obtained above.

*Example 7.13*

Three reservoirs A, B and C are interconnected as shown in figure 7.18. The pipe from A to a junction D is 300 mm diameter and 16 km in length. The pipe from D to B is 225 mm diameter and 9.6 km long and that from D to C is 150 mm diameter and 8 km long. The ends of all pipes are submerged beneath the free water surfaces. Calculate the flow in each pipe. Neglect minor losses and assume a friction factor f = 0.01 for all pipes.

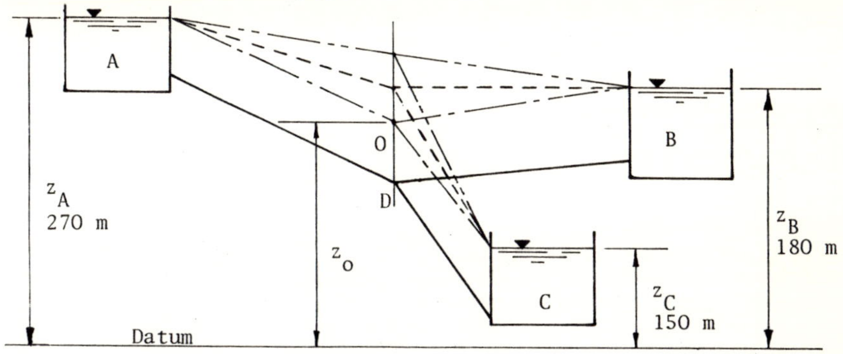


Figure 7.18

This problem may be solved by use of the energy line. Neglecting minor losses, the head loss in each pipe is given by the Darcy equation 7.2.

$$H_L = \frac{4fL\bar{u}^2}{2gD} = \left(\frac{4fL}{2g(\pi/4)^2D^5}\right)\dot{V}^2 = K\dot{V}^2$$

The value of K may be evaluated for each pipe.

$$H_{L_1} = \left(\frac{4 \times 0.01 \times 16 \times 10^3}{2 \times 9.81 \times (\pi/4)^2 \times 0.3^5}\right)\dot{V}^2 = 2.176 \times 10^4\dot{V}^2$$

$$H_{L_2} = \left(\frac{4 \times 0.01 \times 9.6 \times 10^3}{2 \times 9.81 \times (\pi/4)^2 \times 0.225^5}\right)\dot{V}^2 = 5.5 \times 10^4\dot{V}^2$$

$$H_{L_3} = \left(\frac{4 \times 0.01 \times 8 \times 10^3}{2 \times 9.81 \times (\pi/4)^2 \times 0.15^5}\right)\dot{V}^2 = 3.48 \times 10^5\dot{V}^2$$

Energy lines may be drawn for the three possible flow cases and the energy and continuity equations applied. The junction of the energy lines, denoted by O, is at elevation $z_0$ above datum. The value of $z_0$ is determined by the flow case.

Case 1 Flow AD, DB, DC (Energy line (i) ——--—— )

AD $z_A - z_0 = 2.176 \times 10^4\dot{V}_1^2$ (i)

DB $z_0 - z_B = 5.5 \times 10^4\dot{V}_2^2$ (ii)

DC $z_0 - z_C = 3.48 \times 10^5\dot{V}_3^2$ (iii)

$\dot{V}_1 = \dot{V}_2 + \dot{V}_3$ (iv)

Case 2 Flow AD, DC. Zero flow in DB (Energy line (ii) - - - - )

AD $z_A - z_0 = 2.176 \times 10^4\dot{V}_1^2$ (v)

DC $z_0 - z_C = 3.48 \times 10^5\dot{V}_3^2$ (vi)

$$\dot{V}_1 = \dot{V}_3 \qquad \text{(vii)}$$

Case 3 Flow AD, BD, DC (Energy line (iii) ——- ——- - —)

AD $z_A - z_o = 2.176 \times 10^4 \dot{V}_1^2$ (viii)

BD $z_B - z_o = 5.5 \times 10^4 \dot{V}_2^2$ (ix)

DC $z_o - z_C = 3.48 \times 10^5 \dot{V}_3^2$ (x)

$$\dot{V}_1 + \dot{V}_2 = \dot{V}_3 \qquad \text{(xi)}$$

Consider case 2 ($z_o = z_B = 180$ m). Equations (v) and (vi) give

$$\dot{V}_1 = \left(\frac{270 - 180}{2.176 \times 10^4}\right)^{\frac{1}{2}} = 0.0643 \text{ m}^3/\text{s}$$

$$\dot{V}_3 = \left(\frac{180 - 150}{3.48 \times 10^5}\right)^{\frac{1}{2}} = 0.00928 \text{ m}^3/\text{s}$$

Case 2 is not applicable because the calculated values of $\dot{V}_1$ and $\dot{V}_3$ do not satisfy the continuity equation (vii). It may be deduced from the energy lines that Case 1 is the only one that would produce the desired effect of decreasing $\dot{V}_1$, diverting flow $\dot{V}_2$ into B and also increasing $\dot{V}_3$.

There are four equations, (i) to (iv), and four unknowns, $\dot{V}_1$, $\dot{V}_2$, $\dot{V}_3$ and $z_o$, but algebraic solution is very difficult. It is preferable to use an iterative method in which a value is chosen for $z_o$ and the values of $\dot{V}_1$, $\dot{V}_2$ and $\dot{V}_3$ evaluated from equations (i) to (iv). Substituting values for z we obtain

$$270 - z_o = 2.176 \times 10^4 \dot{V}_1^2$$

$$z_o - 180 = 5.5 \times 10^4 \dot{V}_2^2$$

$$z_o - 150 = 3.48 \times 10^5 \dot{V}_3^2$$

$$\dot{V}_1 = \dot{V}_2 + \dot{V}_3$$

When $\dot{V}_1 = \dot{V}_2 + \dot{V}_3$ the correct value of $z_o$ has been assumed. It is convenient to carry out the iterations in the form of the table given below.

| Assumed $z_o$ | $\dot{V}_1$ | $\dot{V}_2$ | $\dot{V}_3$ | $\dot{V}_2 + \dot{V}_3$ |
|---|---|---|---|---|
| 200 | 0.0567 | 0.01907 | 0.012 | 0.031 |
| 240 | 0.0371 | 0.033 | 0.01608 | 0.0491 |
| 230 | 0.0429 | 0.0301 | 0.0152 | 0.0453 |
| 228 | 0.0439 | 0.02954 | 0.015 | 0.0445 |
| 227 | 0.04445 | 0.02923 | 0.01487 | 0.0441 |
| 227.5 | 0.04419 | 0.02939 | 0.01492 | 0.0443 |

The flows are $\dot{V}_1 = 0.0442$ m$^3$/s, $\dot{V}_2 = 0.0294$ m$^3$/s, $\dot{V}_3 = 0.0149$ m$^3$/s.

*Problems*

1 Glycerine is forced through the narrow space formed between two plane glass plates. The distance between the plates is t and the width of the space is b. Assuming laminar two-dimensional and fully developed flow, (i) derive an expression for the longitudinal pressure gradient dp/dx in terms of the volumetric flow rate $\dot{V}$, the dynamic viscosity μ and the distances t and b, (ii) determine the maximum velocity between the plates if μ = 0.96 Pa s, b = 100 mm, t = 0.76 mm and dp/dx = 836 kN/$m^3$, and (iii) by use of the Laplace equation show why this configuration can be used for the Hele-Shaw analogy for ideal, incompressible two-dimensional flow.

[$dp/dx = 12\mu\dot{V}/t^3b$, 63 mm/s, for laminar flow $\nabla^2 p = 0$]

2 A centrifugal pump is used for pumping oil from a tank at the same level as the pump inlet to the bottom of a header tank, whose liquid level is 20 m above the pump inlet. The shaft driving the pump impeller is 20 mm diameter and is supported by the motor bearings. The shaft passes through a simple metal-to-metal seal, 40 mm long in the pump casing, the clearance being 0.1 mm. Calculate the leakage rate through the seal when the pump is not operating, there being a non-return valve at the pump inlet. Assume the oil density is 900 kg/$m^3$ and the viscosity is $10^{-2}$ N s/$m^2$. When the above pump is running at 1500 rev/min the casing is so designed that the pressure at the gland is the same as at the pump inlet. Calculate, from first principles, the power lost in the gland.

[$2.32 \times 10^{-6}$ $m^3$/s, 0.62W]

3 A piston 50 mm diameter and 75 mm long moves vertically in an oil dashpot with a small uniform clearance of 1.25 mm. Under its own weight the piston descends at uniform speed through 38 mm in 40 s. When an additional mass of 100 g is placed on top of the piston, it descends uniformly through the same distance in 25 s. Calculate the dynamic viscosity of the oil.

[0.151 Pa s]

4 (a) A piston of diameter, $D_1$, length, L, and radial clearance, C, is moved concentrically along a cylinder of internal diameter, $D_2$, at constant velocity, U, by means of an external force F. The annular space between the piston and cylinder wall is filled with oil having a coefficient of viscosity, μ. Assuming laminar flow, show that F is given by the expression

$$F = 2\pi\mu LU/\ln(1 + 2C/D_1)$$

(b) Show that the error involved in the value of F, by assuming a linear velocity distribution across the annular space, is less than 1% when $2C/D_1$ is less than 2%.
[N.B. $\log_e(1 + x) = x - x^2/2 + x^3/3 - x^4/4 + \dots$ ]

5 Show that the velocity distribution for fully developed laminar flow in a horizontal annulus of outer radius $R_2$ and inner radius $R_1$ is given by

$$u = -\frac{1}{4\mu}\frac{dp}{dx}\left[R_2{}^2 - r^2 - \frac{(R_2{}^2 - R_1{}^2)}{\ln(R_2/R_1)}\ln\frac{R_2}{r}\right]$$

where x is the flow direction and r is any radius.

6 The velocity distribution for turbulent flow in a pipe is given by $V = V_{max}(1 - r/R)^{1/7}$. Working from first principles determine (i) the mean velocity, (ii) the momentum correction factor, and (iii) the kinetic energy correction factor.

[49 $V_{max}/60$, $\beta = 1.02$, $\alpha = 1.06$]

7 (a) Derive an expression for the velocity of flow at any radius when a viscous fluid flows through a uniform circular pipe under laminar conditions. Hence determine the radius at which a pitot tube must be placed in order to record the mean velocity, $\bar{u}$, in the pipe.

(b) Calculate the error involved if the kinetic energy per unit mass in the above pipe is taken to be $\bar{u}^2/2$.

[$u = (\Delta p/L)(D^2 - 4r^2)/16\mu$, $r = D/(2\sqrt{2})$, $\alpha = 2$, 50%]

8 Water flows through a horizontal 400 mm by 400 mm square duct at the rate of 1 $m^3/s$. There is a right-angle bend in the duct with inner wall radius 200 mm. A mercury manometer is connected to tappings situated in the inner and outer walls of the bend. Determine the pressure difference recorded by the manometer, assuming frictionless flow and a free vortex velocity distribution.

[463 mm]

9 A diffuser is fitted between two pipes of diameters 100 mm and 150 mm respectively. Water flows at a rate of 0.05 $m^3/s$ through the diffuser and it is found that the pressure recovery factor is 0.66. Determine (i) the value of the factor k in the pressure loss term $\frac{1}{2}\rho\bar{u}_1{}^2\, k\,(1 - A_1/A_2)^2$ and (ii) the diffuser efficiency.

[0.9, 83%]

10 The flow of water between two reservoirs is to be controlled by a gate valve situated near to the lower reservoir. If the difference in reservoir water levels is 30 m and the pipe joining the reservoirs is 1 m diameter, 1000 m long and with a mean friction factor of 0.007, obtain an expression for the volume flow rate $\dot{V}$ in terms of k, the coefficient in the valve head loss term $h_L = k\bar{u}^2/2g$. Ignore reservoir exit and entry losses.

By use of the values of k given below comment upon the valve's sensitivity as a flow controller. What would be the effect of (i) a similar valve in series and (ii) a similar or smaller valve in parallel, as a means of improving the controllability of the flow without reducing the maximum water flow rate?

| Valve opening: | full | $\frac{3}{4}$ | $\frac{1}{2}$ | $\frac{1}{4}$ |
|---|---|---|---|---|
| k: | 0.2 | 1.15 | 5.6 | 24 |

[$\dot{V} = [30/(2.31 + 0.082k)]$, small valve in parallel required]

11 Fluid of constant density flows through an abrupt enlargement from a horizontal pipe of radius $R_1$ to a concentric horizontal pipe of radius $R_2$. At exit from the smaller pipe the velocity at a distance r from the axis is given by $u = u_{max}(1 - r/R_1)^{1/7}$. At a downstream section 2, the velocity is given by $u = u_{max}(1 - r/R^2)$. Derive an expression for the head loss between the exit from the smaller pipe and section 2. Neglect friction effects.

$[H_L = u_1^2[0.66(R_1^4/R_2^4) - 2.04(R_1^2/R_2^2) + 1.06]/2g]$

12(a) Water is drawn off from a horizontal water main at a uniform rate q $m^3\ s^{-1}\ m^{-1}$ along the length of the pipe. Obtain an expression for the overall pressure head loss along the pipe of length L, diameter d, friction factor f and volume flow rate at inlet $\dot{V}_1$. Hence show that this loss is one-third that with a constant flow rate $\dot{V}_1$.

(b) A horizontal ring main, 2000 m in perimeter, distributes water at a rate of $250 \times 10^{-6}\ m^3\ s^{-1}\ m^{-1}$. The ring is supplied by a single horizontal pipe, 1000 m long, from a pump. Calculate the pump pressure if the minimum pressure level in the system is to be 40 m water gauge. The diameter and friction factor of the ring main and supply pipe are 0.5 m, 0.01, 1 m and 0.01 respectively.

[43.025 m]

13 Two reservoirs, whose water surface levels differ by 100 m, are connected by a pipe system. A single pipe 200 mm diameter leaves the upper reservoir and, after a straight run of 100 m, it branches into two pipes, each 100 mm diameter. One pipe 50 m long runs straight to the lower reservoir. The second pipe 100 m long rises over a barrier at a point 60 m along the pipe and then runs to the lower reservoir. The top of the barrier is 10 m above the water surface in the lower reservoir. Determine the discharge between the reservoirs and the minimum pressure in the pipe passing over the barrier. Neglect minor losses in the pipe system and assume f = 0.01 throughout.

[0.123 $m^3$/s, 2.1 bar]

14 (a) Oil is to be pumped from a tank A to a higher tank B, with a difference in surface level of 30 m, through 50 m of 100 mm diameter rough pipe ($k_s/D = 0.002$). Determine the pumping power required to maintain a flow of 0.25 $m^3$/s.

(b) The pump is now stopped and it is found that oil leaks back through the pipe against the pump internal resistance which is equal to 28 m head of oil. Assuming that leakage occurs under laminar flow conditions for which f = 16/Re, determine the leakage rate in $m^3$/s. $\nu_{oil}$ = 300 $mm^2$/s, $\rho_{oil}$ = 950 kg/$m^3$.

[2 MW, 0.0032 $m^3$/s]

15 Determine the total ventilation pressure and the fan power required when air flows at the rate of 150 $m^3$/s through the mine shown in figure 7.19.(i) when the rectangular airway has steel girders on timber legs and f = 0.031 (mining data), and (ii) when the rectangular airway is smooth concrete lined and f = 0.0062 (mining data). For case (ii), determine the percentage difference between the fan power obtained using the friction factor, f, from mining data, and

that obtained using a value of f from figure 7.2 with $k_s = 1.3$ mm for concrete ducting. Assume a natural ventilation pressure of 25 mm $H_2O$, assisting the flow, and a constant density 1.2 $kg/m^3$. $\mu_a = 1.8 \times 10^{-5}$ Pa s.

[889 kW, 162 kW, 89.6 kW, 45%]

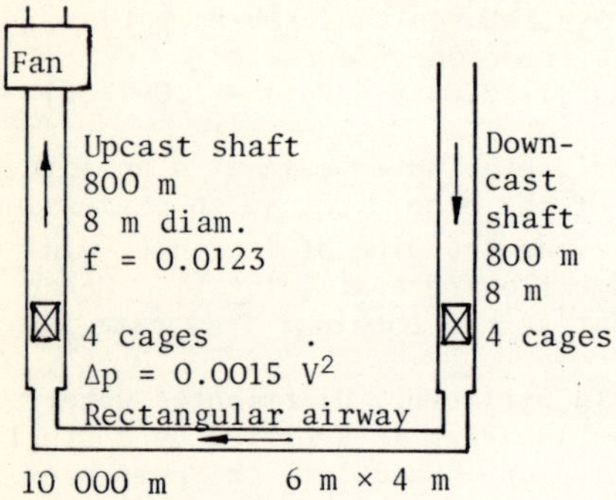


Figure 7.19

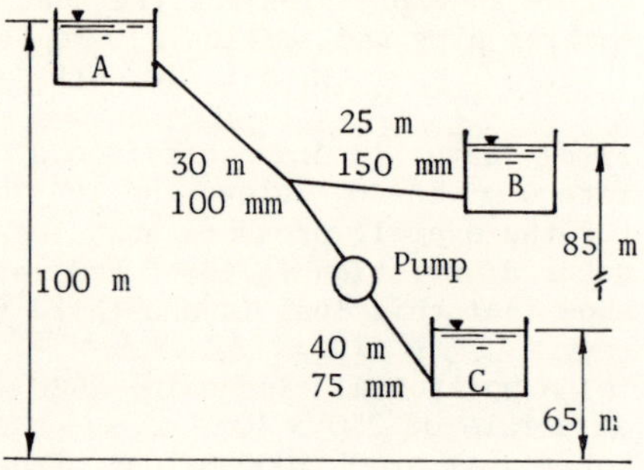


Figure 7.20

16 (a) Two water storage reservoirs are to be linked together by 300 m of single pipeline. If the maximum flow between the reservoirs is to be 0.1 $m^3/s$ when the difference in their water levels is 100 m, calculate the size of pipe required, the pipe friction factor and the velocity of the water. Neglect minor losses and assume turbulent flow through the pipeline for which $f = 0.079\ Re^{-0.25}$. $\nu = 10^{-5}\ m^2/s$.

(b) The flow between the reservoirs is to be increased by 50% by laying a second pipe in parallel with, and adjacent to, the original pipe for part of the distance. If the second pipe is identical in size and friction factor to the first pipe, calculate the length of the new pipe.

[220 mm, 0.0051, 2.63 m/s, 2200 m]

17 Two reservoirs, having a constant difference in surface level of 40 m, are connected by two parallel pipes of diameters 500 mm and 250 mm, respectively. Both pipes are 1 km long. Due to a burst in the 500 mm pipe, near the lower reservoir, the last 100 m section of 500 mm pipe is removed and the total flow over this length is diverted through the 250 mm pipe. Determine the reduction in flow, in $m^3/s$, resulting from the diversion. Neglect all losses other than pipe friction and assume f = 0.01 throughout.

[2.28 $m^3/s$, 1 $m^3/s$, 1.28 $m^3/s$]

18 Water is pumped at the rate of $m^3/s$ from a tank C to two higher tanks A and B as shown in figure 7.20. The pipes have lengths and diameters as specified. Neglecting all losses other than those due to pipe friction, determine (i) the pumping power, and (ii) the discharge to or from each of the tanks A and B. Assume f = 0.01. If the pump is now stopped, determine the flow in each pipe. Neglect the resistance of the pump.

$\left[\begin{array}{l}\dot{V}_A = 0.0328\ m^3/s,\ \dot{V}_B = 0.0628\ m^3/s,\ 22\ kW \\ \dot{V}_A = 0.0383\ m^3/s,\ 0.01917\ m^3/s,\ 0.01913\ m^3/s\end{array}\right]$

# 8 FLOW WITH A FREE SURFACE

Gas flow in a closed duct is always bounded by the solid surfaces of the duct but liquid flow may also take place when the uppermost boundary is the free surface of the liquid itself. Flow with a free surface usually implies flow in *open channels* which may be natural, such as streams and rivers, or artificial, such as canals or irrigation ditches. However, the principles are equally applicable to flow in pipes, sewers and tunnels not running full. The common feature of all these flows is the constant pressure (usually atmospheric) exerted on all points of the free surface.

## 8.1 CHARACTERISTICS OF FLOW IN OPEN CHANNELS

Flow in free surface situations occurs because of the influence of gravity. Hence, the slope of the channel or duct is an important consideration.

Flow in an open channel may be uniform or non-uniform, steady or unsteady. The various types of flow are shown in figure 8.1 in which the slopes of the channel are greatly exaggerated.

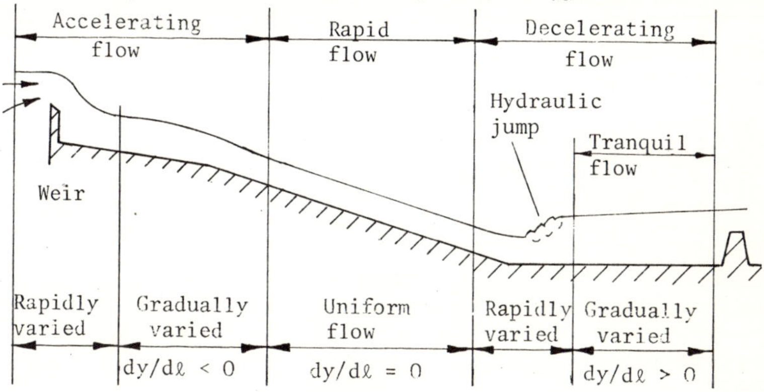


Figure 8.1

*Uniform* flow occurs if the depth does not change along the length of a duct of constant cross-section. Thus, uniform flow is characterised by the liquid surface being parallel to the channel bottom.

*Non-uniform* or *varied* flow occurs if the liquid surface is not parallel to the channel bottom. The change in depth may occur rapidly or gradually.

*Steady* flow occurs if the velocity and depth at a particular point in the channel do not vary with time. Most channel flows are steady apart from isolated phenomena such as the passage of a surge wave.

The flow can also be *laminar* or *turbulent* but it is almost invariably turbulent in cases of practical interest.

The final aspect of interest in channel flow is in connection with the velocity of flow which can be greater or less than the propagation of a surface disturbance. Thus, the flow can be *rapid* or *tranquil*.

### 8.1.1 The Bernoulli Equation

The Bernoulli equation for duct flow (equation 7.1) may be modified for channel flow as shown in figure 8.2. In a channel, the water surface coincides with the piezometric line. The energy line lies one velocity head above the free surface, $S_w$, and the slope of the energy line is called the *hydraulic slope*, $S = H_L/\ell$, where $H_L$ is the dissipation or loss of mechanical energy and $\ell$ is the distance measured along the channel bottom. Thus, between sections 1 and 2 in channel flow

$$y_1 + \alpha_1 \frac{\bar{u}_1^2}{2g} + z_1 = y_2 + \alpha_2 \frac{\bar{u}_2^2}{2g} + z_2 + H_L \qquad (8.1)$$

Because of the irregularities in the velocity profiles in channel flow, it may be necessary to introduce values for the kinetic energy correction factor, $\alpha$, in certain flows. In straight, rectangular channels, $\alpha \simeq 1$.

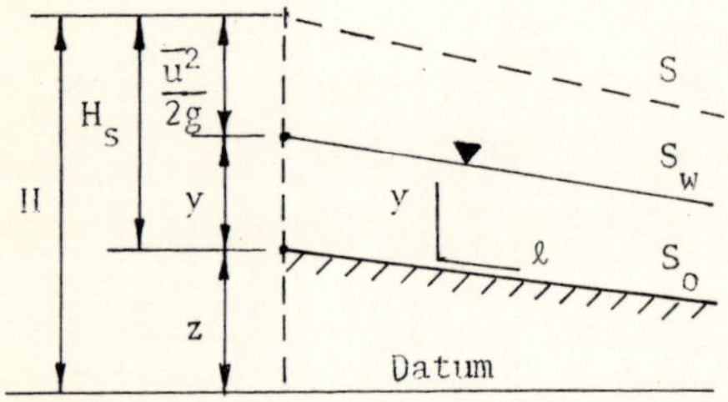


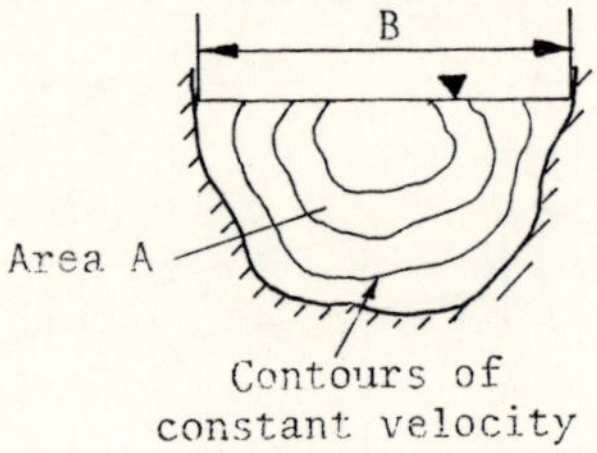


Figure 8.2

The general expression for the profile of the water surface is

$$\frac{dy}{d\ell} = \frac{S_o - S}{1 - \dfrac{\bar{u}^2 B}{gA}}$$

where $S_o$ is the slope of the channel bottom, B is the top width of the flow cross-section and A is the cross-sectional area.

For the special case of uniform flow, $dy/d\ell = 0$ and $S_o = S$. Thus the energy line, liquid surface and channel bottom are parallel for uniform flow.

## 8.2 UNIFORM FLOW

Flow in artificial channels is assumed to be steady and uniform if the cross-section is constant in shape and area, and the liquid surface is parallel to the channel bed. Flow in natural streams is rarely uniform and the equations for uniform flow give results which are only approximations.

### 8.2.1 The Chezy Equation

The basic formula for the determination of uniform (or *normal*) flow is the *Chezy* equation

$$\bar{u} = C\surd(R_h S_o) \qquad (8.3)$$

where $R_h$ is the hydraulic radius or hydraulic mean depth

$$R_h = \frac{\text{Area of cross-section}}{\text{Wetted perimeter}} = \frac{A}{P} \qquad (8.4)$$

and $S_o$ is the slope of the channel bed.

The Chezy constant C is a function of $R_h$, Reynolds number and surface roughness. Note that C has dimensions $L^{1/2}T^{-1}$. A simple formula for C is the *Strickler* formula

$$C = n^{-1}R_h{}^{1/6} \qquad (8.5)$$

where n is a constant varying from 0.01 for very smooth channels to 0.035 for canals and rivers in bad condition.

### 8.2.2 Optimum Shape of Cross-section

The discharge $\dot{V} = A\bar{u}$ and, from equation 8.3, $\bar{u}$ is a function of $R_h$. It follows that for a channel of given roughness, slope and area the maximum discharge occurs when $R_h$ is a maximum and the wetted area a minimum. For sections whose sides do not slope inwards the best section is a semi-circle and the best trapezoidal section is half a regular hexagon. Of sections whose sides do slope inwards, the circular duct is of special interest because drains and sewers frequently run partly full. It is of interest to note that the maximum discharge for a circular duct running nearly full is greater than one which runs completely full.

## 8.3 VELOCITY OF PROPAGATION OF A SURFACE WAVE

Any temporary disturbance of a liquid free surface produces the propagation of waves from the point of disturbance. For gravity waves in shallow water, the velocity of propagation in an irregular cross-section channel is given by

$$c = \surd(g\bar{y}) \qquad (8.6)$$

where $$\bar{y} = \frac{\text{Area of cross-section}}{\text{Width of liquid surface}} = \frac{A}{B} \qquad (8.7)$$

## 8.4 SPECIFIC HEAD

The concept of *specific head*, introduced by *Bakhmeteff* (1912) is useful for the solution of many problems in open channel flow. The specific head, $H_s$, (sometimes called specific energy) is referred to the channel bottom as datum and should not be confused with total head, H, which is referred to an arbitrary horizontal datum (figure 8.2).

$$\text{Thus } H = y + \frac{\bar{u}^2}{2g} + z \tag{8.8}$$

$$\text{but } H_s = y + \frac{\bar{u}^2}{2g} \tag{8.9}$$

These expressions are based on the fact that the hydrostatic law for pressure variation applies i.e. the streamlines are straight and parallel.

The total head must always decrease in any real flow but the specific head may increase or decrease in non-uniform flow. In the case of uniform flow $H_s$ remains constant. Substituting $\bar{u} = \dot{V}/A$ in equation 8.9 we obtain

$$H_s = y + \frac{1}{2g}\left(\frac{\dot{V}}{A}\right)^2 \tag{8.10}$$

For a constant discharge, the minimum value of $H_s$ occurs at a critical velocity, $\bar{u}_c$, when $\partial H_s/\partial y = 0$. Hence

$$\bar{u}_c = \sqrt{(g\bar{y})} \tag{8.11}$$

The critical condition is important because it separates two types of flow. The *super-critical* or *rapid* type of flow occurs when $\bar{u} > \bar{u}_c$ and the *sub-critical* or *tranquil* type of flow occurs when $\bar{u} < \bar{u}_c$. From equations 8.6 and 8.11 it follows that the critical velocity, $\bar{u}_c$, corresponds to the velocity of propagation of a surface wave. Thus, for rapid flow a surface disturbance cannot be propagated upstream against the flow whereas for tranquil flow it can. The ratio $\bar{u}/\sqrt{(g\bar{y})}$ is a form of Froude number. Hence, for critical conditions $\bar{u} = \sqrt{(g\bar{y})}$ and Fr = 1. For rapid flow Fr > 1 and for tranquil flow Fr < 1.

If $\dot{V}$ is maintained constant for flow in a particular section channel, $H_s$ may be plotted against $\bar{y}$ (figure 8.3a). At the minimum value of $H_s$ corresponding to the critical velocity, the depth has a particular value called the critical depth $\bar{y}_c$. For any other value of $H_s$ there are two alternative depths: a value $\bar{y} < \bar{y}_c$ corresponding to rapid flow and a value $\bar{y} > \bar{y}_c$ corresponding to tranquil flow.

If $H_s$ is maintained constant for flow in a particular section channel, $\dot{V}$ may be plotted against $\bar{y}$ (figure 8.3b). It can be shown that $\dot{V}_{max}$ will correspond to the critical conditions of velocity $\bar{u}_c$ and depth $\bar{y}_c$.

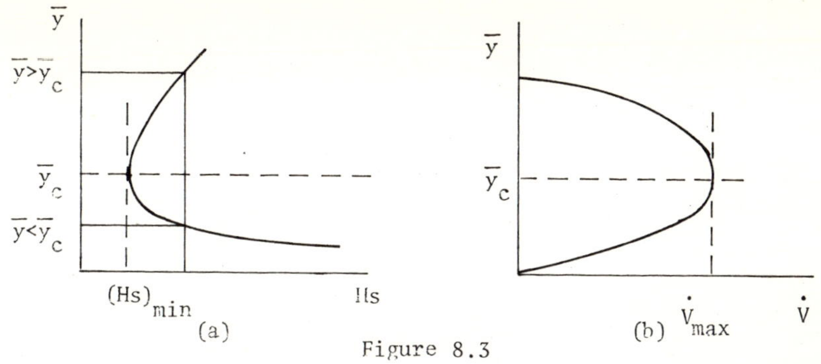


Figure 8.3

Thus, at the critical depth, $H_s$ is minimum for a given discharge and $\dot{V}$ is maximum for a given specific head. The actual relationship between $H_s$ and $\bar{y}_c$ depends upon the section shape.

## 8.5 BED SLOPE

Uniform flow at the critical velocity may be produced in a channel having a critical slope evaluated from the Chezy formula

$$\bar{u}_c = \sqrt{(g\bar{y}_c)} = C\sqrt{(R_h S_c)} \tag{8.12}$$

A slope less than the critical is known as *mild* and a slope greater than the critical is *steep*.

Flow can change from tranquil to rapid or vice versa by virtue of a change in slope. The transition tranquil-rapid can be gradual but the transition rapid-tranquil is always abrupt and called a *hydraulic jump* (figure 8.1).

## 8.6 SURFACE PROFILE

The form of the surface profile for varied flow is given by equation 8.2. Surface profiles are categorised in terms of bed slope, $S_o$, as follows: $S_o > S_c$ (steep), $S_o = S_c$ (critical), $S_o < S_c$ (mild), $S_o = 0$ (horizontal), $S_o < 0$ (adverse). The profile also depends upon the water depth relative to both normal ($y_o$) and critical ($y_c$) depths. If $y > y_o > y_c$, the profile is of type 1, if $y_o > y > y_c$ it is type 2 and if $y_o > y_c > y$ it is type 3. There are twelve possible surface profiles, three of which, based on a mild slope, are illustrated in figure 8.4.

From equation 8.2 a special case occurs when $dy/d\ell = \infty$. This is the condition for flow at the critical depth, $\bar{u}^2/g\bar{y} = 1$. Under this condition a hydraulic jump occurs and the surface profile is steeply inclined but not vertical. The discrepancy is due to the fact that in a jump the flow is rapidly varied, not gradually varied as assumed in the derivation of equation 8.2.

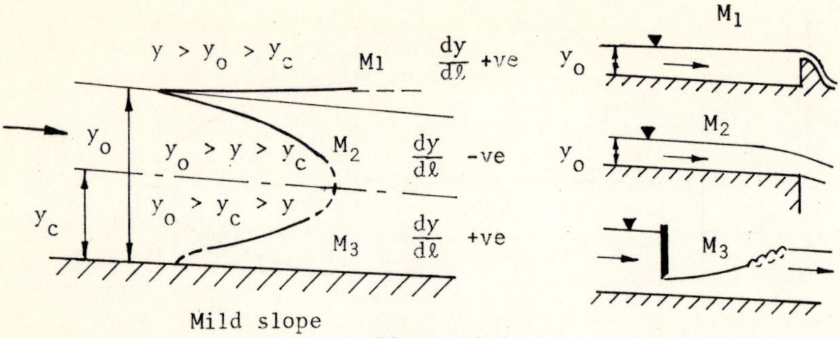


Figure 8.4

*Example 8.1*

The flow of rain-water down the inclined surface of a roof may be treated as laminar flow in a wide, shallow, rectangular channel. Derive expressions for the pressure, p, and velocity, u, at any distance, y, from the surface.

Assuming the shear stress at the water-air interface to be zero, determine the volumetric flow rate of a 2 mm thick film of water flowing down a 2 m wide surface inclined at 30° to the horizontal. $\nu = 10^{-6}$ m$^2$/s.

Refer to figure 8.5 in which the x and y components of the flow are taken parallel and perpendicular to the flow, respectively. The flow is assumed to be steady, uniform and laminar with a shear stress $\tau_s$ at the liquid-air interface.

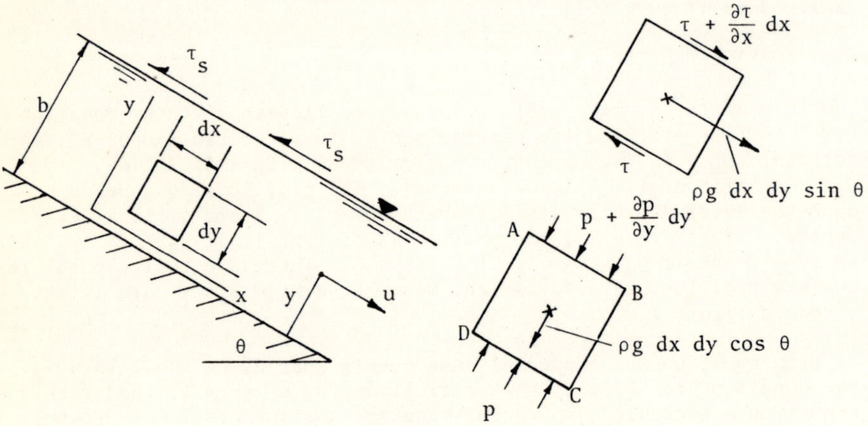


Figure 8.5

Consider the forces acting on a small, rectangular element ABCD. For steady flow, the momentum equation 4.19, applied to forces in the y direction, gives

$$\Sigma F_y = p\,dx - \left(p + \frac{\partial p}{\partial y}\,dy\right) dx - \rho g\,dx\,dy\cos\theta = 0$$

$$\frac{\partial p}{\partial y} = -\rho g\cos\theta$$

$$p = -\rho g y\cos\theta + A$$

The boundary conditions are $p = p_a$, when $y = b$. Therefore

$$p = \rho g\cos\theta\,(b - y) + p_a \qquad \text{(i)}$$

Applying the momentum equation to forces in the x direction we have

$$\Sigma F_x = \rho g\,dx\,dy\sin\theta + \left(\tau + \frac{\partial \tau}{\partial y}\,dy\right) dx - \tau\,dx = 0$$

$$\frac{\partial \tau}{\partial y} = -\rho g\sin\theta$$

$$\tau = -\rho g y\sin\theta + B$$

The boundary conditions are $\tau = -\tau_s$ when $y = b$

$$\tau = \rho g(b - y)\sin\theta - \tau_s \qquad \text{(ii)}$$

For laminar flow $\tau = \mu(du/dy)$. Therefore

$$\mu\frac{du}{dy} = \rho g(b - y)\sin\theta - \tau_s$$

$$u = \frac{\rho g}{\mu}(by - \tfrac{1}{2}y^2)\sin\theta - \frac{\tau_s y}{\mu} + C$$

The boundary conditions are $u = 0$ when $y = 0$. Therefore

$$u = \frac{g}{\nu}(by - \tfrac{1}{2}y^2)\sin\theta - \frac{\tau_s y}{\mu} \qquad \text{(iii)}$$

The volumetric flow, $\dot{V}$, in a channel of width, w, is given by

$$\dot{V} = w\int_0^b u\,dy = w\int_0^b [g(by - \tfrac{1}{2}y^2)/\nu\sin\theta - \tau_s y/\mu]\,dy$$

$$= \frac{wg\sin\theta\,b^3}{3\nu} - \frac{w\tau_s\,b^2}{2\mu} \qquad \text{(iv)}$$

For zero shear stress at the liquid-air interface, equation (iv) reduces to

$$\dot{V} = \frac{wg\sin\theta\,b^3}{3\nu} = \frac{2 \times 9.81 \times \sin 30 \times 2^3 \times 10^{-9}}{3 \times 10^{-6}}$$

$$= 0.0262\ \text{m}^3/\text{s}$$

*Example 8.2*

Using the Chezy formula (a) show that for maximum discharge in a trapezoidal channel the channel should be proportioned to give a hydraulic mean depth equal to half the central depth of flow, and (b) show that the most efficient of all trapezoidal sections is half a regular hexagon.

Refer to figure 8.6 which shows a trapezoidal section, symmetrical about the centre line.

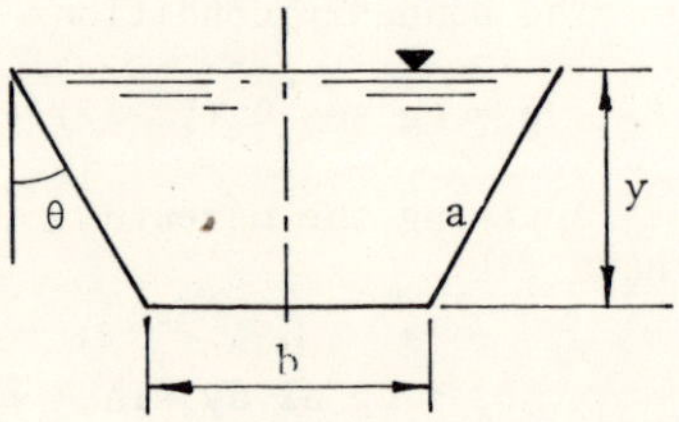

Figure 8.6

Using the Chezy equation 8.2, the discharge is

$$\dot{V} = AC\sqrt{(R_h S_o)}$$

$$= C\sqrt{(A^3 S_o/P)}$$

If the Chezy constant, C, and the slope, $S_o$, are constant and A is a stated constant, the maximum discharge will occur when the depth, y, is critical and P a minimum.

The area is given by

$$A = by + y^2 \tan\theta \qquad \text{(i)}$$

$$b = Ay^{-1} - y\tan\theta$$

Now $$P = b + \frac{2y}{\cos\theta} = Ay^{-1} + y\left(\frac{2}{\cos\theta} - \tan\theta\right) \qquad \text{(ii)}$$

Wetted perimeter, P, will be a minimum when dP/dy = 0

$$\frac{dP}{dy} = -\frac{A}{y^2} + \left(\frac{2}{\cos\theta} - \tan\theta\right) = 0$$

$$A = y^2\left(\frac{2}{\cos\theta} - \tan\theta\right) \qquad \text{(iii)}$$

Substituting for A from (i) and rearranging

$$b = 2y\left(\frac{1}{\cos\theta} - \tan\theta\right) \qquad \text{(iv)}$$

and $$P = b + \frac{2y}{\cos\theta} = 2y\left(\frac{2}{\cos\theta} - \tan\theta\right) \qquad \text{(v)}$$

From equations (iii) and (v)

$$R_h = \frac{A}{P} = \frac{y^2\left(\dfrac{2}{\cos\theta} - \tan\theta\right)}{2y\left(\dfrac{2}{\cos\theta} - \tan\theta\right)} = \frac{y}{2} \qquad \text{(vi)}$$

Thus, for maximum efficiency, a trapezoidal channel should be proportioned to make the hydraulic mean depth equal to half the critical depth of flow. It should be noted that for a rectangular section, $\theta = 0$ and equation (iii) gives $A = 2y^2$. But $A = yb$ therefore $b = 2y$.

Equation (v) shows that P is a minimum when $(2/\cos\theta - \tan\theta)$ is a minimum for all values of y. This condition occurs when $dP/d\theta = 0$. Differentiating

$$\frac{dP}{d\theta} = \frac{2\sin\theta}{\cos^2\theta} - \sec^2\theta = 0$$

$$\theta = 30°$$

Substituting $\theta = 30°$ in equation (iv) we obtain

$$b = 2y\left(\frac{1}{\cos 30} - \tan 30\right) = 1.15y$$

or $$b = \frac{y}{\cos 30} = a$$

Referring to figure 8.6, it can be seen that for this condition to be satisfied the trapezoidal section must be equal to half a regular hexagon.

*Example 8.3*

(a) Show that for a circular duct of diameter, D, and fixed slope, S, the velocity of flow will be a maximum when the depth of liquid, y, in the duct is about 0.81D. Use the Chezy formula and assume a constant value of C.

(b) A circular drain pipe has a diameter of 500 mm and a slope of 1 in 400. If the Chezy constant is 50, determine the maximum velocity of flow and the corresponding discharge.

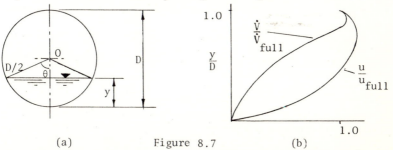

(a) Figure 8.7 (b)

(a) Refer to figure 8.7a in which the liquid surface subtends an angle 2θ at the centre. For any depth, y, the area of the cross-section of liquid is given by

$$A = \left(\frac{D}{2}\right)^2 \theta - 2\left(\tfrac{1}{2} \times \frac{D}{2} \times \sin\theta \times \frac{D}{2} \times \cos\theta\right)$$

$$A = \tfrac{1}{4}D^2(\theta - \tfrac{1}{2}\sin 2\theta) \qquad \text{(i)}$$

The wetted perimeter, P, is

$$P = \frac{D}{2} \times 2\theta = D\theta \qquad \text{(ii)}$$

The Chezy formula gives

$$\bar{u} = C\surd(R_h S)$$

and $\bar{u}$ will be a maximum for the value of $\theta$ which makes $R_h$ a maximum. This occurs for $dR_h/d\theta = 0$.

$$\text{Now} \quad R_h = \frac{A}{P} = \frac{0.25D^2(\theta - 0.5\sin 2\theta)}{D\theta} = 0.25D\left(1 - \frac{\sin 2\theta}{2\theta}\right) \qquad \text{(iii)}$$

$$\text{and} \quad \frac{dR_h}{d\theta} = 0.25D\left(0 - \frac{2\theta \times 2 \times \cos 2\theta - 2 \times \sin 2\theta}{4\theta^2}\right) = 0$$

$$2\theta = \tan 2\theta$$

This equation may be solved by a variety of methods but a table of $2\theta$ (radians) against $\tan 2\theta$ soon yields the solution $\theta = 128.75°$.

From figure 8.7, the depth of flow, y, is given by

$$y = 0.5D - 0.5D\cos\theta = 0.5D(1 - \cos\theta) \qquad \text{(iv)}$$

The maximum velocity occurs when $\theta = 128.75°$. Therefore

$$y = 0.5D(1 - \cos 128.75) = 0.813D$$

(b) For any depth, y, the velocity and discharge are

$$\bar{u} = C\surd(R_h S) = C\surd[0.25D(1 - \sin 2\theta/2\theta)S]$$

$$\dot{V} = \bar{u}A = C\surd[0.25D(1 - \sin 2\theta/2\theta)S] \times 0.25D^2(\theta - 0.5\sin 2\theta)$$

The maximum value of $\bar{u}$ occurs when $\theta = 128.75°$ giving

$$\bar{u}_{max} = 50\surd[0.25 \times 0.5(1 - \sin 257.5/4.5)/400] = 0.975 \text{ m/s}$$

$$\dot{V} = 0.975 \times 0.25 \times 0.5^2(2.25 - 0.5\sin 257.5) = 0.167 \text{ m}^3/\text{s}$$

It should be noted that the maximum discharge does not occur at the maximum velocity condition (figure 8.7b).

*Example 8.4*

Derive expressions for the critical velocity, $\bar{u}_c$, and the critical depth, $y_c$, in a rectangular channel. Hence show that for critical flow $H_s = 3y_c/2$.

Water flows at the rate of 1.5 m$^3$/s in a channel 3 m wide. If the depth of flow at a particular instant is 0.5 m, determine whether the flow is rapid or tranquil and hence determine the alternative depth.

Consider a channel of width, B, and water depth, y. The specific head, $H_s$ is defined

$$H_s = y + \frac{\bar{u}^2}{2g} \qquad \text{(i)}$$

$$H_s = y + \frac{1}{2g}\left(\frac{\dot{V}}{By}\right)^2$$

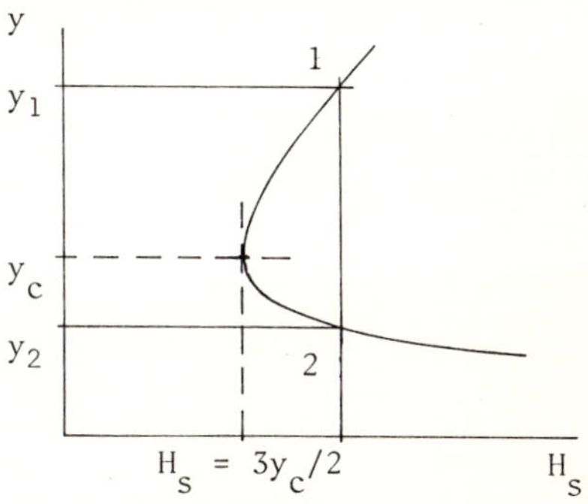

Figure 8.8

If the discharge, $\dot{V}$, is a constant, critical flow occurs when $H_s$ is a minimum i.e. when $dH_s/dy = 0$

$$\frac{dH_s}{dy} = 1 - \frac{\dot{V}^2}{g \times B^2 \times y_c^3} = 0$$

$$y_c = \left(\frac{\dot{V}^2}{gB^2}\right)^{1/3} \qquad \text{(ii)}$$

From continuity, the critical velocity, $\bar{u}_c$, is given by

$$\bar{u}_c = \frac{\dot{V}}{By_c}$$

Substituting for $\dot{V}$ from equation (ii) we obtain

$$\bar{u}_c = \frac{By_c^{3/2}g^{1/2}}{By_c} = \sqrt{(gy_c)} \qquad \text{(iii)}$$

This result is in accordance with equation 8.11, with $\bar{y}_c = y_c$ for a rectangular channel. Substituting for $\bar{u}_c$ in equation (i)

$$H_s = y_c + \frac{gy_c}{2g} = 3y_c/2 \qquad \text{Q.E.D.}$$

Substituting the given values in equation (ii)

$$y_c = \left(\frac{1.5^2}{9.81 \times 3^2}\right)^{1/3} = 0.294 \text{ m}$$

The actual depth of flow is 0.5 m which is greater than the critical depth. The flow is tranquil as confirmed by figure 8.8. For constant $H_s$, there are two alternative depths, $y_1$ and $y_2$. Therefore

$$y_1 + \frac{\bar{u}_1^2}{2g} = y_2 + \frac{\bar{u}_2^2}{2g}$$

$$y_1 + \frac{\dot{V}^2}{2B^2y_1^2g} = y_2 + \frac{\dot{V}^2}{2B^2y_2^2g}$$

$$0.5 + \frac{1.5^2}{2 \times 3^2 \times 0.5^2 \times 9.81} = y_2 + \frac{1.5^2}{2 \times 3^2 \times y_2^2 \times 9.81}$$

$$0.551\ y_2^2 = y_2^3 + 0.0127$$

By successive approximations, or by plotting a graph of $0.551\ y_2^2$ and $y_2^3 + 0.0127$ versus $y_2$, we obtain the solution

$$y_2 = 0.186 \text{ m}$$

*Example 8.5*

(a) Show that the bed slope required to produce uniform flow at the critical depth, $y_c$, in a rectangular channel is given by $S_c = 9.81 y_c n^2 R_h^{-4/3}$, where the Chezy coefficient $C = n^{-1} R_h^{1/6}$. Hence determine the critical slope of a rectangular channel, of width 1 m, to maintain a flow of 1.2 $m^3/s$. Assume n = 0.02.

(b) Sketch the nature of flow (i) when a mild slope changes to a steep slope, and (ii) when a steep slope changes to a mild slope.

(a) In example 8.4, it was established that the critical velocity, $\bar{u}_c$, for flow in a rectangular channel is

$$\bar{u}_c = \sqrt{(g y_c)} \qquad \text{(i)}$$

The Chezy formula for critical flow conditions is

$$\bar{u}_c = C\sqrt{(R_h S_c)} \qquad \text{(ii)}$$

Equations (i) and (ii), together with the given relationship $C = n^{-1} R_h^{1/6}$, yield

$$S_c = 9.81 y_c n^2 R_h^{-4/3} \qquad \text{Q.E.D.}$$

In example 8.4, expressions for $y_c$ and $R_h$ were derived

$$y_c = \left(\frac{\dot{V}^2}{gB^2}\right)^{1/3} = \left(\frac{1.2^2}{9.81 \times 1^2}\right)^{1/3} = 0.528 \text{ m}$$

and $$R_h = \frac{B y_c}{B + 2y_c} = \frac{1 \times 0.528}{1 + 2 \times 0.528} = 0.256 \text{ m}$$

$$S_c = 9.81 \times 0.528 \times 0.02^2 \times 0.256^{-4/3} = 1/80$$

(b)(i) The transition from tranquil flow on a mild slope to rapid flow on a steep slope is shown in figure 8.9a. For a gradual change in slope, there is only a small loss in energy due to the disturbance in the flow but for an abrupt change there is increased disturbance and loss. The change in specific head follows the $y$-$H_s$ curve.

(ii) The transition from rapid to tranquil flow is marked by a highly turbulent region in which air entrainment occurs. The surface rises rapidly and unevenly to the sub-critical depth and, on the surface itself, rollers are formed which repeatedly dash themselves against the oncoming current. A large amount of energy is dissipated. This phenomenon is known as a *hydraulic jump* or *standing wave*. The physical explanation for the hydraulic jump is that the change in specific head from a super-critical depth $y_1$ to a sub-critical depth $y_2$ cannot occur by following the y-$H_s$ curve without the addition of specific head from outside. This is not available due to the uniform slope of the channel in the region of the jump and so a discontinuity occurs with the depth jumping from $y_1$ to $y_2$ as shown in figure 8.9b. The loss in specific head due to turbulence is ultimately dissipated to thermal energy. The hydraulic jump is a useful energy dissipator and is used to reduce high flow velocities and prevent excessive scouring of the channel walls.

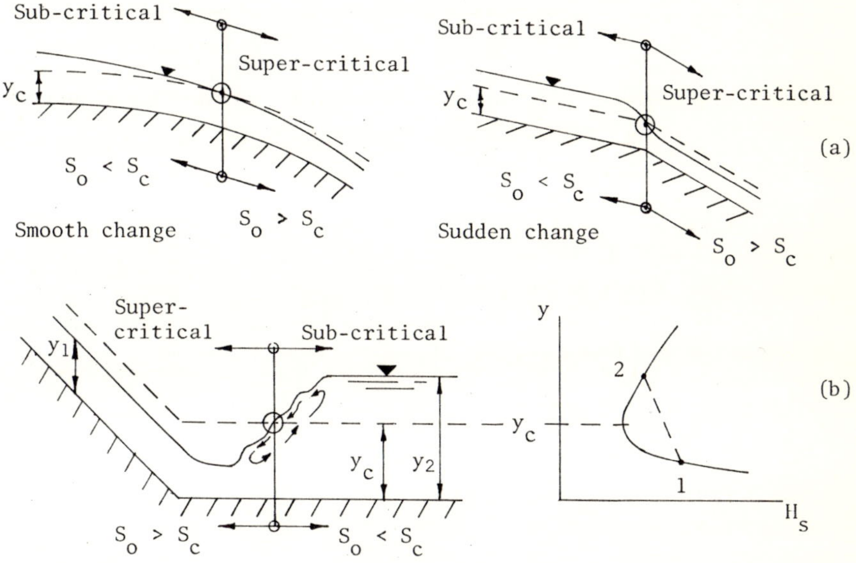


Figure 8.9

*Example 8.6*

(a) Show that for a two-dimensional hydraulic jump formed on a horizontal surface

$$y_2/y_1 = \tfrac{1}{2}[\sqrt{(1 + 8Fr^2)} - 1]$$

where $y_1$ and $y_2$ are the depths of flow upstream and downstream of the jump and Fr is the Froude number $\bar{u}/\sqrt{(gy)}$.

(b) In a particular jump, the depth increases from 0.5 m to 1.2 m. Determine the loss in energy per unit mass of water.

(a) Refer to figure 8.10 which shows a section through a two-dimensional jump. A section through a jump in a rectangular channel of width, B, would look the same. Consider the control volume enclosing the jump. For steady flow, the only forces acting on the control volume are wall viscous forces and hydrostatic pressure forces. It is reasonable to neglect wall viscous forces therefore the momentum equation 4.19 gives

$$\rho g \times \tfrac{1}{2}y_1 \times By_1 - \rho g \times \tfrac{1}{2}y_2 \times By_2 = \rho\dot{V}(\bar{u}_2 - \bar{u}_1)$$

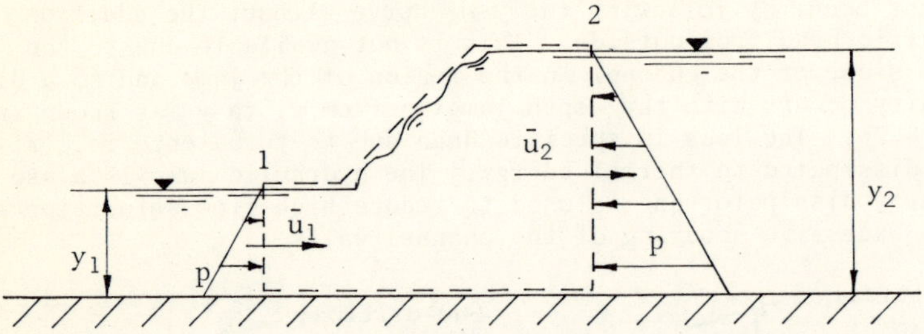


Figure 8.10

Substituting $\bar{u}_2 = \dot{V}/By_2$ and $\bar{u}_1 = \dot{V}/By_1$ and rearranging, we obtain

$$y_2^2 - y_1^2 = \frac{2}{g}\left(\frac{\dot{V}}{B}\right)^2 \frac{(y_2 - y_1)}{y_1 y_2}$$

and $$y_2^2 + y_1 y_2 - \frac{2}{gy_1}\left(\frac{\dot{V}}{B}\right)^2 = 0 \qquad \text{(i)}$$

The solution to this quadratic equation is

$$y_2 = -\frac{y_1}{2} + \left[\frac{y_1^2}{4} + \frac{2}{gy_1}\left(\frac{\dot{V}}{B}\right)^2\right]^{\frac{1}{2}}$$

or $$\frac{y_2}{y_1} = \frac{1}{2}\left(1 + \frac{8\bar{u}_1^2}{gy_1}\right)^{\frac{1}{2}} - \frac{1}{2}$$

But a Froude number using depth y as the characteristic length is $Fr = \bar{u}_1/\surd(gy_1)$. Therefore we may state

$$y_2/y_1 = \tfrac{1}{2}[\surd(1 + 8Fr^2) - 1] \qquad \text{Q.E.D}$$

For $y_2/y_1 > 1$, $Fr > 1$ and $\bar{u}_1 > \surd(gy_1)$. It follows that in a hydraulic jump, $\bar{u}_1$ is supercritical.

(b) The energy dissipated in the jump per unit mass of water, is given by

$$\Delta e_s = g(H_{s1} - H_{s2}) = g(y_1 - y_2) + \tfrac{1}{2}(\bar{u}_1^2 - \bar{u}_2^2)$$

or $$\Delta e_s = g(y_1 - y_2) + \frac{1}{2y_1^2 y_2^2}\left(\frac{\dot{V}}{B}\right)^2 (y_2^2 - y_1^2) \qquad \text{(ii)}$$

Eliminating $(\dot{V}/B)^2$ between equations (i) and (ii)

$$\Delta e_s = \frac{g(y_2 - y_1)^3}{4y_1 y_2} = \frac{9.81 \times (1.2 - 0.5)^3}{4 \times 1.2 \times 0.5} = 1.41 \text{ J/kg}$$

Note that $y_2 > y_1$, otherwise $\Delta e_s$ is negative, implying an increase in $H_s$ (and mechanical forms of energy) across the jump.

*Example 8.7*

A broad channel carries a flow of 2.75 $m^3/s$ per metre breadth, the depth of flow being 1.5 m. Calculate the rise in floor level required to produce critical flow conditions. What is the corresponding fall in the surface level?

Refer to figure 8.11 which shows two-dimensional flow with a rise in floor level, D, produced (say) by a rectangular section block. (This configuration is used in a broad-crested weir for flow measurement.) The specific head upstream is

$$H_s = y_1 + \bar{u}_1^2/2g \qquad \text{(i)}$$

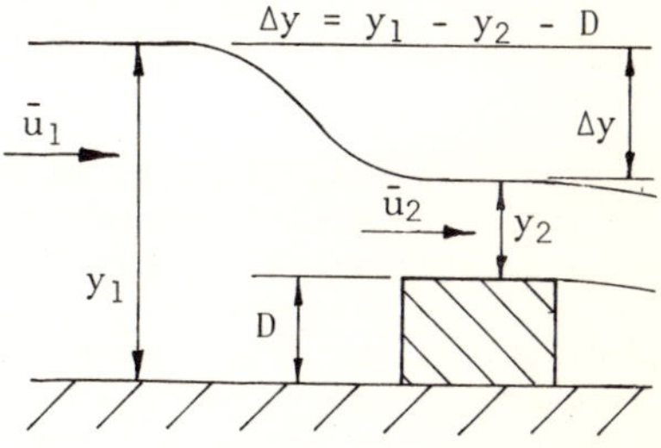


Figure 8.11

The specific head relative to the upper face of the block is

$$(H_s)_b = H_s - D = y_2 + \bar{u}_2^2/2g$$

$$\bar{u}_2 = \sqrt{[2g(H_s - y_2 - D)]} \qquad \text{(ii)}$$

Consider the flow through a rectangular channel of breadth B with a discharge coefficient $C_d$

$$\dot{V} = C_d B \bar{u}_2 y_2 = C_d B y_2 \sqrt{[2g(H_s - y_2 - D)]}$$

For critical flow conditions over the block $y_2 = y_c$ and, from example 8.4, $y_c = 2(H_s)_b/3 = 2(H_s - D)/3$. Therefore

$$\dot{V} = 2C_d B(H_s - D)\sqrt{[2g(H_s - D)/3]}/3 = 1.7C_d B(H_s - D)^{3/2} \qquad \text{(iii)}$$

Now for unit breadth of channel, equation (i) gives

$$H_s = y_1 + \frac{\bar{u}_1^2}{2g} = y_1 + \frac{\dot{V}^2}{2gy_1^2}$$

$$= 1.5 + \frac{2.75^2}{2 \times 9.81 \times 1.5^2} = 1.671 \text{ m}$$

For two-dimensional flow and unit breadth, and $C_d = 1$, equation (iii) gives

$$\dot{V} = 1.7(H_s - D)^{3/2}$$

$$D = H_s - (\dot{V}/1.7)^{2/3} = 1.671 - (2.75/1.7)^{2/3} = 0.293 \text{ m}$$

The fall in water level, $\Delta y$, is given by

$$\Delta y = y_1 - (y_c + D) = y_1 - [2(H_s - D)/3 + D]$$

$$= y_1 - \frac{2H_s}{3} - \frac{D}{3} = 1.5 - \frac{2 \times 1.671}{3} - \frac{0.293}{3} = 0.288 \text{ m}$$

*Example 8.8*

(a) Describe the action of a venturi flume, used for measuring the discharge of a stream, with and without the formation of a hydraulic jump. Derive an expression for the discharge in each case.

(b) A venturi flume with a level bed is 12 m wide and has a throat width of 6 m. Calculate the free-flow discharge when the upstream depth is 1.5 m. Assume $C_d = 0.95$ and solve by successive approximations.

(a) A venturi flume is an artificial, streamlined contraction in a channel which, by causing a change in velocity and depth, facilitates the measurement of discharge. Only a small *afflux* (difference in depth upstream and downstream) is necessary. The advantages of the venturi flume are a robust construction and a clear passage unlikely to be clogged or damaged by floating debris.

Assuming no loss in specific head between upstream and throat sections

$$H_s = y_1 + \frac{u_1^2}{2g} = y_2 + \frac{u_2^2}{2g}$$

$$u_2 = \sqrt{[2g(H_s - y_2)]}$$

$$\dot{V} = B_2 y_2 \sqrt{[2g(H_s - y_2)]}$$

Introducing a discharge coefficient $C_d$

$$\dot{V} = B_2 y_2 C_d \sqrt{[2g(H_s - y_2)]} \qquad \text{(i)}$$

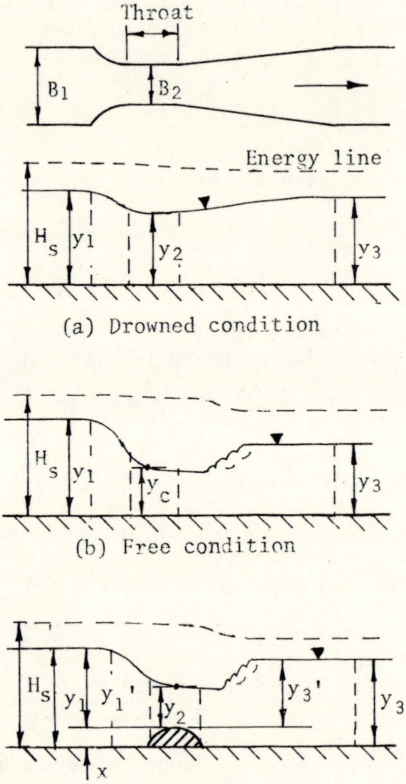


(a) Drowned condition

(b) Free condition

(c) Flume with hump

Figure 8.12

If the flow is sub-critical throughout, the flume operates in the *drowned* condition and there is no hydraulic jump. At the throat the surface is unstable and $y_1 - y_2$ cannot be accurately determined. A flume is therefore designed to operate with super-critical or *free* flow, the proportions being such that a hydraulic jump is formed just downstream with the critical depth, $y_c = y_2$ at the throat. Now in example 8.4 it was shown that for critical flow in a rectangular channel

$$y_c = \frac{2}{3} H_s \qquad \text{(ii)}$$

and $$\frac{u_c^2}{gy_c} = 1 \qquad \text{(iii)}$$

Combining the continuity equation with equations (ii) and (iii) and again introducing a discharge coefficient, $C_d$, we obtain

$$\dot{V} = u_c y_c B_2 = 1.705 C_d B_2 H_s^{3/2} \qquad \text{(iv)}$$

In practice, for free flow to occur, the *submergence ratio* ($y_3/y_1$) must be appreciably less than unity. The upper limiting value of this ratio is called the *modular limit* and is taken as about 0.75. It is sometimes found impossible to obtain free flow with a level bed flume and low discharges and so a streamlined hump is introduced in the flume bed (figure 8.12c).

(b) As a first approximation the velocity of approach, $u_1$, is neglected. Therefore

$$\dot{V} = 1.705 C_d B_2 y_1^{3/2} = 1.705 \times 0.95 \times 6 \times 1.5^{3/2} = 17.85 \text{ m}^3/\text{s}$$

The velocity of approach may now be estimated

$$u_1 = \frac{\dot{V}}{B_1 y_1} = \frac{17.85}{12 \times 1.5} = 0.992 \text{ m/s}$$

Substituting values in equation (iv) we obtain

$$\dot{V} = 1.705 \times 0.95 \times 6 \times (1.5 + 0.992^2/2 \times 9.81)^{3/2} = 18.76 \text{ m}^3/\text{s}$$

A second correction gives

$$\dot{V} = 18.85 \text{ m}^3/\text{s}$$

*Example 8.9*

Water flows at the rate of 8 $m^3/s$ in a channel of rectangular cross-section, 5 m wide, with a slope of 1 in 4000. The Chezy constant is 80. A dam is built across the channel with the result that the depth just upstream of the dam is 3 m. How far upstream will the depth be within 50 mm of the normal depth? Solve (i) using the assumption that the slope at the mean depth applies to the whole backwater curve and (ii) by dividing the range of depth into three equal parts and using a numerical integration.

The normal depth, $y_o$, occurs when the flow is uniform. For uniform flow the Chezy equation 8.3 may be applied.

$$u = C\sqrt{(R_h S_o)}$$

or $$\frac{\dot{V}}{By_o} = C\left(\frac{By_o S_o}{B + 2y_o}\right)^{\frac{1}{2}}$$

$$\frac{8}{5y_o} = 80\left[\frac{5y_o}{(5 + 2y_o)4000}\right]^{\frac{1}{2}}$$

Solution by successive approximations gives $y_o$ = 1.35 m. The required upstream depth is 1.35 + 0.05 = 1.4 m.

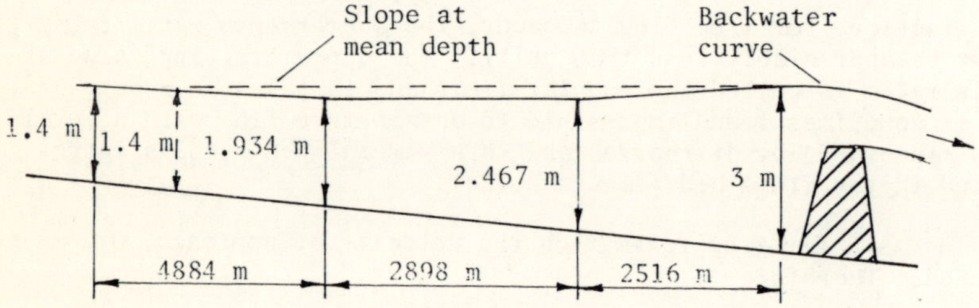


Figure 8.13

(i) The mean depth of water is (1.4 + 3)/2 = 2.2 m. At the mean depth

$$R_h = \frac{5 \times 2.2}{5 + 2 \times 2.2} = 1.17 \text{ m}$$

and $\bar{u} = \dot{V}/By = 8/5 \times 2.2 = 0.727$ m/s

For non-uniform flow, from the Chezy equation

$$\bar{u} = C\surd(R_h S)$$

where S is the energy line slope

$$S = \frac{0.727^2}{80^2 \times 1.17} = 7.06 \times 10^{-5}$$

From equation 8.2

$$\frac{dy}{d\ell} = \frac{S_o - S}{1 - \frac{\bar{u}^2 B}{gA}} = \frac{2.5 \times 10^{-4} - 7.06 \times 10^{-5}}{1 + 0.727^2/(9.81 \times 2.2)} = 1.84 \times 10^{-4} \text{ m/m}$$

Assuming that this slope is maintained constant from a depth of 3 m at the dam to 1.4 m upstream, the distance involved is given by

$$\text{distance} = \frac{3 - 1.4}{1.84 \times 10^{-4}} = 8695 \text{ m}$$

(ii) For numerical integration the range of depth 3 m to 1.4 is divided into three equal parts. Starting at the dam and working upstream, the calculations may be set out in tabular form

| *y(m)* | *Average y(m)* | *A*($m^2$) | *P(m)* | $R_h$*(m)* | $\bar{u}$*(m/s)* | $\bar{u}^2/gy$ |
|---|---|---|---|---|---|---|
| 3-2.467 | 2.733 | 13.67 | 10.47 | 1.31 | 0.585 | 0.0128 |
| 2.467-1.934 | 2.2 | 11 | 9.4 | 1.17 | 0.727 | 0.0245 |
| 1.934-1.4 | 1.667 | 8.34 | 8.33 | 1.0 | 0.959 | 0.0562 |

| $1-\bar{u}^2/gy$ | $S=(\bar{u}^2/C^2R_h)10^4$ | $(S_o-S)10^4$ | $d\ell/dy$ | $\Delta y$ | $\Delta \ell$ |
|---|---|---|---|---|---|
| 0.9872 | 0.409 | 2.091 | 4721 | 0.533 | 2516 |
| 0.9755 | 0.706 | 1.794 | 5438 | 0.533 | 2898 |
| 0.9438 | 1.437 | 1.063 | 8878 | 0.533 | 4732 |
| | | | | $\Sigma\Delta\ell$ = | 10 146 |

Thus the total distance upstream is 10 146 m [cf. 8695 m obtained by method (i)]. The accuracy can obviously be improved by dividing the range of depth into a greater number of parts than the three chosen.

*Problems*

1 A roof sloping 20° from the horizontal is to be covered with a layer of asphalt of thickness 3 mm. During hot weather there is a tendency for the softened asphalt to flow down the slope. What must be the effective kinematic viscosity of the asphalt in $m^2/s$ if the top surface must not move more than 0.025 mm relative to the bottom surface in a 10 h period of time?

[$4.34 \times 10^4$ $m^3/s$]

2 Water flows with a free surface through a conduit whose section is shown in figure 8.14. Calculate the depth of flow for which the velocity is a maximum.

[1.704 m]

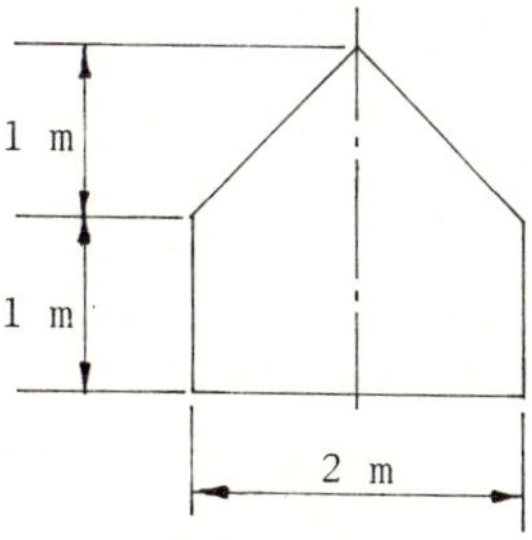


Figure 8.14

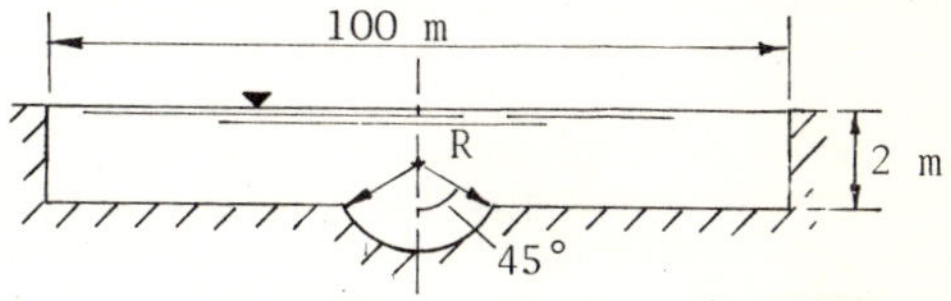


Figure 8.15

3 The cross-section of a river channel is to be modified in such a way that the normal winter discharges are contained within a main central channel, while exceptional flows spill over into a flood plain on either side, bounded by banks constructed parallel with the river. The central channel approximates to a circular segment, as shown in figure 8.15, and is to carry water at the rate of 10 $m^3/s$. The flood plain has banks 2 m high and 100 m apart. If the bed slope is 1/1000 and the Chezy constant is 30, determine

(i) the radius of the central channel, and (ii) the approximate maximum flow rate that can be contained if the water level is to be 0.5 m below the top of the flood banks in order to allow for waves. Determine also, from first principles, whether the flood flow is sub-critical or super-critical.

[5.96 m, 172 $m^3/s$, 0.67 m sub-critical]

4 Derive an expression for the volume flow rate of a fluid flowing uniformly along an open channel, in terms of the cross-sectional area of the flow A, the wetted perimeter of the channel P, the gradient of the channel base S, the friction factor f, and the specific gravitational force g.

A trapezoidal channel with sides at 45° to the base is to be dug to carry a flow of 2.56 $m^3/s$ with a gradient of 1 in 50. The friction factor is 0.1. Determine the dimensions of the channel which will have the minimum cross-sectional area capable of carrying the given flow.

[$\dot{V} = A\surd(2g/f)\surd(AS/P)$, y = 1 m, B = 0.828 m]

5 Derive an expression for the critical depth of flow in a triangular section channel of included angle θ, whose sides are equally inclined to the vertical axis, when the volumetric flow rate is $\dot{V}$.

A channel of triangular section, with included angle 60°, carries water at the rate of 25 $m^3/s$. Determine the critical depth and the value of the Chezy coefficient if the slope of the channel required to maintain critical depth is 0.001.

[$y = [\dot{V}^2/(g \tan \theta/2)]^{1/3}$, 4.8 m, 54.4]

6 A circular pipe of diameter 500 mm and length 20 m has a slope of 1 in 200. Assuming the Chezy formula $v = 60\surd(RS)$ to apply for all conditions of flow, determine the rate of energy dissipation (i) when the flow in the pipe is running at 0.95D which is the optimum depth for maximum discharge, and (ii) when the pipe is running full. Determine also the pumping power required to increase the flow rate in the pipe running full to the maximum discharge at condition (i).

[303 W, 289 W, 31.6 W]

7 For the same specific energy, calculate the alternative depths of flow in a channel which is 6 m wide and the depth of stream 6 m. The rate of flow is 12 000 ℓ/s.

[6 m, 0.187 m]

8 A sluice is placed across a channel 5 m wide and the depth of discharge under the sluice is 1 m. If the depth upstream of the sluice is 5 m, determine the flow rate in the channel. An obstruction downstream causes a hydraulic jump to be formed after the sluice Working from first principles, derive an expression relating the depths upstream and downstream of the jump to the Froude number. Hence, determine the depth downstream of the jump and the power dissipated in the jump. Neglect the loss of energy at the sluice.

[45.2 m, 3.61 m, 546 kW]

9 Water flows at 5 $m^3/s$ under a wide sluice gate into a long rectangular channel 3 m wide. The slope of the channel is 0.001. A hydraulic jump is formed with the ratio of upstream to downstream depths 0.5. Determine the value of the constant n in the empirical formula $\bar{u} = n^{-1}R_h^{2/3}S^{1/2}$.

[0.0318]

10 Water flows through a horizontal rectangular tunnel 2.5 m high and 4 m wide. The flow is controlled by a sluice which is raised to produce a depth and velocity of flow downstream of 1.25 m and 10 m/s, respectively. Further downstream a hydraulic jump is formed which reaches the roof of the tunnel. Determine the absolute pressure at the roof of the tunnel and the power dissipated in the jump. Atmospheric pressure is 1 bar.

[115.8 kPa, 473 kW]

11 A level channel 0.5 m wide has a block of wood fastened across it to form a broad-crested weir, the face being 75 mm above the bed. The depth of water upstream from the weir is 150 mm. Determine the rate of flow, making one approximation for the velocity of approach in the channel, assuming critical conditions to exist over the weir. Work from first principles.

[0.0185 $m^3/s$]

12 A venturi flume with a level bed is 13 m wide and 1.6 m deep upstream and has a throat width of 6.5 m. Calculate the discharge if $C_d = 0.94$. Successive approximations or any other suitable method may be used to solve the equation.

[22.12 $m^3/s$]

13 A venturi flume 1.3 m wide at the entrance and 0.6 m at the throat has a horizontal bottom. Neglecting hydraulic losses in the flume, calculate the flow if the depths are 0.6 m and 0.55 m at entrance and throat respectively. A hump is now installed at the throat of height 200 mm so that a standing wave is formed beyond the throat. Assuming the same flow as before, show that the increase in upstream depth is nearly 80 mm.

[0.355 $m^3/s$]

14 The bed of a river 16 m wide has a uniform slope of 1 in 14 000 and the Chezy constant C = 90. A dam across the river has a spillway across the top in the form of a broad-crested weir, the height of the sill being 4 m above the river bed. Assuming that the whole of the river flow passes over the spillway, $C_d = 0.9$, find the afflux at the dam (i.e. the increase in river depth caused by the dam), when the depth of the river well upstream is 1.6 m. State fully the assumptions made.

[3.5 m]

15 A rectangular channel 1.3 m wide has a uniform slope of 1 in 1600 and normal flow with a constant depth of 0.6 m occurs when $\dot{V} = 0.6$ $m^3/s$. The lowering of a sluice when this quantity is flowing increases the depth just upstream of the sluice to 1 m. Assuming that $S = u^2/C^2R$, determine how far upstream from the sluice the depth will be 0.8 m.

[Example 8.9 method (i) 491 m, (ii) 498 m]

# 9 UNSTEADY FLOW

Previous chapters have been devoted to steady flow situations in which properties such as velocity, pressure and density, at a particular point, do not vary with time. In this respect, turbulent flow is treated as steady because the small-scale velocity fluctuations involved do not change the time average bulk mean velocity at a point. In unsteady flow, however, mean values do vary with time and methods of solution of problems are more complex, frequently involving numerical solution. In this chapter some practical problems amenable to simple analytical solution will be considered.

## 9.1 PULSATING FLOW

A special case of unsteady flow, known as *pulsating* flow, occurs if there are *periodic* fluctuations in the mean velocity and pressure at any point in the flow. Pulsating flow occurs most severely in the inlet and exhaust ducts of reciprocating engines and the suction and delivery pipes of reciprocating pumps and compressors. It is not proposed to analyse pulsating flow in detail here but it is necessary to point out an inherent error in the use of steady flow measurement devices.

The true mass flow rate at any section of area A is, for incompressible flow

$$\dot{m} = (\rho A/t) \int u \, dt = \rho A \bar{u}_t \qquad (9.1)$$

where $\bar{u}_t$ is the time mean average velocity.

The flow through a meter of the pressure differential type (e.g. venturi meter) is determined from

$$\dot{m} = K\surd(\Delta p)$$

where K is a meter constant.

For turbulent, pulsating flow the pressure drop due to area change is

$$(\Delta p)_{true} = C(\bar{u}_t)^2 \qquad (9.2)$$

and the mean pressure drop across the flow meter, taken from the manometer is

$$\overline{\Delta p} = C\overline{u_t^2} \qquad (9.3)$$

Equations 9.2 and 9.3 do not give the same result therefore it is inherently inaccurate to use the mean manometer reading to find mass flow rate in a turbulent pulsating flow.

However, for laminar, pulsating flow the pressure drop due to area change is

$$(\Delta p)_{true} = C\bar{u}_t = \overline{\Delta p} \tag{9.4}$$

and it is accurate to use the mean manometer reading.

This is the basis of viscous flow meters in which the original turbulent flow in the pipe is changed to laminar flow in a large number of small passages.

## 9.2 QUASI-STEADY FLOW

If the mean velocity at any point steadily increases or decreases with time but the rate of change is low enough for acceleration forces to be neglected then a quasi-steady analysis may be adopted. This analysis assumes that, at any instant, steady flow exists and steady flow equations may be used e.g. the Bernoulli equation. These equations, together with the continuity equation, will yield expressions for the mass flow rate at any instant of time.

The flow between a reservoir and a sink, during a given time interval, may be estimated if the manner of change of the pressure difference is defined. This presents difficulty in the case of subsonic flow of a gas but for the flow of a liquid, in which the change of surface level is directly related to flow, a simple analytical treatment is possible.

## 9.3 FLOW WITH SIGNIFICANT ACCELERATION FORCES

If a volume of liquid in a pipeline is accelerated at a moderate rate due to the slow opening or closing of a valve then inertia effects must be considered. In the simplest case, the volume of liquid is assumed to be accelerated as a whole due to a net force acting on the volume. The net force is the difference between the net piezometric pressure force $(p_1^* - p_2^*)$ A and the resisting force due to pipe and fittings resistance. The use of piezometric pressure eliminates the need to consider the weight of the liquid column.

In some situations with a long pipe, the inertia head can be very high and the actual value of $p_2$ may fall below the vapour pressure e.g. in the suction pipe of a reciprocating pump.

## 9.4 FLOW WITH SIGNIFICANT ELASTIC FORCES

If changes in velocity occur extremely rapidly due to instantaneous opening or closing of a valve then elastic forces in the fluid become important for both liquids and gases. Changes in pressure and velocity do not now take place instantaneously throughout the body of fluid, as in section 9.3, but are propagated by pressure waves.

This phenomenon, which can lead to noise and vibration, is known as *water hammer*, although its effects are not restricted to water; any compressible fluid may be involved.

Consider a pipeline conveying fluid at velocity u. If a valve at the end of the pipeline is suddenly closed, the fluid near the valve is brought to rest, and the pressure increases above that of the moving fluid. The junction between the stationary and moving fluid then moves away from the valve with celerity c, leaving a high pressure region behind it. Thus, a pressure wave moves along the pipe to the open end, as the body of fluid in the pipe is brought to rest. This pressure wave is repeatedly reflected from the open end of the pipe, and then the valve, and each time the wave passes a particular point the pressure changes to a new value of positive or negative gauge pressure. Eventually, the pressure transient dies away due to fluid friction. It is impossible in practice to close a valve instantaneously, but if the closure time is less than the time taken for a pressure wave to travel from the valve to the open end of the pipe, and back, closure may be treated as instantaneous.

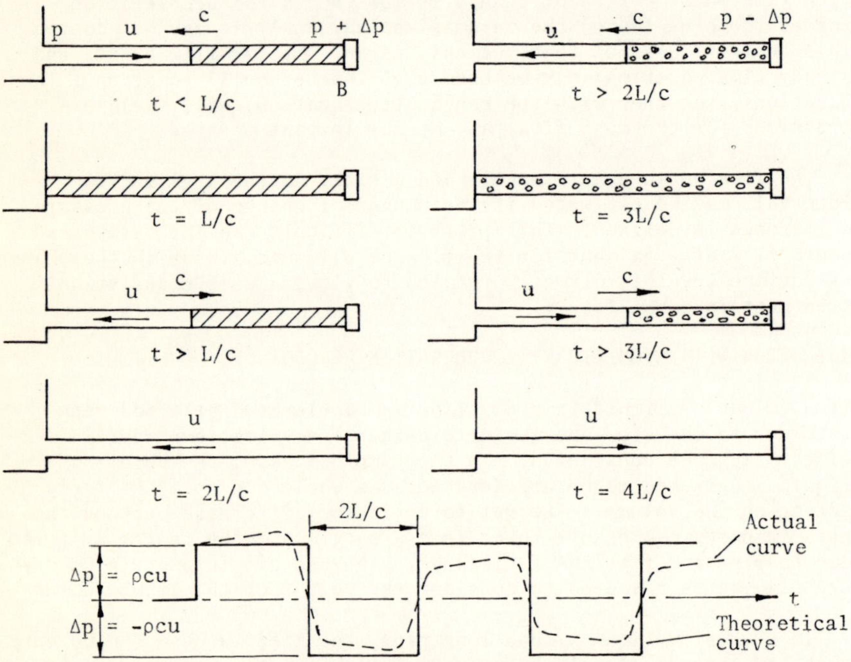


Pressure transient at valve B

Figure 9.1

The pressure rise, $\Delta p$, associated with the water hammer effect is

$$\Delta p = \rho u c \qquad (9.5)$$

This pressure rise can be large enough to fracture the pipe. In hydro-electric installations, surge tanks are frequently used to minimise the water hammer effect.

*Example 9.1*

(a) Derive expressions for the velocity distribution and the volumetric flow rate for two-dimensional laminar flow between parallel surfaces.

(b) Air flows through a viscous flow meter which consists of 100 parallel rectangular ducts normal to the flow, each 150 mm by 1 mm. The length of the meter is 200 mm. Neglecting end effects and assuming laminar flow through each duct determine the mean pressure difference recorded by an alcohol manometer (S.G. = 0.8) for a pulsating flow in which the volumetric flow rate varies between 0.15 $m^3/s$ and 0.175 $m^3/s$. Assume $\rho$ = 1.2 kg/s and $\mu = 1.8 \times 10^{-5}$ N $s/m^2$.

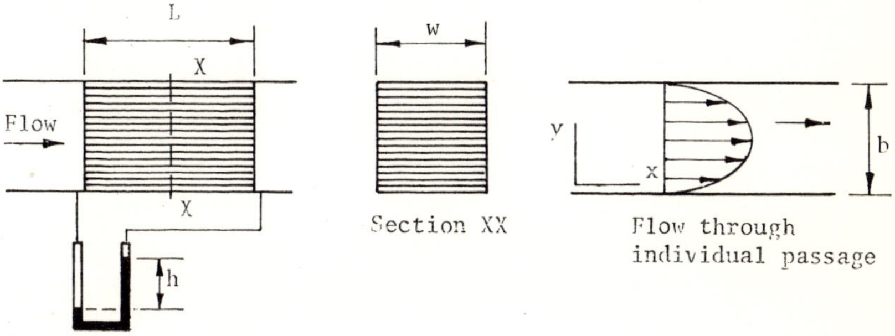


Figure 9.2

(a) The general expression for laminar flow between parallel surfaces was derived in example 7.2

$$u = \frac{1}{2\mu}\left(\frac{dp}{dx}\right) y^2 + Ay + B \qquad \text{(i)}$$

The boundary conditions are u = 0 when y = 0 and u = 0 when y = b. Therefore

$$u = \frac{1}{2\mu}\left(\frac{dp}{dx}\right)(y^2 - yb) \qquad \text{(ii)}$$

This type of flow is called *Poiseuille* or *Hagen-Poiseuille* flow and the velocity distribution is parabolic. The volumetric flow rate for a width, w, is given by

$$\dot{V} = \int_0^b wu\, dy = \frac{w}{2\mu}\left(\frac{dp}{dx}\right)\int_0^b (y^2 - yb)\, dy = -\frac{w}{12\mu}\left(\frac{dp}{dx}\right) b^3 \qquad \text{(iii)}$$

From the continuity equation

$$\dot{V} = \bar{u}(wb) \qquad \text{(iv)}$$

Therefore from equations (iii) and (iv)

$$\bar{u} = -\frac{1}{12\mu}\left(\frac{dp}{dx}\right) b^2 \qquad \text{(v)}$$

The negative sign in the expressions for velocity and volumetric flow rate indicates that the pressure gradient dp/dx is negative in the flow direction.

(b) The viscous flow meter consists of N rectangular ducts each of width, w, and breadth, b. Assuming that $dp/dx = \Delta p/L$ for flow through a duct, equation (iii) gives

$$\frac{\dot{V}}{N} = -\frac{w}{12\mu}\left(\frac{dp}{dx}\right)b^3$$

$$\frac{\Delta p}{L} = -\frac{12\mu\dot{V}}{wb^3N}$$

$$\Delta p = -\left(\frac{12\mu L}{wb^3N}\right)\dot{V}$$

The pressure drop across the meter fluctuates between $\Delta p_1$ and $\Delta p_2$ as the flow fluctuates between $\dot{V}_1$ and $\dot{V}_2$. Thus

$$\Delta p_1 = \frac{12 \times 1.8 \times 10^{-5} \times 0.2 \times 0.15}{0.15 \times 1^3 \times 10^{-9} \times 10^2} = 432 \text{ Pa}$$

$$\Delta p_2 = \frac{432 \times 0.175}{0.15} = 504 \text{ Pa}$$

In section 9.1 it was shown that for laminar flow the mean $\Delta p$ gives a correct mean $\dot{V}$. Therefore, neglecting the response of the manometer itself (i.e. wave action in the manometer connecting leads and manometer liquid damping) the mean difference in liquid level in the manometer is

$$\bar{h} = \frac{\overline{\Delta p}}{\rho g} = \frac{(504 + 432)/2}{800 \times 9.81} = 59.6 \text{ mm}$$

The assumption of laminar flow should be checked (i.e. Re < 2000).

$$Re = \frac{\rho\bar{u}D_e}{\mu} = \frac{\rho}{\mu}\left(\frac{\Delta pb^2}{12\mu L}\right)\left(\frac{4wb}{2(w + b)}\right)$$

$$= \frac{1.2 \times 504 \times 1^2 \times 10^{-6} \times 4 \times 0.15 \times 1 \times 10^{-3}}{1.8^2 \times 10^{-10} \times 12 \times 0.2 \times 2 \times 0.151} = 1545$$

*Example 9.2*

Water flows over a 60° triangular notch from a storage tank in which the surface area is 20 $m^2$. Find the time for the surface level to fall from 500 mm to 100 mm above the bottom of the notch. Derive any formulae used. Assume a constant discharge coefficient of 0.6.

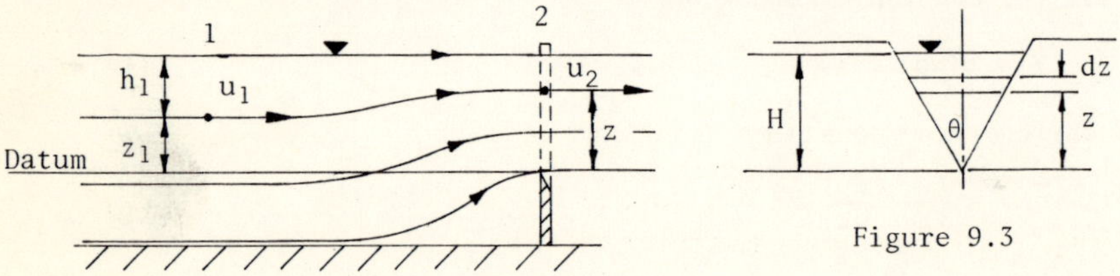


Figure 9.3

The first step is to derive an expression for idealised flow through the notch under steady conditions (refer to example 4.6). At section 2, in the plane of the notch, the pressure is atmospheric throughout and the velocity, $u_2$, at any point is found by applying the Bernoulli equation along a streamline.

$$\frac{p_a}{\rho g} + h_1 + \frac{u_1^2}{2g} + z_1 = \frac{p_a}{\rho g} + \frac{u_2^2}{2g} + z_2$$

Neglecting the velocity of approach, $u_1$, substituting $H = h_1 + z_1$ and $z_2 = z$, we obtain

$$u_2 = \sqrt{[2g(H - z)]}$$

From continuity, the discharge through an element of area $2z \tan\theta \, dz$ is given by

$$u_2 \, dA = \sqrt{[2g(H - z)]} \; 2z \tan\theta \, dz$$

and the total discharge through the notch is

$$\dot{V} = 2 \tan\theta \sqrt{(2g)} \int_0^H \sqrt{(H - z)} z \, dz = 2.35 \tan\theta \, H^{2.5} \qquad \text{(i)}$$

In practice, a discharge coefficient, $C_d$, must be introduced

$$\dot{V} = 2.35 C_d \tan\theta \, H^{2.5} \qquad \text{(ii)}$$

Let the surface area of the tank be A. In a time, dt, the level of water in the tank will fall by dH and the volume will reduce by dV. From continuity

$$dV = -A \, dH$$

Also $dV = \dot{V} \, dt = 2.35 C_d \tan\theta \, H^{2.5} \, dt$

$$dt = \frac{-A \, dH}{2.35 C_d \tan\theta \, H^{2.5}}$$

$$t_2 - t_1 = -\frac{A}{2.35 C_d \tan\theta} \int_{H_1}^{H_2} H^{-2.5} \, dH$$

$$= \frac{20}{2.35 \times 0.6 \times \tan 30} \int_{0.1}^{0.5} H^{-2.5} \, dH = 473 \text{ s}$$

*Example 9.3*

Two open vertical cylindrical tanks of diameters 3 m and 2 m are mounted with their axes vertical. They are connected by a single pipe 100 mm diameter and 50 m long. Minor losses in the pipe may be taken as 2.5 times the velocity head and the friction factor as 0.01. If the water level in the 3 m diameter tank is initially 5 m above that in the smaller tank, determine the drop in level in a time of 5 min.

Applying the energy equation 7.11 between points on the water surfaces in tanks A and B we obtain

$$\frac{p_a}{\rho g} + 0 + z_A = \frac{p_a}{\rho g} + 0 + z_B + \frac{4fL\bar{u}^2}{2gD} + 2.5\frac{\bar{u}^2}{2g}$$

$$\bar{u} = \left(\frac{2g}{4fL/D + 2.5}\right)^{\frac{1}{2}} H^{\frac{1}{2}} \quad \text{(i)}$$

Applying the continuity equation to flow in the pipe

$$\dot{V} = A\bar{u} = \tfrac{1}{4}\pi D^2\left(\frac{2g}{4fL/D + 2.5}\right)^{\frac{1}{2}} H^{\frac{1}{2}} \quad \text{(ii)}$$

Applying the continuity equation to flow in tanks A and B

$$\dot{V} = A_A\left(-\frac{dz_A}{dt}\right) = A_B\left(\frac{dz_B}{dt}\right)$$

$$\frac{dz_B}{dt} = -\frac{A_A}{A_B}\frac{dz_A}{dt} \quad \text{(iii)}$$

Also $\frac{dH}{dt} = \frac{d}{dt}(z_A - z_B) = \frac{dz_A}{dt} - \frac{dz_B}{dt}$ (iv)

From equations (ii), (iii) and (iv)

$$\dot{V} = -\frac{dH}{dt}\frac{A_A}{(1 + A_A/A_B)} = \tfrac{1}{4}\pi D^2\left(\frac{2g}{4fL/D + 2.5}\right)^{\frac{1}{2}} H^{\frac{1}{2}}$$

$$-H^{-\frac{1}{2}}\,dH = \frac{(1 + A_A/A_B)\pi D^2}{4A_A}\left(\frac{2g}{4fL/D + 2.5}\right)^{\frac{1}{2}} dt$$

$$2(H_1^{\frac{1}{2}} - H_2^{\frac{1}{2}}) = \frac{(1 + A_A/A_B)\pi D^2}{4A_A}\left(\frac{2g}{4fL/D + 2.5}\right)^{\frac{1}{2}}(t_2 - t_1)$$

$$2(5^{\frac{1}{2}} - H_2^{\frac{1}{2}}) = \frac{(1 + 9/4)\pi \times 0.1^2}{4 \times \frac{1}{4}\pi \times 3^2}\left(\frac{2 \times 9.81}{4 \times 0.01 \times 50/0.1 + 2.5}\right)^{\frac{1}{2}} \times 300$$

$H_2 = 3$ m

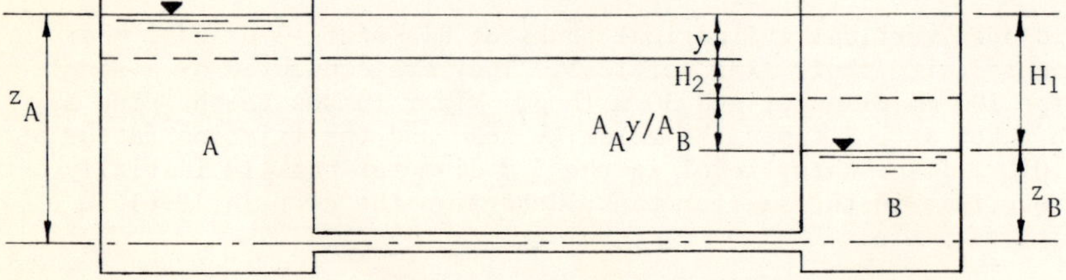


Figure 9.4

Refer to figure 9.4. As the level in tank A drops by an amount y, the level in tank B rises by $(A_A/A_B)y$.

$$H_1 = y + H_2 + (A_A/A_B)y$$

$$y = \frac{H_1 - H_2}{1 + A_A/A_B} = \frac{5 - 3}{1 + 9/4} = 0.615 \text{ m}$$

*Example 9.4*

Water is discharged from a 25 mm diameter orifice in the vertical side of the tank shown in figure 9.5. The tank has two parallel sides 1.2 m apart and the base is rectangular. The water level above the centre of the orifice is observed to fall from 1.5 m to 0.5 m in a time of 10 min. Determine the coefficient of discharge. The free jet initially falls 250 mm to a flat surface in a horizontal distance of 1.2 m from the vena contracta. Determine the discharge from the tank as the point of impact with the flat surface moves from its initial distance of 1.2 m to 1.0 m from the vena contracta. Assume a coefficient of contraction of 0.67.

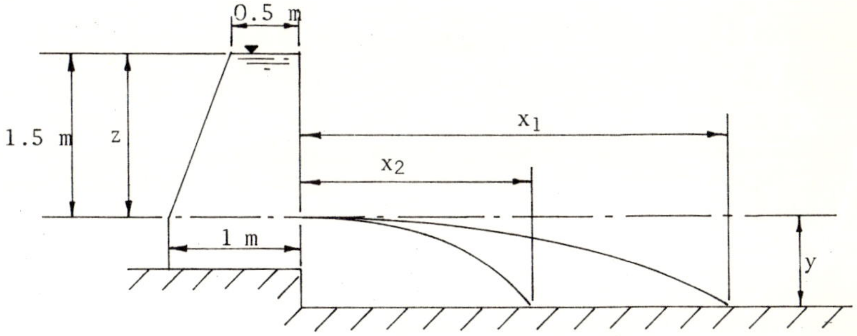


Figure 9.5

At any surface level, z, relative to the centre line of the orifice, the area of the water surface, $A_s$, is given by

$$A_s = 1.2(1 - z/3) \tag{i}$$

and the discharge, $\dot{V}$, from the tank is

$$\dot{V} = A_s\left(-\frac{dz}{dt}\right) = (0.4z - 1.2)\frac{dz}{dt} \tag{ii}$$

Applying the Bernoulli equation between a point on the water surface and a point at exit from the orifice, we have

$$\frac{p_a}{\rho g} + 0 + z = \frac{p_a}{\rho g} + \frac{u_t^2}{2g} + 0$$

$$u_t = \surd(2gz) \tag{iii}$$

Introducing a discharge coefficient, $C_d$, the actual discharge is

$$\dot{V} = C_d A_o u_t = C_d A_o \surd(2gz) \qquad \text{(iv)}$$

From equations (ii) and (iv)

$$\dot{V} = C_d A_o \surd(2gz) = (0.4z - 1.2)\,\frac{dz}{dt}$$

$$C_d\, dt = \frac{(0.4z^{0.5} - 1.2z^{-0.5})\, dz}{A_o \surd(2g)}$$

$$C_d\,(t_2 - t_1) = \frac{1}{A_o \surd(2g)} \left[\frac{2 \times 0.4z^{1.5}}{3} - 2 \times 1.2z^{0.5}\right]_{z_1}^{z_2}$$

$$C_d \times 10 \times 60 = \frac{1}{\frac{1}{4}\pi \times 0.025^2 \surd(2 \times 9.81)} \left[0.267z^{1.5} - 2.4z^{0.5}\right]_{1.5}^{0.5}$$

$$C_d = 0.65$$

From the coordinates of the water jet at the two points of impact, velocities of the jet at the vena contracta are given by

$$u_1 = x_1\left(\frac{g}{2y}\right)^{\frac{1}{2}} = 1.2 \times \left(\frac{9.81}{2 \times 0.25}\right)^{\frac{1}{2}} = 5.32 \text{ m/s}$$

$$u_2 = x_2\left(\frac{g}{2y}\right)^{\frac{1}{2}} = 1 \times \left(\frac{9.81}{2 \times 0.25}\right)^{\frac{1}{2}} = 4.43 \text{ m/s}$$

From equation (iii) the theoretical velocity at the initial head of $z_1 = 1.5$ m, is

$$u_{t_1} = \surd(2gz_1) = \surd(2 \times 9.81 \times 1.5) = 5.43 \text{ m/s}$$

The coefficient of velocity, $C_v$, is defined by

$$C_v = \frac{u_1}{u_{t_1}} = \frac{5.32}{5.43} = 0.98$$

$$u_{t_2} = \frac{u_2}{C_v} = \frac{4.43}{0.98} = 4.52 \text{ m/s}$$

$$z_2 = \frac{u_{t_2}{}^2}{2g} = \frac{4.52^2}{2 \times 9.81} = 1.04 \text{ m}$$

From equation (i)

$$A_{s_2} = 1.2(1 - 1.04/3) = 0.783 \text{ m}^2$$

$$A_{s_1} = 1.2(1 - 1.5/3) = 0.6 \text{ m}^2$$

The volume of water discharged from the tank is

$$V = \left(\frac{A_{s_2} + A_{s_1}}{2}\right)(z_1 - z_2) = \left(\frac{0.783 + 0.6}{2}\right)(1.5 - 1.04)$$

$$= 0.318 \text{ m}^3$$

*Example 9.5*

Water flows from a reservoir through a pipe 100 m long and 75 mm diameter. A valve at the end of the pipe is slowly closed in such a way that the water velocity is reduced uniformly to zero in a time of 2 s. If the water level in the reservoir remains constant at 8 m above the pipe outlet determine (i) the velocity of flow 1 s after commencement of valve closure, and (ii) the maximum pressure at the valve during closure. Assume f = 0.008 and neglect compressibility of the water.

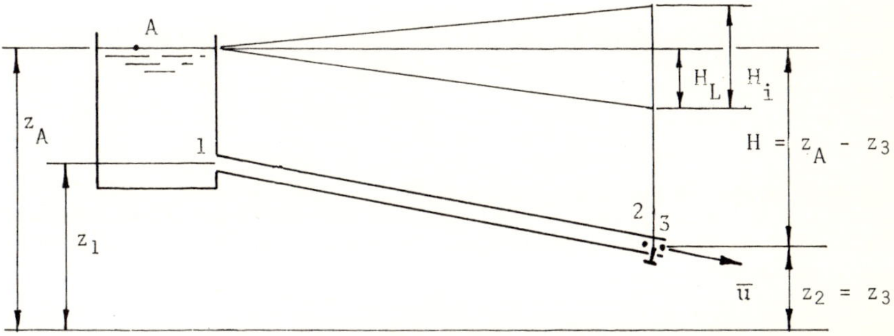


Figure 9.6

In this problem we are concerned with the deceleration of the liquid column in the pipe from an initial steady velocity $u_o$.

Applying the energy equation 7.11 between the liquid surface and exit from the valve, we obtain

$$\frac{p_a}{\rho g} + 0 + z_A - \frac{4fL\bar{u}_o^2}{2gD} = \frac{p_a}{\rho g} + \frac{\bar{u}_o^2}{2g} + z_3$$

$$\bar{u}_o = \left(\frac{2gH}{\frac{4fL}{D} + 1}\right)^{\frac{1}{2}} = \left(\frac{2 \times 9.81 \times 8}{\frac{4 \times 0.008 \times 100}{0.075} + 1}\right)^{\frac{1}{2}} = 1.9 \text{ m/s}$$

The velocity is reduced uniformly to zero. Therefore after 1 s

$$\bar{u} = \left(\frac{t_{total} - t}{t_{total}}\right)\bar{u}_o = \frac{1 \times 1.9}{2} = 0.95 \text{ m/s}$$

At any instant during valve closure the net force which tends to decelerate the liquid column is equal to the net piezometric force $(p_1^* - p_2^*)A$ and the resisting force due to pipe and fitting resistance $\Delta p_L A$.

$$(p_1^* - p_2^*)A - \Delta p_L A = \rho AL\left(-\frac{du}{dt}\right) \qquad \text{(i)}$$

$$[(p_{g_1} + \rho g z_1) - (p_{g_2} + \rho g z_2)] - \rho g\frac{4fL\bar{u}^2}{2gD} = -\rho L\frac{du}{dt}$$

Neglecting changes in pressure at entry to the pipe

$$[\rho g(z_A - z_1) + \rho g z_1 - p_{g_2} - \rho g z_2] - \rho g \frac{4fL\bar{u}^2}{2gD} = -\rho L \frac{d\bar{u}}{dt} \quad \text{(ii)}$$

$$\frac{p_{g2}}{\rho g} = (z_A - z_2) - \frac{4fL\bar{u}^2}{2gD} + \frac{L}{g}\frac{d\bar{u}}{dt} \quad \text{(iii)}$$

It is convenient to use head terms in conjunction with figure 9.6

$$\frac{p_{g2}}{\rho g} = H - H_L + H_i \quad \text{(iv)}$$

The maximum pressure at the valve occurs at the instant of complete closure when $\bar{u} = 0$. From equation (iii)

$$p_{g2} = \rho g H + \rho L \frac{d\bar{u}}{dt} = 10^3 \times 9.81 \times 8 + 10^3 \times 100 \times \frac{1.9}{2} = 174 \text{ kPa}$$

It is possible to use a modified form of the Bernoulli equation to obtain the same result. Applying the Bernoulli equation between the liquid surface and a point just before the valve and including the inertia head

$$\left(\frac{p_a}{\rho g} + 0 + z_A\right) - \frac{4fL\bar{u}^2}{2gD} = \left(\frac{p_{g2}}{\rho g} + \frac{p_a}{\rho g} + \frac{\bar{u}^2}{2g} + z_2\right) + \frac{L}{g}\left(-\frac{d\bar{u}}{dt}\right)$$

$$\frac{p_{g2}}{\rho g} = (z_A - z_2) - \frac{4fL\bar{u}^2}{2gD} - \frac{\bar{u}^2}{2g} + \frac{L}{g}\frac{d\bar{u}}{dt} \quad \text{(v)}$$

Thus, the expression obtained is the same as equation (iii) with the exception of the term $\bar{u}^2/2g$. This velocity term does not appear in equation (iii) because it was assumed that the pressure acting on the liquid column at the pipe entry is equal to the hydrostatic pressure just outside. In fact, there is a drop in pressure equal to $\bar{u}^2/2g$ as liquid enters the pipe and to be strictly correct this should have been included.

The Bernoulli approach to this type of problem is frequently preferred because it appears simpler than the full momentum analysis shown earlier. It must be emphasised, however, that, used in this way, the Bernoulli equation is a true momentum equation rather than an energy equation expressing mechanical energy dissipation.

*Example 9.6*

(a) Briefly explain the effects of fitting large air vessels on to the suction and delivery pipes of a reciprocating pump.

(b) Determine the maximum speeds at which a double-acting reciprocating pump could be run with (i) no air vessels fitted, and (ii) a large air vessel fitted to the suction side close to the pump. The pump piston has a diameter of 0.15 m and moves with simple harmonic motion through a stroke of 0.45 m. The pump is situated 3.6 m above the lower reservoir, to which it is connected by a pipe

6 m long by 0.1 m diameter with a friction factor of 0.006. The barometric pressure head is 9.8 m of water, and the minimum allowable water pressure head is 1.8 m.

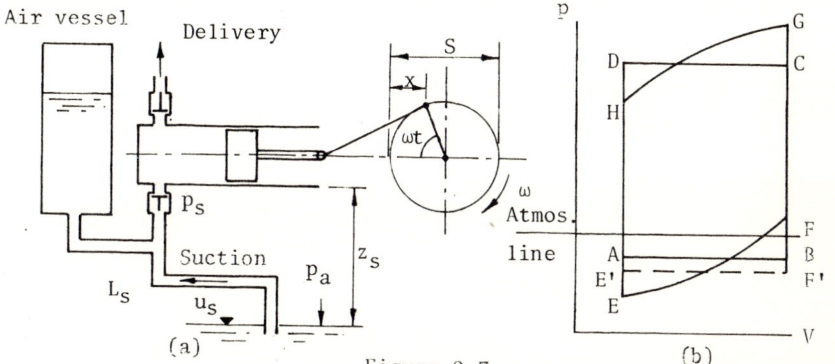


Figure 9.7

(a) The ideal cylinder pressure-volume diagram for a reciprocating pump is a rectangle ABCD (figure 9.7b). Due to the inertia and friction of the liquid in the suction and delivery pipes, the actual diagram is EFGH. The effect of liquid inertia in the suction pipe is to reduce the pressure at inlet by AE and the speed of the pump is limited by the pressure at E. This pressure must not be allowed to fall below the value at which dissolved gases are liberated from the liquid.

Pulsations in either the suction or delivery pipes may be largely eliminated by connecting a large closed air vessel to the pipe at a point close to the pump. The acceleration head and friction loss are then restricted to the short length of pipe between the cylinder and the vessel. With an air vessel connected to the suction pipe, as shown in figure 9.7a, a practically steady velocity is produced in the pipe. This gives a constant friction loss, AE', much less than the previous maximum value and, because E' corresponds to a higher pressure value than E, the pump may operate at a much higher speed.

(b)(i) Applying the Bernoulli equation between the reservoir surface and the pipe connection to the pump we have

$$\frac{p_a}{\rho g} + 0 + 0 - \frac{4fL_s\bar{u}_s^{\,2}}{2gd} = \frac{p_s}{\rho g} + \frac{\bar{u}^2}{2g} + z_s + \frac{L_s}{g}\frac{du}{dt} \qquad \text{(i)}$$

For simple harmonic motion of the piston the displacement, x, is

$$x = \tfrac{1}{2}S(1 - \cos \omega t) = \tfrac{1}{2} \times 0.45(1 - \cos \omega t)$$

The acceleration of the piston (area A) is

$$\frac{dx}{dt} = 0.225\omega^2 \cos \omega t$$

The acceleration of the water in a pipe of area a is

$$\frac{d\bar{u}}{dt} = \frac{A}{a} \times 0.225\omega^2 \cos \omega t = \frac{0.225D^2\omega^2 \cos \omega t}{d^2} \qquad \text{(ii)}$$

Combining equations (i) and (ii) we obtain

$$\frac{p_a}{\rho g} = \frac{p_s}{\rho g} + \frac{\bar{u}_s^{\,2}}{2g} + z_s + \frac{4fL_s\bar{u}_s^{\,2}}{2gd} + \frac{0.225\ L_sD^2\omega^2 \cos \omega t}{d^2g}$$

The minimum pressure occurs at the beginning of the suction stroke when the flow and the friction loss are zero (t = 0). Therefore

$$\frac{p_a}{\rho g} = \frac{p_s}{\rho g} + z_s + \frac{0.225L_sD^2\omega^2}{d^2g}$$

$$9.8 = 1.8 + 3.6 + \frac{0.225 \times 6 \times 0.15^2\omega^2}{0.1^2 \times 9.81}$$

$$\omega = 3.77 \text{ rad/s}$$

(ii) When an air vessel is fitted, there is no acceleration of the liquid in the pipe. Equation (i) reduces to

$$\frac{p_a}{\rho g} = \frac{p_s}{\rho g} + \frac{\bar{u}_s^{\,2}}{2g}\left(1 + \frac{4fL_s}{d}\right) + z_s \qquad \text{(iii)}$$

For a double-acting pump the velocity of flow in the pipe is

$$\bar{u}_s = 2\left(\frac{\omega}{2\pi}\right)\frac{AS}{a} = \frac{2 \times 0.15^2 \times 0.45}{2 \times \pi \times 0.1^2}\,\omega = 0.322\omega \qquad \text{(iv)}$$

Substituting values and combining equations (iii) and (iv)

$$9.8 = 1.8 + \frac{0.322^2\omega^2}{2 \times 9.81}\left(1 + \frac{4 \times 0.006 \times 6}{0.1}\right) + 3.6$$

$$\omega = 18.47 \text{ rad/s}$$

*Example 9.7*

A valve at the outlet of a pipe conveying water is suddenly closed. If only the elasticity of the water is taken into account show that the pressure rise produced by valve closure is given by $\Delta p = \rho uc$, where u is the initial mean velocity of flow in the pipe and c is the velocity of propagation of a pressure wave in water.

A pipe 850 m long contains water flowing at 2 m/s. If a valve at the outlet end of the pipe is closed in a time of 1 s determine the pressure rise in the pipe. Show that the maximum pressure rise depends upon the total time of closure and in this case is not affected by the momentary rate of closure. $K_{water} = 2.14\ GN/m^2$.

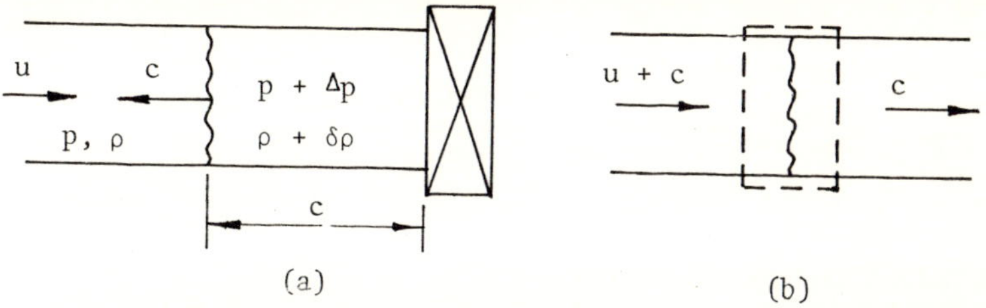


Figure 9.8

Consider the sudden closing of a valve in a pipeline (figure 9.8). The liquid near the valve is brought to rest and as the pressure wave moves through the liquid with celerity c the pressure and density increase to $p + \Delta p$ and $\rho + \delta\rho$, respectively.

In unit time the pressure wave moves a distance c from the valve and the increase in density of the stationary fluid of volume Ac is due to an increase in mass due to fluid inflow $\rho Au$. Therefore

$$\delta\rho \; Ac = \rho Au \tag{i}$$

From equation 1.22, the bulk modulus of the liquid is

$$K = \rho\left(\frac{\Delta p}{\delta\rho}\right)$$

Substituting for $\delta\rho$ in (i)

$$c = \frac{uK}{\Delta p} \tag{ii}$$

Now give the whole system a velocity c to the right (figure 9.8b) so that the wave is stationary relative to coordinate axes moving with the wave. Applying the momentum theorem to the control volume

$$\Sigma F = \dot{m}(u_2 - u_1)$$

$$-\Delta p \; A = \rho A(u + c)[c - (u + c)]$$

$$\Delta p = \rho(u + c)u \tag{iii}$$

Substituting for c from equation (ii)

$$\Delta p = \rho\left(1 + \frac{K}{\Delta p}\right)u^2$$

Now K has a very high value for liquids, therefore $\frac{K}{\Delta p} >> 1$ and

$$\Delta p = \rho\frac{K}{\Delta p}\,u^2 = u\surd(\rho K) \tag{iv}$$

From equations (ii) and (iv)

$$c = \sqrt{(K/\rho)} \qquad \text{(v)}$$

Thus the velocity with which a wave is propagated relative to the pipe, c, is equal to the speed of sound in an infinite fluid. The velocity relative to the liquid is (u + c) but since c >> u the relative and absolute pressure wave celerities are virtually the same. Combining equations (iv) and (v)

$$\Delta p = \rho uc \qquad \text{Q.E.D.}$$

As the liquid is brought to rest, energy is stored in the compressed liquid as strain energy. The strain energy per unit volume is

$$E_s = \frac{(\Delta p)^2}{2K} \qquad \text{(vi)}$$

Substituting for $\Delta p$ from equation (iv) we have

$$E_s = \rho u^2/2 \qquad \text{(vii)}$$

But $\rho u^2/2$ is the kinetic energy per unit volume of the liquid moving with initial velocity u. Thus all of the original kinetic energy of the liquid is transformed to strain energy and none is degraded into heat by the pressure transient. In practice, however, some degradation will occur due to pipe friction.

Strictly speaking, $\Delta p = \rho uc$ is obtained for instantaneous valve closure. No valve may be closed instantaneously in practice but if closure is completed before the initial pressure wave is reflected back to the valve (i.e. in a time less than 2L/c) then the whole of the kinetic energy is converted into strain energy and closure may be treated as instantaneous. In these circumstances only the total time of closure is important and not the rate of closure. If, however, t >> 2L/c then the method of example 9.5 must be adopted to determine the pressure rise. For the given pipe

$$\frac{2L}{c} = \frac{2L}{\sqrt{(K/\rho)}} = \frac{2 \times 850}{\sqrt{(2.14 \times 10^9/10^3)}} = 1.165 \text{ s}$$

The actual closure time of 1 s is less than 2L/c therefore the closure may be treated as instantaneous. Substituting values in equation (iv)

$$\Delta p = 2\sqrt{(10^3 \times 2.14 \times 10^9)} = 2.92 \text{ MPa}$$

*Example 9.8*

Show that when a valve is closed at the end of a pipe conveying liquid, the velocity at which a pressure wave is propagated is given by

$$c' = \left[\frac{1}{\rho(1/K + D)xE}\right]^{\frac{1}{2}}$$

if circumferential strain in the pipe is taken account of. Hence, determine the pressure rise in a straight steel pipe, 1000 m long, 300 mm diameter and 20 mm wall thickness, when a valve at the outlet end is closed in a time of 1.3 s. At the instant of valve closure, water is passing through the pipe at a velocity of 2.5 m/s. $K_{water} = 2.14$ GPa, $E_{steel} = 200$ GPa.

Assume that as the water in the pipe is brought to rest by instantaneous valve closure, the whole of the kinetic energy of the water is converted into strain energy of the water and circumferential strain energy of the pipe material (refer to example 9.7 for justification).

$$\begin{matrix}\text{Strain energy} \\ \text{in water per} \\ \text{unit mass of} \\ \text{water}\end{matrix} + \begin{matrix}\text{Strain energy} \\ \text{in pipe per} \\ \text{unit mass of} \\ \text{water}\end{matrix} = \begin{matrix}\text{Kinetic energy} \\ \text{per unit mass} \\ \text{of water}\end{matrix} \qquad \text{(i)}$$

Now the circumferential stress in a pipe of diameter, D, and wall thickness, x, is given by $\sigma = (\Delta p)D/2x$ and the strain energy in the pipe material is $(\sigma^2/2E) \times$ volume. Thus, the strain energy in a pipe of length L is given by

$$(E_s)_L = \left(\frac{\Delta pD}{2x}\right)^2 \frac{\pi DxL}{2E} \qquad \text{(ii)}$$

The pipe length L occupied by unit mass of water is given by

$$1 = \rho V = \rho \tfrac{1}{4}\pi D^2 L$$

$$L = \frac{4}{\rho\pi D^2} \qquad \text{(iii)}$$

The strain energy in the pipe wall per unit mass of water is obtained by substituting equation (iii) in (ii)

$$(E_s)_L = \left(\frac{\Delta pD}{2x}\right)^2 \frac{\pi Dx}{2E}\left(\frac{4}{\rho\pi D^2}\right) = \frac{(\Delta p)^2 D}{2\rho xE} \qquad \text{(iv)}$$

From example 9.7, the strain energy in the liquid per unit mass is $(\Delta p)^2/2\rho K$ [equation (vi)] and the kinetic energy per unit mass of liquid is $u^2/2$ [equation (vii)]. Substituting in equation (i) we obtain

$$\frac{(\Delta p)^2}{2\rho K} + \frac{(\Delta p)^2 D}{2\rho xE} = \frac{u^2}{2}$$

$$\Delta p = \rho u \left(\frac{1}{\rho(1/K + D/xE)}\right)^{\frac{1}{2}} \qquad \text{(v)}$$

Comparing equation (v) with $\Delta p = \rho uc$ (equation 9.5) obtained for the pressure rise when pipe expansion is neglected, it can be seen that the pressure wave celerity in a pipe with circumferential expansion is effectively

$$c' = \left(\frac{1}{\rho(1/K + D)xE}\right)^{\frac{1}{2}} \qquad \text{(vi)}$$

$$= \left(\frac{1}{10^3[1/(2.14 \times 10^9) + 0.3/(0.02 \times 200 \times 10^9]}\right)^{\frac{1}{2}} = 1360 \text{ m/s}$$

$$\frac{2L}{c'} = \frac{2 \times 1000}{1360} = 1.47 \text{ s}$$

The actual closure time of 1.3 s is less than 2L/c' therefore the valve closure may be treated as instantaneous

$$\Delta p = \rho uc' = 10^3 \times 2.5 \times 1360 = 3.4 \text{ MPa}$$

*Problems*

1 Derive an expression for the pressure drop in a fluid flowing through a circular pipe under laminar flow conditions.

It is required to measure the mass flow rate of air entering an internal combustion engine under pulsating conditions. The flow is known to be turbulent and therefore a viscous flow meter is constructed, consisting of a number of parallel circular tubes. The flow area of the meter is to be equal to that of the intake pipe which is 100 mm diameter. Assuming that the mean Reynolds number of flow in the intake pipe is $2 \times 10^4$ and that the mean velocity, density and viscosity of air are maintained in the viscous flow meter, determine (i) the number and diameter of tubes required for the meter to ensure the same flow and a mean meter Reynolds number of $10^3$, and (ii) the actual mass flow rate of air entering the engine when the difference in level recorded by an alcohol manometer fluctuates between 0.8 mm and 1.2 mm alcohol. The manometer leads are connected to tapping points 150 mm apart on the meter. Neglect the response of the manometer and leads. Assume $\rho_{air} = 1.25$ kg/m$^3$, $\mu_{air} = 18 \times 10^{-6}$ N s/m$^2$, $\rho_{alcohol} = 800$ kg/m$^3$.

[400, 5 mm, 0.0225 kg/s]

2 Show that the discharge over a rectangular notch of width b is given by $V = (2/3)C_d b\sqrt{(2g)}H^{3/2}$. Water flows over a sharp-edged rectangular notch 1 m wide from a reservoir in which the water surface area is 100 m$^2$. The original water level is 750 mm above the sill of the notch. Determine the fall in level in a time of 5 min. Assume $C_d = 0.6$.

[682 mm]

3 Water is discharged from an orifice, 60 mm diameter, in the base of a vertical, conical tank. The tank is 10 m high and the base is 5 m diameter. Determine the time for the tank to empty if the initial water level is 6 m above the bottom. Assume a discharge coefficient of 0.65.

[1.6 h]

4 A horizontal, cylindrical tank is 3 m diameter and 4 m long. The tank is initially half full of water. Determine the diameter of an

orifice, situated in the base of the tank, needed to empty the tank in one hour. Assume $C_d = 0.6$.

[48.8 mm]

5 A vertical plate divides a tank of cross-sectional area 4.4 $m^2$ into two compartments with an area ratio of 3 to 1. The water level in the larger compartment is initially 5 m above that in the smaller one. What time will be required to raise the level in the smaller compartment by 3 m if flow between the compartments takes place through a 50 mm diameter orifice in the base of the vertical plate? Assume $C_d = 0.6$.

[391 s]

6 A tank of 4 $m^2$ cross-sectional area, discharges water at the rate $\dot{V}_o$ through an orifice 50 mm diameter and there is a steady inflow of $\dot{V}_i = 0.01$ $m^3/s$. How long will it take for the water surface to rise from 1 m to 3 m above the centre of the orifice? What is the maximum possible head of water above the orifice under these conditions? $C_d = 0.6$.

[58.26 min, 3.7 m]

7 A straight pipe 200 m long and 100 mm diameter is connected to a reservoir in which the water level is 10 m above the pipe outlet. If a valve at the pipe outlet is now opened determine the time taken for the velocity in the pipe to reach one half of the maximum velocity. Neglect compressibility and minor losses and assume f = 0.01.

[1.75 s]

8 A vertical tank of rectangular cross-section is 3 m high and has two parallel sides 2 m apart and two sides whose distance apart varies linearly from 4 m at the base to 2 m at the top. A pipe 10 m long is attached to the base of the tank and the outlet of the pipe is 3 m below the bottom of the tank. If the depth of water in the tank falls from 2.5 m to 1.5 m in 15 min, determine the diameter of the pipe. If, at the instant that the water depth in the tank is 1.5 m, a valve at the end of the pipe is closed in a time which may be treated as instantaneous, determine the pressure rise just before the valve. Assume a pipe entry loss equal to one velocity head and a pipe friction factor of 0.008. The bulk modulus and density of water may be assumed equal to 2.14 $GN/m^2$ and 1000 $kg/m^3$ respectively.

[48.8 mm, 4.7 MPa]

9 Derive a formula for the pressure rise in a fluid flowing in a pipe when the valve at the end from which the fluid escapes is closed (a) slowly, and (b) very quickly. Water flows through a steel pipe 200 mm bore and 6 mm thick at 2.4 m/s, pipe length 120 m. Calculate the maximum pressure if (a) flow is reduced uniformly to zero in 5 s, and (b) valve is closed in a time which may be treated as instantaneous. K for water = 2.14 $GN/m^2$; E for steel = 200 $GN/m^2$.

[126.6 kPa, 3 MPa]

# BIBLIOGRAPHY

The author has included problems and ideas from a variety of sources and it would be impossible to list them all here. The following books have proved particularly useful in the preparation of this text and the student is referred to them for further reading and derivation of the relationships quoted in the theory sections.


1 Engineering Sciences Data Unit, *Internal Flow, Vols 1, 2, 4, 5* (E.S.D.U. London)

2 Chartered Institution of Building Services, *CIBS Guide C4*

3 Fox, J.A., *Introduction to Engineering Fluid Mechanics*, 2nd ed. (The Macmillan Press, London, 1977)

4 Lay, J.E., *Thermodynamics* (Pitman & Sons Ltd., London, 1964)

5 Massey, B.S., *Mechanics of Fluids*, 3rd ed. (Van Nostrand Co. Ltd. London, 1975)

6 Mayhew, Y.R. and Rogers, G.F.C., *Thermodynamic and Transport Properties of Fluids*, SI units (Basil Blackwell, Oxford 1968)

7 Rogers, G.F.C. and Mayhew, Y.R., *Engineering Thermodynamics, Work and Heat Transfer*, 2nd ed. (Longmans, Green & Co., 1967)

8 Shepherd, D.G., *Elements of Fluid Mechanics* (Harcourt, Brace & World, New York, 1965)

9 Vallentine, H.R., *Applied Hydrodynamics*, 2nd ed. (Butterworths, London, 1967)

10 Vennard, J.K. and Street R.L., *Elementary Fluid Mechanics*, 5th ed. (John Wiley & Sons Inc., London, 1976)

11 Webber, N.B., *Fluid Mechanics for Civil Engineers*, SI ed. (Chapman & Hall Ltd., London, 1971)